Springer Tracts in Mechanical Engineering

About this Series

Springer Tracts in Mechanical Engineering (STME) publishes the latest developments in Mechanical Engineering - quickly, informally and with high quality. The intent is to cover all the main branches of mechanical engineering, both theoretical and applied, including:

- Engineering Design
- Machinery and Machine Elements
- Mechanical structures and stress analysis
- Automotive Engineering
- Engine Technology
- Aerospace Technology and Astronautics
- Nanotechnology and Microengineering
- Control, Robotics, Mechatronics
- MEMS
- Theoretical and Applied Mechanics
- Dynamical Systems, Control
- Fluids mechanics
- Engineering Thermodynamics, Heat and Mass Transfer
- Manufacturing
- Precision engineering, Instrumentation, Measurement
- Materials Engineering
- Tribology and surface technology

Within the scopes of the series are monographs, professional books or graduate textbooks, edited volumes as well as outstanding PhD theses and books purposely devoted to support education in mechanical engineering at graduate and postgraduate levels.

More information about this series at http://www.springer.com/series/11693

Hung Nguyen-Schäfer

Rotordynamics of Automotive Turbochargers

Second Edition

Springer

Hung Nguyen-Schäfer
EM-motive GmbH
A Joint Company of Daimler and Bosch
Ludwigsburg
Germany

ISSN 2195-9862 ISSN 2195-9870 (electronic)
Springer Tracts in Mechanical Engineering
ISBN 978-3-319-17643-7 ISBN 978-3-319-17644-4 (eBook)
DOI 10.1007/978-3-319-17644-4

Library of Congress Control Number: 2015938734

Springer Cham Heidelberg New York Dordrecht London

Printed on acid-free paper

Springer International Publishing AG Switzerland is part of Springer Science+Business Media
(www.springer.com)

Preface to the Second Edition

In this second edition, all chapters are revised in which typos and errors are corrected; and comments and feedbacks are included. In addition, a few new topics on automotive turbochargers are added in this book.

First, the new CO_2 emission laws for passenger and light duty commercial vehicles in the European Union countries (EU) from 2012 to 2021 are discussed and some possible measures that should be carried out to comply with the new emission laws. Second, the turbo compound is used in normal and hybrid operations of commercial vehicles. Third, the new technique of three-stage turbochargers is applied for the first time to the passenger vehicle of BMW M550d. Fourth, the two-phase lubrication Reynolds equation with the bearing filling grade that depends on the bubble fraction of the oil mixture is used in the bearing computation. Fifth, the failure mechanisms of the bearings due to wears and fatigue are analyzed and displayed.

Furthermore, the new Chap. 10 deals with designing the platforms of automotive turbochargers using physical similarity laws. The platform design is based on the market analysis of the engine requirements and the strategy for the type series of the platform. Then using the similarity laws among the turbomachines, downsized engines, and classical mechanics, the power characteristics and geometrical sizes of the turbocharger type series are computed for the platforms of passenger and commercial vehicles.

Rotordynamics of automotive turbochargers is a special case and the most difficult application in turbomachinery; therefore, it involves the broadly interdisciplinary field of applied physics and mechanical engineering. Hence, some mathematical backgrounds of vector calculus, differential equations, and bifurcation theory are required. This all-in-one book of turbochargers is intended for scientific and engineering researchers, practitioners working in the rotordynamics field of automotive turbochargers, and graduate students in applied physics and mechanical engineering.

I would like to thank Dr. Jan-Philip Schmidt and Mrs. P. Jantzen at Springer in Heidelberg for the helpful and cooperative work on the publishing of this book.

In addition, I am grateful to Ms. Sivajothi Ganerarathinam at SPS in Chennai, India for the best support of producing this book.

Finally, my special thanks go to my wife for her understanding patience and endless support during the revisal of the second edition in my leisure and vacation time.

Ludwigsburg, Germany Hung Nguyen-Schäfer

Preface to the First Edition

This book has arisen from my many years of experience in the automotive industry, as a development engineer and as a senior expert in rotordynamics of automotive turbochargers. It is intended for senior undergraduates and graduates in mechanical engineering, research scientists, and practicing engineers who work on the rotordynamics of automotive turbochargers. It could be also used as a rotordynamic textbook in colleges and universities, and practical handbook of rotordynamics in the automotive turbochargers.

The topic of rotordynamics of automotive turbochargers is a widely interdisciplinary working field, first involving *rotordynamics* to study dynamics of rotating machines at very high rotor speeds as well as to balance the rotor. Second, it involves *thermodynamics* and *turbo matching* to compute working conditions of the turbochargers. Third, it involves *fluid and bearing dynamics* to compute the acting loads in the bearings at various operating conditions, and to design the hydrodynamic oil–film bearings. Lastly, it involves *applied tribology* to reduce bearing friction and wears of the journal and bearings. In order to understand the rotordynamic phenomena, readers are assumed to have some mathematical requisite backgrounds of modeling and simulating nonlinear rotordynamics of turbochargers. The author tries to keep the mathematics requirement as simple as possible in this book; however, without any mathematical background, it is quite difficult to comprehend and thoroughly understand the rotordynamic behaviors of turbochargers.

Exhaust gas turbochargers used in passenger, commercial vehicles, and off-road engines have some important discrepancies with the heavy turbomachines applied to power plants, chemical, and aeroplane industries. The automotive turbochargers are much smaller compared to industrial turbomachines. Therefore, they generally work at very high rotor speeds in various dynamically operating conditions, such as highly transient rotor speeds, variable pressures, high temperatures of exhaust gas, as well as unsteady-state mass flow rates of the intake air and exhaust gas. The industrial turbomachines are larger and heavier, and often operate at a nearly stationary condition. Due to the large compressor and turbine wheels, they operate at relatively low rotational speeds from 3,000 rpm (Europe) or 3,600 rpm (US) in power plants for electrical frequency of 50 or 60 Hz; up to about 15,000 rpm in chemical industries and aeroplanes. By contrast, exhaust gas turbochargers mostly work at high rotor speeds from 150,000 to 350,000 rpm in the automotive applications. Therefore, the unbalanced force is much larger than the rotor weight, leading to nonlinear characteristics of the oil–film bearings used in the automotive turbochargers. For this reason, nonlinear rotordynamics is usually applied to the turbochargers to study and compute the nonlinear rotor responses of the harmonic, sub-, and supersynchronous vibrations.

Moreover, turbocharger engineers in the industry have to confront many problems at once, namely good quality, feasibility, form tolerances in mass production, time to market (TTM), highly innovative products, and product price. The last one is an important concern for a company. No matter how good the products are, nobody can afford them because they are very expensive. Then, the question is, how long can the company survive without selling any product or always selling products at a loss. Parallel to the product price, turbochargers must be qualitative and innovative in terms of high efficiency, best low-end-torque, working at high temperatures of the exhaust gas, less or no wear of bearings, and low airborne noises. They should come to the market as soon as possible since the first bird gets the worm; i.e., despite highly innovative products, the time to market (TTM) is always shorter because the competitors never sleep. Additionally, the turbochargers should work in all operating conditions while they are produced at a possibly wide range of form tolerances in mass-production; e.g., radial and thrust bearings with large form tolerances since producing them with narrow ones increases the production cost, leading to rise in the product price.

All these boundary conditions make the turbocharger development in the industry much more difficult, especially in the nonlinear rotordynamics of turbochargers. Therefore, development engineers of turbochargers need to have deep understanding of rotordynamics and bearing systems containing radial and thrust bearings applied to automotive turbochargers. Furthermore, issues of rotor balancing and tribology in the bearings have to be coped with, so that the produced turbochargers work in any case at the given industrial development conditions. Customer requirements of the automotive turbochargers are very high, in terms of good rotordynamic stability, low airborne noises, less or no wear of the bearings at high oil temperatures, and an acceptable product price.

Despite all careful efforts, there would be some unpredictable errors in this book. I would be very grateful to get your feedbacks and hints of errors. For this reason, readers of this book need to have a thorough analysis before applying it to their individual applications, and take their own responsibilities for possible damages.

I like to thank the board of directors of Bosch Mahle Turbo Systems (BMTS), Dr. Knopf, Dr. Prang, and Mr. Jennes for their support and for allowing me to use some pictures of BMTS in this book. Especially, I learned a great deal from working with Dr. Engels on turbocharging. In addition, I am indebted to my colleagues at BMTS who supported me in technical discussions, and provided help in this book: Dr. Haiser; Schnaithmann; Ahrens, Kothe, and Kleinschmidt; Lemke and Kreth; Di Giandomenico (Bosch).

For fruitful discussions of the computation of nonlinear rotordynamics, I would like to acknowledge Dr. Schmied at Delta JS, Switzerland.

In addition, I like to thank Dr. Jan-Philip Schmidt at the Springer Publisher in Heidelberg for the good and helpful cooperation during the publishing of this book.

Finally, my special thanks go to my brother, Richard Nguyen at First American in Santa Ana, California for carefully reading this book with constructive critiques.

Stuttgart, Germany Hung Nguyen-Schäfer

Contents

About the Author

Dr. Hung Nguyen-Schäfer is a senior technical manager in development of electric machines for hybrid and electric vehicles at EM-motive GmbH, a joint company of Daimler and Bosch in Germany. He received the B.Sc. and M.Sc. in mechanical engineering with nonlinear vibrations in fluid mechanics from the University of Karlsruhe (KIT), Germany in 1985; and a Ph.D. degree in nonlinear thermo- and fluid dynamics from the same university in 1989. He joined Bosch Company and worked as a technical manager on many development projects. Between 2007 and 2013, he was in charge of rotordynamics, bearings, and design of platforms for automotive turbochargers at Bosch Mahle Turbo Systems in Stuttgart.

He is also the author of a professional engineering book *Aero and Vibroacoustics of Automotive Turbochargers, Springer Berlin-Heidelberg (2013)*; in addition, coauthor of a mathematical engineering book *Tensor Analysis and Elementary Differential Geometry for Physicists and Engineers, Springer Berlin-Heidelberg (2014)*.

Chapter 1
Turbocharging Concepts

1.1 Introduction

The average CO_2 emission for new passenger vehicles in the European Union countries (EU) is limited to 120 g/km from 2012 (65 % produced vehicles) to 2015 (100 %). This emission limit is reduced to the target of 95 g/km from 2020 (95 % produced vehicles) and 2021 (100 %). According to the emission law, the average emission of the produced passenger vehicles (PV) for a car manufacturer must be reduced by about 21 % from 2015 to 2020 (see Fig. 1.1).

The average CO_2 emission for the produced light-duty commercial vehicles (LDCV) is limited to 175 g/km from 2014 (70 % manufactured vehicles) to 2017 (100 %). Furthermore, it must be reduced from 175 g/km to 147 g/km (100 % manufactured vehicles) from 2020; i.e., nearly 16 % less CO_2 emission in 3 years, as shown in Fig. 1.2.

If car manufacturers do not fulfill the emission limit, they have to pay a high fine for the excessive amount of g/km CO_2 for each produced vehicle. The fine amount could be between 1000 and 1500 Euros for a new high-power vehicle. Therefore, the reduction of CO_2 emission has recently become one of the main tasks of car manufacturers in the EU.

In order to reduce carbon dioxide (CO_2) and nitrogen oxides (NO_x) exhausted by passenger and commercial vehicles and to improve the fuel consumption of the engines, many measures have been already carried out, e.g., high-pressure direct injection (HPDI), exhaust gas recirculation (EGR), variable valve train (VVT), variable compression ratio (VCR), turbocharged downsized engines (TC), and hybrid and electric vehicles (HEV) [3].

From 2012 more hybrid and electric PVs have been produced in the EU to reduce the average emission according to the current emission law. However, such vehicles remain at a very small number because they are still very expensive and the driving range is limited by 100–120 km, except Tesla's electric vehicles at

H. Nguyen-Schäfer, *Rotordynamics of Automotive Turbochargers*,
Springer Tracts in Mechanical Engineering, DOI 10.1007/978-3-319-17644-4_1

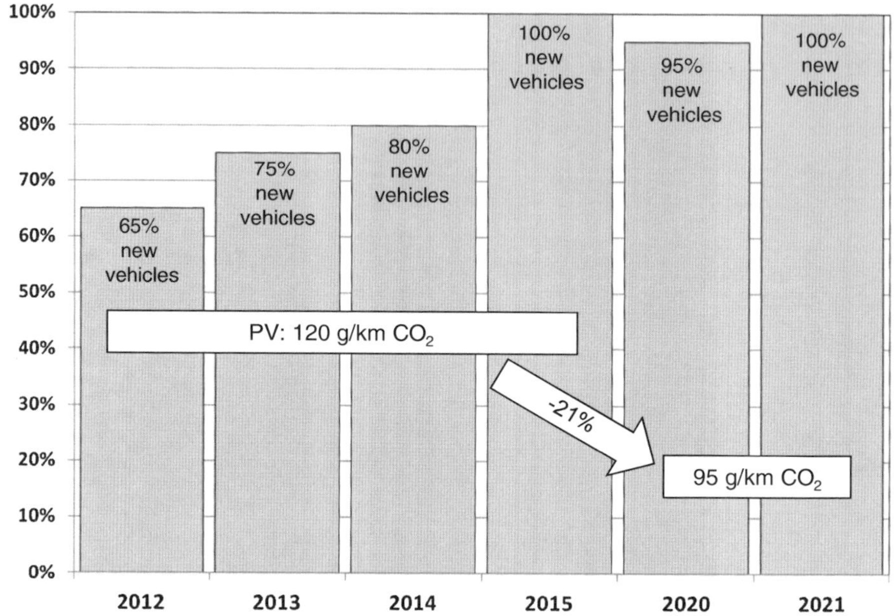

Fig. 1.1 Average CO_2 emission limit for passenger vehicles

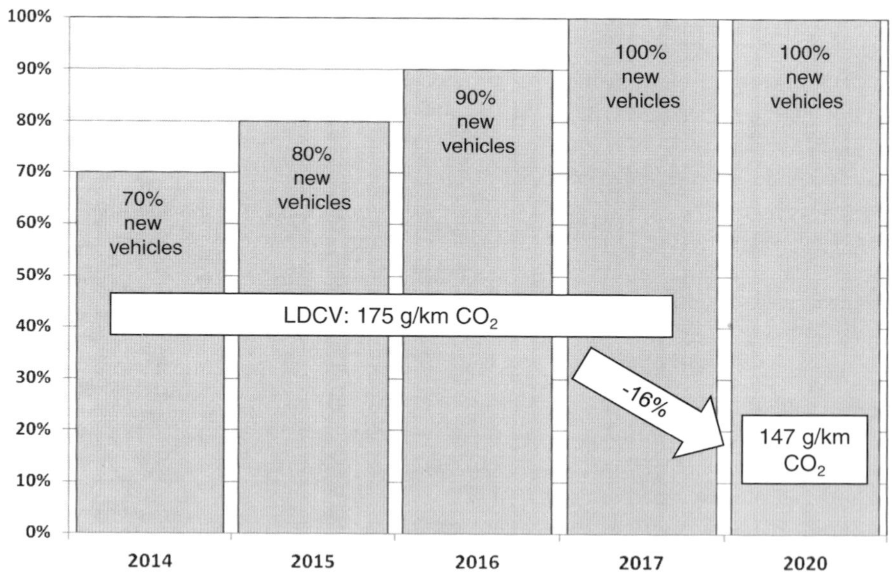

Fig. 1.2 Average CO_2 emission limit for light duty commercial vehicles

400–500 km range. It will be expected that the HEV will rise from 2020 in order to reduce about 21 % emission to 95 g/km CO_2 emission.

Furthermore, Toyota (in 2015), Nissan (2016), and Daimler (2017) will apply the technology of fuel cell vehicles (FCV) to zero-emission vehicles (ZEV).

In this chapter, we mainly consider the turbocharged downsized engines in which rotordynamics and bearings play an important role in automotive turbochargers. Two important aspects are *downsizing engines* by reducing the number of cylinders or volumetric size of cylinders and *turbocharging*. Engines with fewer numbers of cylinders or small cylinder volumes induce less friction between the pistons and cylinders. Additionally, the total weight of the vehicle is also reduced due to small engines leading to less driving friction. Evidently, small engines need less fuel consumption; in turn, they produce less engine power. Small engines consume less fuel and therefore produce less carbon dioxide (CO_2) as well as nitrogen oxides (NO_x). From the point of view of energy and air pollution, they have done a good job to sustain our energy resources and to keep the environment clean and less polluted.

However, it should be made more fun driving, especially with a large acceleration; it needs more power in small engines. Hence, the specific power defined as the engine power per cylinder volume (kW/liter) must be improved. Therefore, we have to capture the unused and cost-free energy source in vehicles to boost the small engines.

Figure 1.3 shows the average energy flux in a passenger vehicle. Only about 10–15 % of the input fuel energy is really used for driving power. The rest of the fuel energy is needed for thermal cooling (27–33 %), thermal exhaust gas (30–35 %), friction or load change loss (10–14 %), and miscellaneous supply powers of control devices, electric motors, and entertainment devices (2–4 %). The other energy losses (anergy) cannot be gained to empower the engine, except the thermal exhaust gas (partial exergy) of nearly 30–35 % of the fuel energy. We could use a part of this energy to increase the specific power of the engine leading to improving the total engine power and the engine efficiency. Without using hybrid technology, the maximum efficiency of a gasoline combustion engine is about 35 %, while that of a turbocharged diesel engine is nearly 40 %.

After combustion of fuel in the engine, a large energy of the exhaust gas still remains in the form of enthalpy at the engine outlet because the temperature of the exhaust gas is quite high (e.g., diesel 820–850 °C and gasoline 950–1050 °C). Without turbocharging, the enthalpy energy escapes from the engine to the environment. Why do we not capture it to boost the engine for improving its specific power? To do that, an exhaust-gas turbocharger is necessary to collect the exhaust gas in the turbine and compress the intake air in the compressor to a high pressure for the combustion process. This procedure is called turbocharging with which more intake air is compressed in the engine. As a result, more engine power is generated.

The exhaust-gas turbocharger consists of the core unit CHRA (center housing and rotating assembly), turbine, compressor, and actuator. Both compressor and turbine wheels are mounted and fixed on the rotor shaft, whose bearing system includes two radial bearings and thrust bearing supports in radial and axial

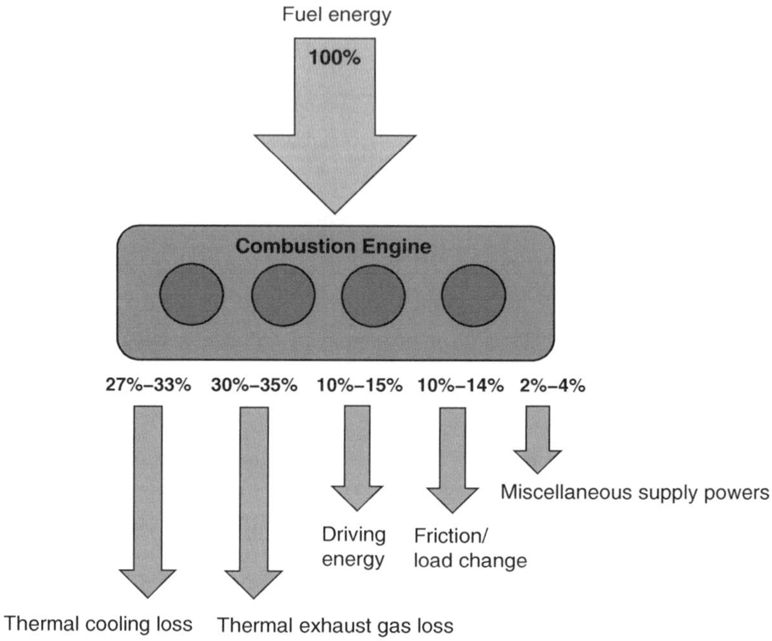

Fig. 1.3 Average energy flux of a passenger vehicle

directions, respectively (see Fig. 6.2). The exhaust gas expands in the turbine wheel and generates the rotational kinetic energy. The turbine energy drives the rotor and therefore compresses the intake air to a high pressure by means of the compressor.

The concept of using exhaust-gas turbochargers takes advantages of improving the specific power of engines and reducing CO_2 and NO_x. First, turbocharged downsized engines have nearly the same power as conventional engines (without downsizing and turbocharging) but much less fuel consumption leading to strong reduction of carbon dioxide and nitrogen oxides. Second, turbocharged engines without downsizing have more power than conventional engines but quite little fuel consumption for those who need more fun while driving (cf. Table 1.1). The first case is an important key to reduce the emission of small and middle class vehicles and to satisfy the driving fun as well.

In general, turbocharged downsized engines with the same power could save nearly 10 % fuel consumption by reducing the cylinder volume by 25 %. To abide by the current emission law of automotive vehicles or to boost engines of PV, two-stage and three-stage turbochargers are applied to engines operating at large compression ratios of the charge air (>4). Multi-stage turbochargers consist of many single-stage turbochargers of different volumetric sizes that are sequentially set up in the engine. Some applications of different turbocharger types in the engines are discussed in the next section.

Table 1.1 Comparison of engine types

Engine types	Conventional (without Turbocharging and downsizing)	Turbocharged and downsized	Turbocharged without downsizing
Engine power	0	0	+
Fuel efficiency	0	++	+
Driving fun	0	+	++
Reducing emission	0	++	0/+

1.2 Applications of Turbochargers to Downsized Engines

In automotive applications, single-stage turbochargers are normally used at a compression ratio of the charge air to nearly 2.5. There are two setups of turbochargers with Exhaust Gas Recirculation at high pressure (HP-EGR) and low pressure (LP-EGR) [1]. By means of the EGR to the engine, the burning temperature spike of the fuel–air mixture in the engine is lower leading to reduction of nitrogen oxides (NO_x).

(a) Single-Stage Turbochargers with EGR

Figure 1.4 shows a single-stage turbocharger with high-pressure EGR (HP-EGR) in which the exhaust gas partly returns to the engine with a maximum of 50 % EGR rate before entering the turbine. The EGR rate is defined as the ratio of the recirculation to outflow mass flow rates of the exhaust gas. The EGR valve regulates the EGR mass flow rate so that NO_x emission fulfills the current emission law. Due to expansion energy of exhaust gas, the turbine (T) propels the compressor (C) that compresses the intake air at the pressure ratio to nearly 2.5. The compression process in the compressor brings the inlet air from the ambient condition to higher pressure and temperature. To maintain the compressed charge air with a large mass flow rate for the engine combustion, the intercooler cools it behind the compressor. Otherwise, the mass flow rate of the charge air drops at high temperatures. In case of turbochargers with HP-EGR, the pressure of the recirculated exhaust gas behind the EGR-intercooler must be higher than the charge air pressure at the engine inlet. Hence, the exhaust gas pressure must be high enough to overcome the charge air pressure. Generally, turbines with HP-EGR are designed to be smaller than turbines without HP-EGR, so that the exhaust gas pressure remains higher than the compressed charge air pressure. Therefore, the engine must work against the high exhaust gas pressure at the engine outlet. As a result, the fuel consumption increases compared to the system without HP-EGR, especially at high turbocharging ratios. That is the reason why the HP-EGR valve should be closed as soon as the exhaust gas meets the requirements of the current emission law.

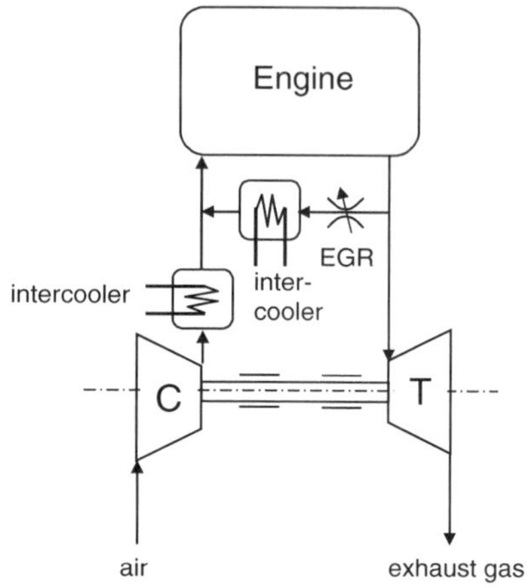

Fig. 1.4 Schematic setup of a single-stage turbocharger with HP-EGR

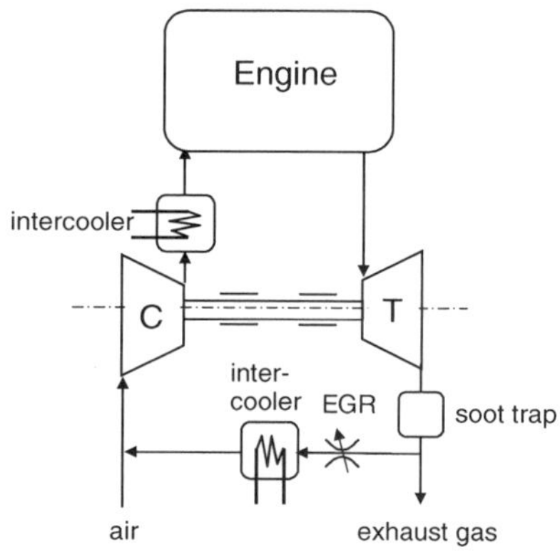

Fig. 1.5 Schematic setup of a single-stage turbocharger with LP-EGR

To overcome this disadvantage, the HP-EGR valve is moved in the low-pressure site at the turbine outlet. The setup of a turbocharger with LP-EGR is shown in Fig. 1.5. In this case, the exhaust gas pressure at the engine outlet is reduced leading to improving the engine operating condition. As a result, the fuel performance becomes more efficient.

However, the LP-EGR turbocharger has two disadvantages. First, the pressure drop in the soot trap limits the mass flow rate of EGR, and hence the EGR rate. Second, the exhaust gas and ambient air mix at the compressor inlet and are compressed in the compressor to higher pressure. Due to pressure increase in the compressor wheel, the exhaust gas condenses in the compressor wheel causing chemical erosion on the surface of the wheel. Additionally, the unburned hard particles in the exhaust gas impact and damage the aluminum compressor wheel at high rotor speeds (erosion), especially the blades at the inlet of the compressor wheel.

To prevent the compressor wheel from such damages, it is usually coated by a NiP (Nickel Phosphate) layer. However, there are two negative effects of the coating layer. First, its thickness causes a reduction in the air mass flow rate in the compressor wheel and, therefore, reduced engine power. Second, the weight of the compressor wheel is heavier due to the additional layer. As a result, the lifetime of the compressor wheel is shortened by low-cycle fatigue (LCF).

(b) Bi-turbochargers

Bi-turbo (twin-turbo) consists of two small turbochargers of the same volumetric size. Both turbochargers of the bi-turbo are parallel setup and operate at the same time in the entire rotor speed range (parallel bi-turbo). The mass flow rate of the exhaust gas is divided into both turbines of the bi-turbo by opening both shut-off valves. In case of an engine with four cylinders, two of them provide exhaust gas for each turbocharger, as shown in Fig. 1.6. After compression and intercooling, both compressors provide the engine with the total compressed charge air [2, 3].

At the small geometry of turbochargers, the rotor has small mass inertia moment that speeds up quickly to achieve the maximum torque. Therefore, the transient behavior of the parallel bi-turbo is strongly improved in low-end torque (LET). However, a small turbocharger delivers a small mass flow rate of the charge air that alone does not fulfill the required nominal engine power. Hence, both small turbochargers must parallel operate to deliver enough compressed charge air for the nominal engine power.

In case of a sequential bi-turbo that is parallel setup and sequentially operates, only one of the sequential bi-turbo, e.g., the left one operates alone at low engine speed by closing both shut-off valves. The small turbocharger builds up the charge-air pressure much earlier and improves the transient behavior of the turbocharger at low engine speed. At high engine speeds, the other turbocharger (the right one) is turned on by opening both shut-off valves. In this case, both turbochargers operate parallel like the parallel bi-turbo. Hence, the mass flow rate of the charge air increases to empower the engine at high rotor speeds.

An application of a sequential quad-turbo (four single turbochargers) is used for the strongly powered W-engine with 16 cylinders (4 × 4) in the super sports car Bugatti Veyron 16.4 with 1200 hp. The exhaust gas of each of the four cylinders supplies one turbocharger of the sequential quad-turbo.

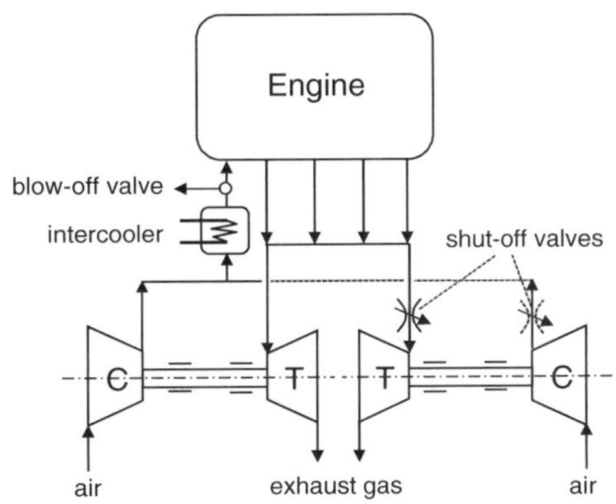

Fig. 1.6 Schematic setup of a bi-/twin-turbo

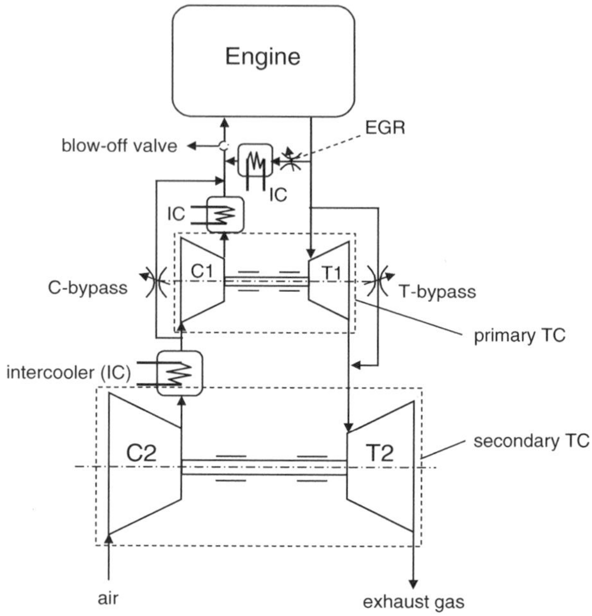

Fig. 1.7 Schematic setup of a regulated two-stage turbocharger

(c) Two-Stage Turbochargers

Two-stage turbochargers are used for HP ratios in which two different volumetric sizes of turbochargers are sequentially setup and operate at the procedure regulated by the bypass valves (see Fig. 1.7). Leaving the engine, the exhaust gas contains a large enthalpy at high pressure and temperature. Therefore, the primary turbocharger (C1, T1) is smaller than the secondary turbocharger (C2, T2). In case of unregulated operating condition in applications of commercial vehicles, the primary turbocharger is about 15 % smaller than the single-stage turbocharger; the secondary turbocharger is 15 % larger than the single-stage one [2].

At low engine speeds, only the primary turbocharger (small one) operates with its small mass inertia moment. Therefore, the transient behavior is significantly improved compared to the secondary turbocharger (large one). During this time, both bypass valves of the primary turbocharger are closed and the waste gate the secondary turbine is fully opened so that the secondary turbocharger is decoupled from the two-stage turbocharger. At the middle engine speeds from about 1500 rpm, both turbochargers operate sequentially by means of regulated bypass valves. As soon as the required charge-air pressure is reached at high engine speeds, only the secondary turbocharger operates since the larger turbine T2 has high efficiency at high rotor speeds. In this case, the primary turbocharger is decoupled from the two-stage turbocharger where both bypass valves of the primary turbocharger are fully opened.

Similar to other types of turbochargers, the compressed charge air should be intercooled before entering the engine in order to maintain the high mass flow rate of the charge air for the required nominal engine power.

The pressure ratio of the two-stage turbocharger is displayed in Fig. 1.8. The compressor pressure ratio in the full load curve results from the product of the pressure ratios of both compressors of the two-stage turbocharger. Due to the small volumetric size of the primary turbocharger, its pressure load curve speeds up faster at low engine speeds in order to receive a good transient response in low-end torque.

The engine torques of the single-stage turbocharger and two-stage turbochargers are compared to each other over the rotor speed, as illustrated in Fig. 1.9. The single-stage turbocharger is larger than the primary but smaller than the secondary turbocharger of the two-stage turbocharger. Therefore, the transient response of the two-stage is better than the single-stage. The advantages of the two-stage turbocharger are better response behavior, HP ratio of the compressed charge air, and higher engine torque in the whole operating speed range.

Fig. 1.8 Pressure ratio of a regulated two-stage turbocharger

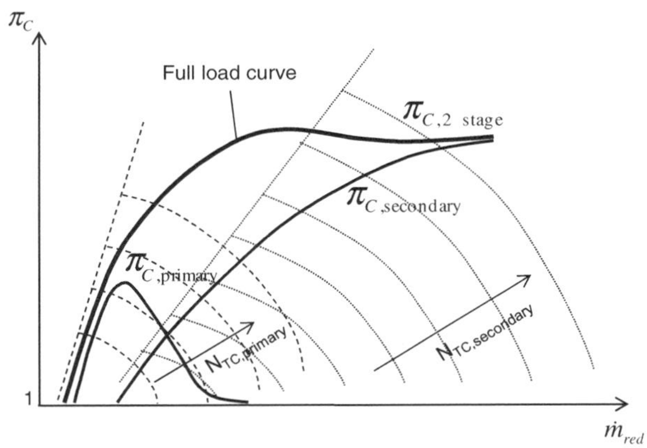

Fig. 1.9 Engine torques of the single- and two-stage turbocharger

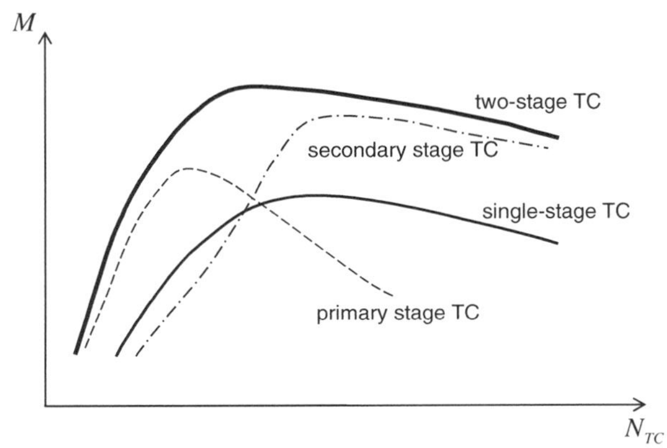

(d) Three-Stage Turbochargers

Three-stage turbochargers are used for extremely high driving torques at low-end speeds and an HP ratio to 3.0. Three different volumetric sizes of turbochargers are sequentially setup and regulated by the bypass valves so that the turbochargers operate sequentially and parallel as well.

The first commercial three-stage turbocharger is applied to the three-liter diesel engine BMW M550d X-drive with 280 kW (nearly 375 HP) at 4000 rpm and a maximum driving torque of 740 Nm at 2000 rpm. The tri-turbocharger consists of two HP primary and supplementary turbochargers (TC1 and TC3) using variable turbine geometry (VTG) and one low-pressure (LP) secondary turbocharger (TC2) using waste gate (WG), as displayed in Fig. 1.10.

Due to three-stage turbocharging, the charge-air temperature is very high causing negative impacts on the engine, such as knocking problem, increasing NO_x, reducing the charge-air density leading to reduction of engine power and overall efficiency. To overcome these problems, liquid water is directly injected into the charge air behind the last intercooler of the compressor C1 (see Fig. 1.10). The injected water evaporates using the charge-air enthalpy. As a result, the charge-air temperature strongly drops before entering the engine.

The working principle of the tri-turbocharger is briefly described as follows (cf. Fig. 1.11):

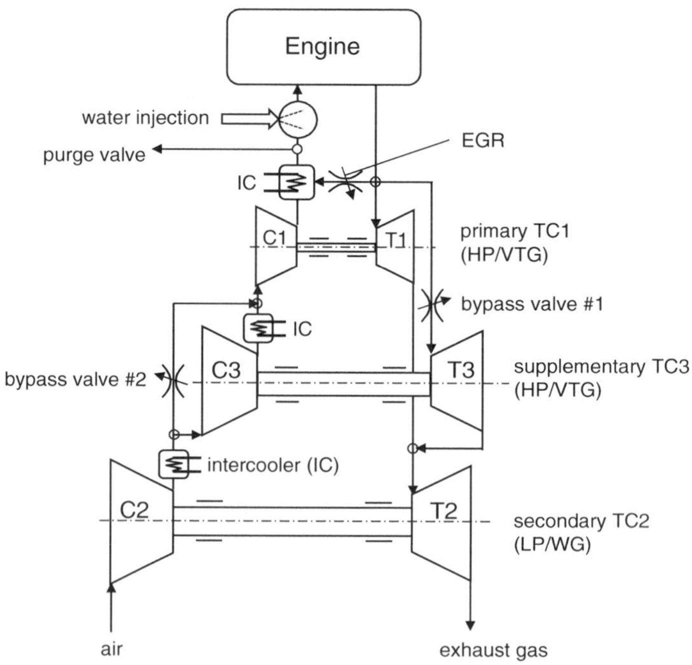

Fig. 1.10 Schematic setup of a regulated three-stage turbocharger

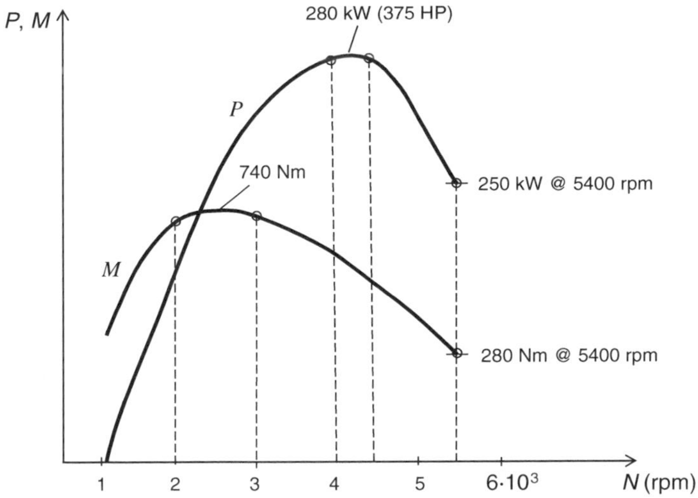

Fig. 1.11 Engine torque and power versus engine speed of BMW M550d

- At low engine speeds to 1500 rpm, the smallest primary turbocharger TC1 (HP/VTG) operates alone where the bypass valve #1 is closed, the bypass valve #2 is opened, and the WG of the secondary TC2 is fully opened. Due to small primary TC1, the driving torque increases much more quickly at low engine speeds leading to very good dynamic response.
- From the engine speeds of 1500–2700 rpm, the working principle is similar to the two-stage turbocharger. The primary (HP/VTG) and secondary (LP/WG) turbochargers TC1 and TC2 operate sequentially in which the bypass valve #1 is closed and the bypass valve #2 is opened. In this case, the waste gate (WG) of the secondary TC2 is regulated to control the charge-air pressure. At engine speed of about 2000 rpm, the engine reaches the maximum torque of 740 Nm, which remains nearly constant until 3000 rpm.
- At engine speed over 2700 rpm, three turbochargers TC1, TC2, and TC3 operate sequentially and parallel where the bypass valve #1 is regulated to control the bypassed mass flow rate of the exhaust gas to the supplementary TC3, the bypass valve #2 is closed, and the WG of the secondary TC2 is regulated. In this case, the mass flow rate of the exhaust gas at high temperature through the bypass valve #1 operates the supplementary TC3. As a result, the compressor C3 additionally increases the charge-air pressure in order to improve the engine power to the maximum power of 280 kW at engine speed from 4000 to 4400 rpm.
- At very high engine speeds over 4000 rpm, the driving torque and engine power are still very high with 280 Nm and 250 kW at engine speed of 5400 rpm. Using this three-stage turbocharger, the BMW M550d X-drive needs only 5 s to accelerate the car from zero to 100 km/h and about 20 s from zero to 200 km/h.

(e) Electrically Driven Turbochargers in Fuel Cell Vehicles

Automotive fuel cells are usually PEMFC type (Proton Exchange Membrane Fuel Cells). Gaseous hydrogen H_2 is supplied to the anode of the membrane electrode assembly (MEA) of the fuel cell. At the anode side, hydrogen molecules H_2 are catalytically split into hydrogen protons H^+ and electrons e^-. The protons cross over the humid membrane to the cathode; the electrons flow over the external load circuit to the cathode side of MEA. Oxygen in the supply air reacts with the crossed-over hydrogen protons and the electrons in the cathode, resulting in exhaust gases, such as N_2, residual O_2, and water steam H_2O.

The transport of electrons between the anode and cathode of the MEA produces electric power at 250–450 V DC. The generated electric power depends on the reaction pressure of the supply air and hydrogen in the fuel cell and other parameters. Therefore, the charge-air pressure should be 1.5–2 bars absolute so that the fuel cell operates in optimum condition. In this case, the single-stage exhaust gas turbocharger compresses ambient air into the pressurized charge air. However, at the low power operating condition, the exhaust gas temperature is relatively low (between 80 and 100 °C) and its mass flow rate is quite small. Hence, an electric motor (EM) is additionally required besides the turbine wheel to drive the compressor at the low engine speeds. At increasing fuel cell power, the mass flow rate of the exhaust gas increases. As a result, the turbine wheel alone generates enough energy to drive the compressor without using the electric motor. In this case, the electric motor is turned off at once in order to spare the generated fuel cell power.

To avoid electric sparks and wear in the coal burst, the electric motor must work with a high alternative current supply voltage (VAC) that is inverted from the direct current voltage (VDC) of the fuel cell by a DC/AC inverter, as shown in Fig. 1.12. The electrically driven turbocharger (EDTC) normally operates at the maximum rotor speed of approximately 120,000 rpm due to material stability of its large compressor and turbine wheels. Additionally, a high electric motor power is required for the transient response τ_{90} (about 0.8 s) of the fuel cell. As a result, the rotor speed is limited by about 120,000 rpm to obtain the material stability and good overall efficiency of the EDTC.

(f) Turbo-Compound

Turbo-compounds consist of a turbocharger and power turbine (PT) that uses the exhaust gas from the turbocharger at a still high temperature, to generate additional power for the engine. Using a transmission gearbox, the power turbine shaft is directly geared to the engine drive shaft to increase the engine torque and power to help increasing the engine efficiency, as shown in Fig. 1.13. Turbo compounds are possibly applied in commercial vehicles to gain some additional driven energy to the engine shaft.

Furthermore, the turbo-compound can be used to generate the electric energy for other applications in commercial vehicles. The power turbine drives a generator (G) that generates the electric energy to charge the electric vehicle battery (Li-Ion EVB) in hybrid commercial vehicles. Using the PWM method (pulse-width modulation), the DC/AC inverter converts two-phase current of the battery direct into three-phase

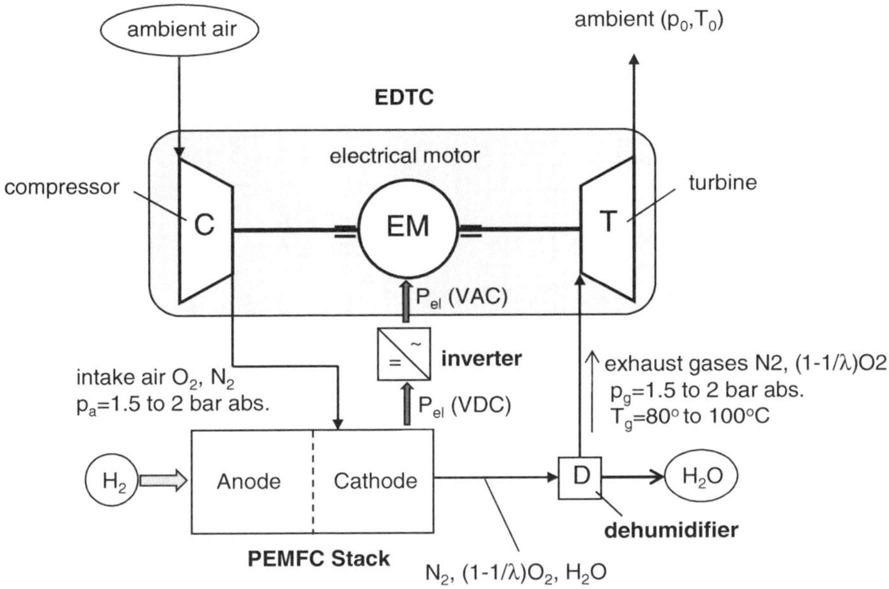

Fig. 1.12 Schematic setup of EDTC in PEM-Fuel Cells

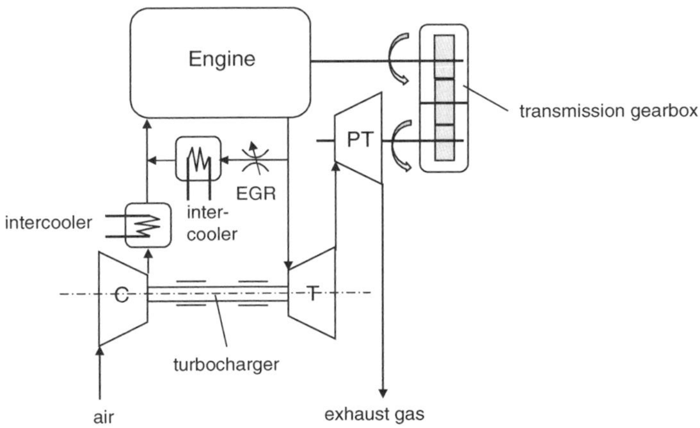

Fig. 1.13 Turbo-compound used in commercial vehicles

current for the electric machine, as shown in Fig. 1.14. The electric machine is used to drive light duty commercial vehicles (LDCV), such as delivery and emergency vehicles, for a short distance within the city. For a long distance on the freeway, the combustion engine operates and charges the Li-Ion EVB for the next trip within the city.

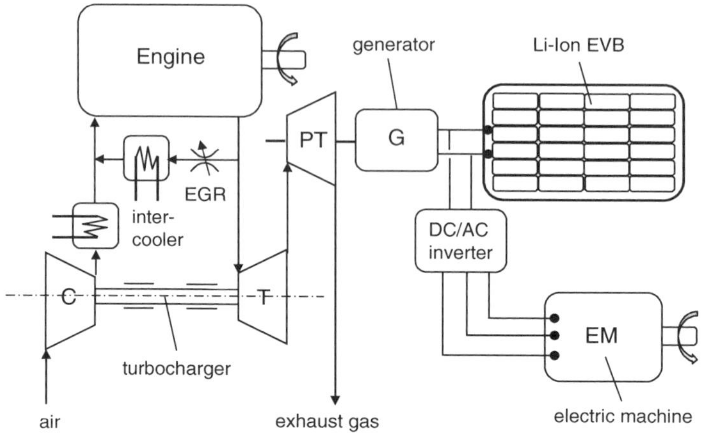

Fig. 1.14 Turbo-compound used in hybrid vehicles

Using the combustion engine and electric machine, hybrid vehicles can contribute to reducing the average emission of the whole vehicle fleet and therefore keep the city environment clean and green.

1.3 Regulation of Charge-Air Pressure

Three kinds of regulation of charge-air pressure are applied in automotive and industrial turbochargers using unregulated, wastegated (WG), and variable turbine geometry (VTG) turbochargers. The pressure ratio of the charge air varies from 2.0 to 2.5 in PV and from 3.5 to 4.0 in commercial vehicles [5].

There is neither waste gate (WG) nor variable turbine geometry (VTG) in the unregulated turbochargers that are mostly used in commercial vehicles. Figure 1.15 shows the unregulated pressures of the exhaust gas and charge air in the engine. At increasing engine speeds, more charge air is required, therefore, the charge air pressure is increased from the ambient pressure to the maximum pressure for the engine combustion. The charge-air pressure is always higher than the exhaust gas pressure in the unregulated turbochargers in the whole engine speed range.

In case of regulation of the charge air pressure, two measures are used in turbochargers: waste gate (WG) and variable turbine geometry (VTG). The pressure behavior of the wastegated turbocharger is shown in Fig. 1.16. As soon as the turbocharger reaches the maximum torque, the waste gate in the turbine is continuously opened to regulate the charge-air pressure remaining nearly constant. At high engine speeds, the exhaust gas pressure increases higher than the charge-air pressure since the flow resistance of the exhaust gas in the catalytic converter (CAT), diesel particle filter (DPF), and muffler increases with the gas mass flow rate.

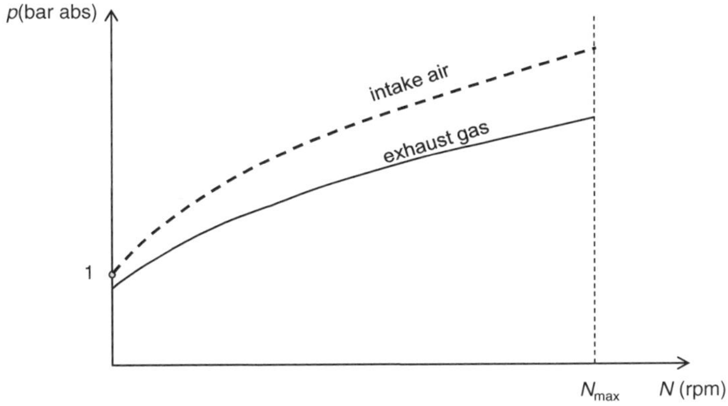

Fig. 1.15 Unregulated pressures versus engine speed

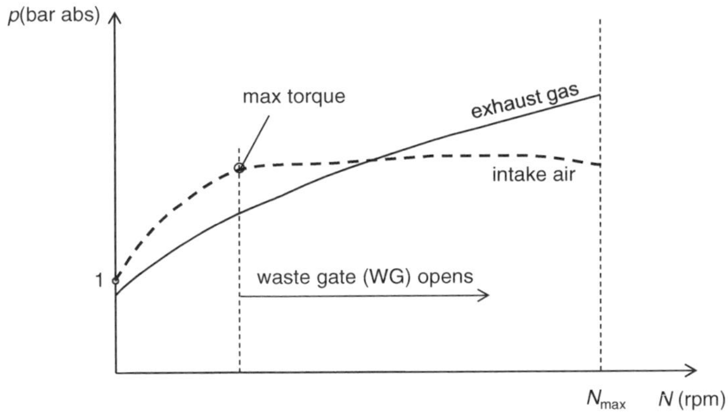

Fig. 1.16 Regulated pressures of a wastegated turbocharger

Figure 1.17 shows the pressure behavior of a VTG turbocharger versus the engine speed. At low engine speeds, the VTG is in the minimum open position (min-flow position). Therefore, the exhaust gas pressure rises higher than the charge-air pressure. After reaching the maximum torque, the VTG continuously further opens at increasing engine speeds in order to regulate the charge-air pressure remaining constant. Shortly after opening the VTG, drops the exhaust gas pressure a little lower than the charge-air pressure. Then, it increases over the charge-air pressure at the high engine speeds. The pressure drop of the exhaust gas results from suddenly opening the VTG. After that, the VTG continuously opens at increasing the mass flow rates of the exhaust gas to the maximum engine speed to keep the charge-air pressure quasi constant. By increasing the engine speeds, more exhaust gas is produced leading to increase of the exhaust gas pressure due to increasing the flow resistance.

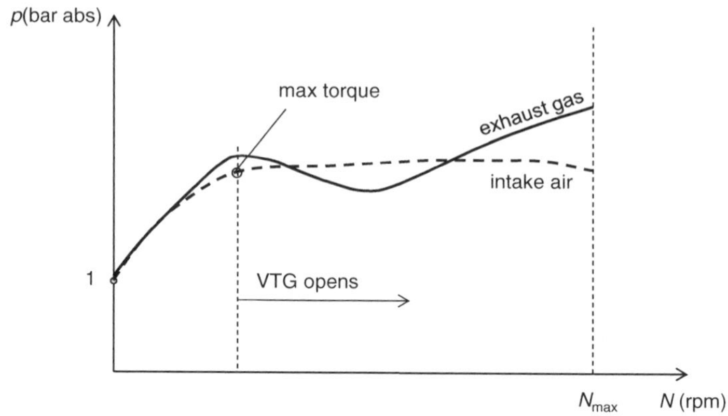

Fig. 1.17 Regulated pressures of a VTG turbocharger

1.4 Required Charge-Air Pressure of Downsized Engines

Depending on the engine power, cylinder volumetric size, and number of cylinders, the charge-air pressure is determined for any engine speed.

The engine power is calculated as [3, 4]

$$P_e = \dot{m}_f \eta_f Q_f = \dot{m}_a \eta_f Q_f \frac{\dot{m}_f}{\dot{m}_a} \tag{1.1}$$

where;

$\dot{m}_f, \dot{m}_a$ are the mass flow rates of fuel and charge air;
η_f is the fuel combustion efficiency;
Q_f is the fuel heating value.

The air-fuel ratio is defined as

$$AFR \equiv \frac{\dot{m}_a}{\dot{m}_f} \tag{1.2}$$

The air–fuel ratio varies from 12 to 18 in applications to SI engines (spark-ignition gasoline engines) and from 18 to 70 to CI engines (compression-ignition diesel engines).

The mass flow rate of the charge air required for the engine results from the air density ρ_a, cylinder total volume V_{cyl}, and engine speed N.

$$\dot{m}_a = \eta_{vol} \rho_a V_{cyl} \frac{N}{n_R} = m_a \frac{N}{n_R} \tag{1.3}$$

within

$V_{cyl} = zV_c$ where z is the number of cylinders,
V_c is the single cylinder volume;
η_{vol} is the volumetric efficiency;
m_a is the real mass of the charge air in the cylinders;
N is the engine speed (in rps: rounds per second);
$n_R = 1$ for two-stroke; = 2 for four-stroke engine.

The λ number (relative air-fuel ratio) is defined as the ratio of the real air mass flow rate to the air mass flow rate required for the stoichiometric combustion.

$$\lambda = \frac{\dot{m}_a}{\dot{m}_{a,stoich}} \tag{1.4}$$

The relative air–fuel ratio can be written in the air–fuel ratio AFR and the stoichiometric air-fuel ratio as

$$\lambda = \frac{AFR}{AFR_{stoich}} \tag{1.5}$$

where

$\lambda = 1$ for a stoichiometric combustion (ideal combustion process);
$\lambda < 1$ for a rich mixture (more unburned hydrocarbons in exhaust gas);
$\lambda > 1$ for a lean mixture (more residual oxygen in exhaust gas).

In a stoichiometric combustion of gasoline with octane fuel (C_8H_{18}), AFR_{stoich} equals 14.7. In applications of SI gasoline engines, the λ number is chosen between 0.95 and 1.05 ($\lambda = 1 \pm 5\ \%$) for optimal specific fuel consumption and for reducing the exhaust gas emission in the three-way catalytic converter.

Substituting Eqs. (1.1, 1.2, and 1.3), the engine power becomes

$$P_e = \left(\eta_{vol}\rho_a V_{cyl}\frac{N}{n_R}\right)\eta_f Q_f \frac{1}{AFR} \tag{1.6a}$$

The specific engine power is defined as the engine power per cylinder volume.

$$P_{e,sp} \equiv \frac{P_e}{V_{cyl}} = \left(\eta_{vol}\rho_a \frac{N}{n_R}\right)\eta_f Q_f \frac{1}{AFR} \tag{1.6b}$$

In case of downsizing engines by means of smaller cylinder volume V_{cyl} (less number of cylinders or smaller volume of each cylinder), the air density must be increased so that the required nominal engine power is achieved according to Eq. (1.6a).

Downsized engines have less friction between the cylinders and pistons. Hence, the fuel mileage MPG (miles per gallon) is improved. In the turbocharged engines without downsizing, the cylinder volume remains unchanged; however, the engine power increases due to high density of the charge air.

The density of charge air is increased by means of turbocharging. Assumed that the charge air is a perfect gas, the charge air pressure results in

$$p_a = \rho_a R_a T_a \tag{1.7}$$

where

ρ_a is the density of charge air;
R_a is the gas constant of charge air;
T_a is the temperature of charge air.

Inserting Eq. (1.7) into Eq. (1.6a), the engine power is calculated at the engine speed N as [4]

$$P_e = \left(\frac{\eta_{vol} p_a V_{cyl}}{R_a T_a} \cdot \frac{N}{n_R} \right) \eta_f Q_f \frac{1}{AFR} \tag{1.8}$$

Using the isentropic process, the charge-air temperature is given in

$$T_a = T_1 \left(1 + \frac{1}{\eta_C} \left[\left(\frac{p_a}{p_1} \right)^{\frac{\kappa_a - 1}{\kappa_a}} - 1 \right] \right) \tag{1.9}$$

where

T_1 is the inlet temperature of the intake air;
p_1 is the inlet pressure of the intake air;
p_a is the charge-air pressure;
η_C is the isentropic compressor efficiency;
κ_a is the isentropic exponent of the charge air ($\approx$ 1.4).

Using the state equation of a perfect gas given in Eq. (1.7), the density ratio of the charge air to the ambient air results as

$$\begin{aligned} \frac{\rho_a}{\rho_1} &= \left(\frac{p_a}{p_1} \right) \left(\frac{T_a}{T_1} \right)^{-1} \\ &= \frac{\pi_C}{1 + \frac{1}{\eta_C} \left[\pi_C^{\left(\frac{\kappa_a - 1}{\kappa_a} \right)} - 1 \right]} \end{aligned} \tag{1.10}$$

where π_C is the compression ratio of the compressor.

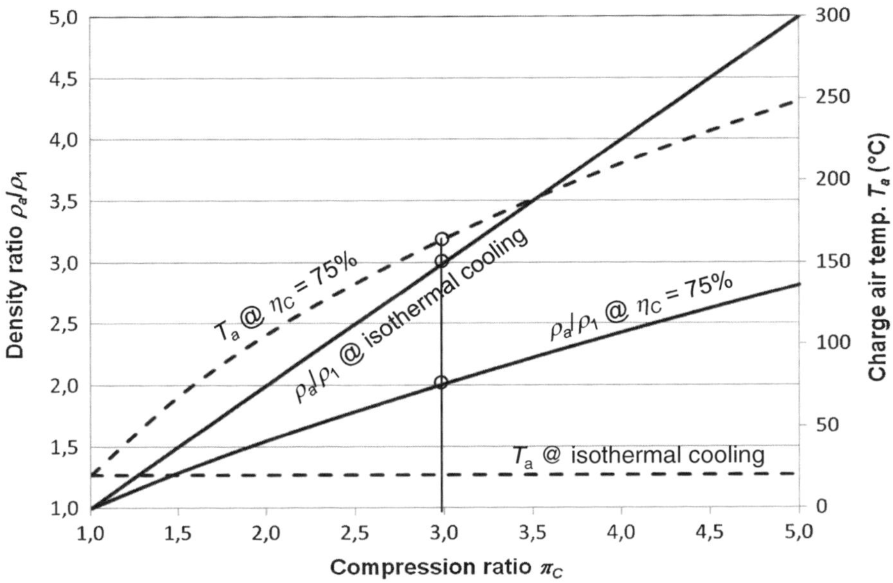

Fig. 1.18 Characteristics of the charge air versus compression ratio

Figure 1.18 shows the density ratio and temperature of the charge air versus compression ratio at ambient temperature of 20 °C. The charge-air density without intercoolers increases to twice the intake-air density at compression ratio of 3 and average compressor efficiency of 75 %. In this case, the charge-air temperature T_a rises from the ambient temperature T_1 of 20 °C to nearly 165 °C.

At lower compressor efficiency, the charge-air temperature is higher and therefore its density is lower leading to smaller mass flow rate of the charge air for the engine. As a result, the engine power is reduced according to Eq. (1.6a). By contrast, at isothermal intercooling after compression, the charge-air density at the compression ratio of 3 is triple the ambient air density; i.e., 50 % more than the charge-air density without intercoolers. In practice, the charge-air density ratio at the compression ratio of 3 is between 2 and 3 with using a charge-air intercooler in the turbocharger (see Sect. 2.5).

The engine torque M_e results from the engine power P_e and rotational speed ω_e.

$$P_e = M_e\omega_e = 2\pi N M_e \tag{1.11}$$

Thus,

$$\begin{aligned} M_e &= \left(\frac{\eta_{vol} p_a V_{cyl}}{2\pi R_a T_a n_R}\right)\eta_f Q_f \frac{1}{AFR} \\ &= \left(\frac{\eta_{vol} \rho_a V_{cyl}}{2\pi n_R}\right)\eta_f Q_f \frac{1}{AFR} \end{aligned} \tag{1.12}$$

Equations (1.8 and 1.12) indicate that the engine power P_e and engine torque M_e do not increase much even at the high charge-air pressure p_a when the charge-air temperature T_a increases at the same time. That is the reason why the charge air must be intercooled behind the compressor, as shown in Sect. 1.2. Note that the charge-air density ρ_a is the key issue in turbocharging of engines, and not the charge-air pressure p_a!

To satisfy the requirement of the engine power and engine torque of downsized engines, the charge-air density must be increased by raising the charge-air pressure and dropping its temperature at the same time according to Eqs. (1.8, 1.9, and 1.10). Therefore, the turbocharger with intercoolers is used for increasing the charge-air mass flow rate in the turbocharger.

The mean effective pressure p_{me} of the engine is used to characterize the engine performance. It is defined as the engine work produced in a cycle per cylinder volume displaced in the same cycle.

$$p_{me} = \frac{W_e}{V_{cyl}} = \frac{P_e n_R}{V_{cyl} N} \tag{1.13}$$

In order to maintain the same engine power of the downsized engine by reducing the cylinder volume, the mean effective pressure p_{me} must be increased by turbocharging.

$$P_e = \frac{p_{me} V_{cyl} N}{n_R} \tag{1.14}$$

To estimate the fuel consumption of the engine, the brake-specific fuel consumption (*bsfc*) is defined as the required fuel mass flow rate per engine power.

$$bsfc = \frac{\dot{m}_f}{P_e} \tag{1.15}$$

The unit for the brake-specific fuel consumption *bsfc* is g/(kW s) or g/kJ.

References

1. Baines, NC.: Fundamentals of Turbocharging. Concepts ETI, Inc., White River Junction (2005)
2. Basshuysen, R., et al.: Gasoline Engine with Direct Injection. Vieweg and Teubner, Berlin (2009)
3. Golloch, R.: Downsizing bei Verbrennungsmotoren. Springer, New York (2005)
4. Heywood, J.B.: Internal Combustion Engine Fundamentals. McGraw-Hill, New York (1988)
5. Japikse, D., Baines, N.C.: Introduction to Turbomachinery. Concepts ETI, Inc. White River Junction (1994)

Chapter 2
Thermodynamics of Turbochargers

2.1 Thermodynamic Characteristics

Some essential thermodynamic characteristics of gases are needed to know in the turbocharging. They are usually applied to thermodynamics of turbochargers where the charge air and exhaust gas are assumed to be compressible perfect gases [1–4].

- The total temperature T_t(K) results from the sum of the static T_s(K) and dynamic temperatures T_{dyn}(K). The static temperature is measured at the wall where the gas velocity equals zero due to the viscous boundary layer.

$$T_t = T_s + T_{dyn} = T_s + \frac{c^2}{2c_p} \tag{2.1}$$

where c is the gas velocity; c_p is the heat capacity at constant pressure.

- The total pressure p_t is calculated from the isentropic gas equation as follows:

$$p_t = p_s \left(\frac{T_t}{T_s}\right)^{\frac{\kappa}{\kappa-1}} = p_s \left(1 + \frac{\kappa - 1}{2} M^2\right)^{\frac{\kappa}{\kappa-1}} \tag{2.2}$$

where

p_s	is the static pressure;
$\kappa = c_p/c_v$	is the isentropic exponent of gas;
M	is the Mach number of gas ($M = c/a$), a is the sound velocity

- The specific total enthalpy h_t results from the sum of the gas specific enthalpy and specific kinetic energy of gas.

$$h_t = h + \frac{c^2}{2} \tag{2.3}$$

H. Nguyen-Schäfer, *Rotordynamics of Automotive Turbochargers*,
Springer Tracts in Mechanical Engineering, DOI 10.1007/978-3-319-17644-4_2

where the gas specific enthalpy h (enthalpy per mass unit, J/kg) is defined as

$$\begin{aligned} h(T) &= c_p(T - T_0) + h(T_0) \\ &= u(T) + \frac{p}{\rho} = u(T) + R_g T \end{aligned} \tag{2.4a}$$

within ρ is the gas density; T_0 is the reference temperature (K); u is the specific internal energy; R_g is the gas constant; $h(T_0) \equiv 0$ at $T_0 = 0$ K (= −273.14 °C).

The specific internal energy u is defined as

$$u(T) - u(T_0) = c_v(T - T_0) = c_v \Delta T \tag{2.4b}$$

where $u(T_0) \equiv 0$ at $T_0 = 0$ K.

Thus, $u(T) = c_v T$; $h(T) = c_p T$; T in K.

2.2 Efficiencies of Compressor and Turbine

The compression process in the compressor is a polytropic process with increasing entropy due to friction and losses in the compressor. Figure 2.1 shows the compression process of the intake air from the state 1 at the compressor inlet (p_1, T_1) to state 2 at the compressor outlet (p_2, T_2).

The compressor efficiency η_C is defined as the ratio of the isentropic total enthalpy change from 1t to 2st to the polytropic total enthalpy change from 1t to 2t. In other words, the compressor needs more energy in the polytropic process (real process) than the possibly minimal required energy of the compressor stage in the isentropic process (ideal process) [2, 5, and 6].

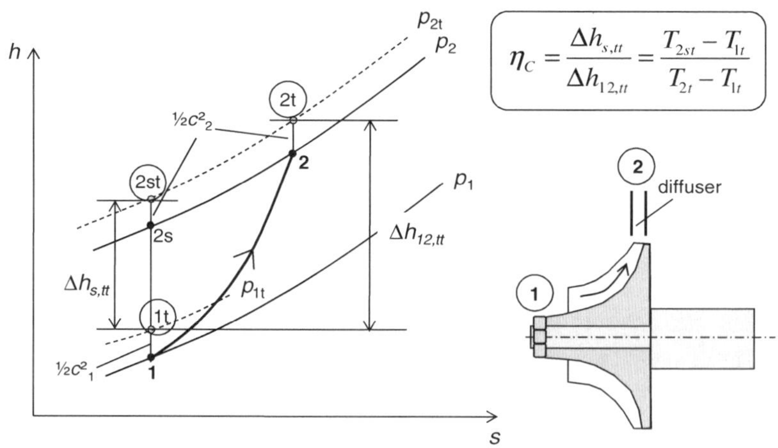

Fig. 2.1 Compression process in the compressor stage in *h-s* diagram

The total–total isentropic efficiency of the compressor stage (further called compressor) consisting of the compressor wheel and diffuser is defined as

$$\eta_C = \frac{\Delta h_{s,tt}}{\Delta h_{12,tt}} = \frac{T_{2st} - T_{1t}}{T_{2t} - T_{1t}} \tag{2.5}$$

The total–total isentropic efficiency is generally used in the compressor since the kinetic energy of gas in the state 2 could be transformed into the pressure energy in the diffuser. The process increases the charge-air pressure.

Using thermodynamic equations for the isentropic process, the compressor efficiency is written in terms of the total pressures and temperatures at the inlet and outlet of the compressor, and the isentropic exponent of the charge air $\kappa_a \approx 1.4$.

$$\eta_C = \frac{\left(\frac{p_{2t}}{p_{1t}}\right)^{\left(\frac{\kappa-1}{\kappa}\right)_a} - 1}{\left(\frac{T_{2t}}{T_{1t}}\right) - 1} \tag{2.6}$$

The compressor efficiency is determined by measuring the total pressures and temperatures at the inlet and outlet of the compressor according to Eq. (2.6). The maximum total–total isentropic efficiency of the compressor η_C is normally between 70 and 80 % at the design point in the compressor performance map.

Analogous to the compressor, the efficiency of turbine results from the polytropic expansion process of the exhaust gas from the state 3 at the turbine inlet (p_3, T_3) to state 4 at the turbine outlet (p_4, T_4), see Fig. 2.2. The turbine efficiency η_T is defined as the ratio of the polytropic total enthalpy change from 3t to 4t to the isentropic total enthalpy change from 3t to 4s. Physically speaking, the turbine delivers less output energy due to friction and losses in the polytropic expansion process than the possibly maximum energy given in the isentropic process.

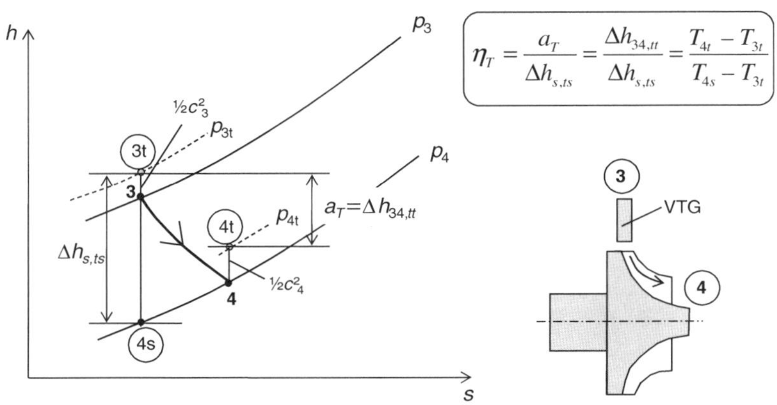

Fig. 2.2 Expansion process in the turbine stage in *h-s* diagram

The total-static isentropic efficiency of the turbine stage (further called turbine) consisting of the turbine wheel and variable turbine geometry (VTG) or waste gate (WG) is defined as [1, 3]

$$\eta_T = \frac{\Delta h_{34,tt}}{\Delta h_{s,ts}} = \frac{T_{4t} - T_{3t}}{T_{4s} - T_{3t}} \tag{2.7}$$

The total-static isentropic efficiency is normally used in the turbine since the kinetic energy of the exhaust gas in the state 4 does not generate any additional power for the turbine.

Applying thermodynamic equations to the isentropic process, the turbine efficiency is expressed in terms of the total pressure and temperature at the turbine inlet and outlet and the isentropic exponent of the exhaust gas $\kappa_g \approx 1.32$.

$$\eta_T = \frac{1 - \left(\frac{T_{4t}}{T_{3t}}\right)}{1 - \left(\frac{p_{4s}}{p_{3t}}\right)^{\left(\frac{\kappa - 1}{\kappa}\right)_g}} \tag{2.8}$$

The turbine efficiency is determined by measuring the total pressure and temperature at the turbine inlet and outlet according to Eq. (2.8). The maximum total-static isentropic efficiency of the turbine η_T is normally between 65 and 70 % at the design point of the turbine performance map [5–7].

2.3 Turbocharger Equations

The turbocharger consists of the turbine, compressor, and core unit (CHRA: center housing and rotating assembly) including the rotor and bearing system. Both turbine and compressor wheels are fixed in the rotor shaft that is supported on the bearing system of the radial and thrust bearings. The rotating shaft including the compressor wheel, turbine wheel, thrust rings, radial bearings, and seal rings is called the *rotor* of the turbocharger.

Due to expanding the exhaust gas of the engine in the turbine, it generates the turbine power that depends on the mass flow rate of the exhaust gas through the turbine and the isentropic enthalpy drop in the turbine. The effective turbine power results as

$$P_T = \eta_T \dot{m}_T |\Delta h_{sT}| \tag{2.9}$$

The isentropic enthalpy drop in the turbine stage is calculated using thermodynamic equations.

$$|\Delta h_{sT}| = c_{p,g}T_3\left[1 - \left(\frac{p_4}{p_3}\right)^{\left(\frac{\kappa-1}{\kappa}\right)_g}\right] \tag{2.10}$$

Inserting Eq. (2.10) into Eq. (2.9), one obtains the effective turbine power in function of the mass flow rate, inlet temperature, and pressure ratio of the turbine.

$$P_T = \eta_T P_{T,ideal} \equiv \eta_T \dot{m}_T c_{p,g} T_3\left[1 - \left(\frac{p_4}{p_3}\right)^{\left(\frac{\kappa-1}{\kappa}\right)_g}\right] \tag{2.11}$$

Due to the friction loss in the bearing system, the required compressor power results from the effective turbine power and mechanical efficiency η_m.

$$P_C = \eta_m P_T = \eta_m \eta_T \dot{m}_T c_{p,g} T_3\left[1 - \left(\frac{p_4}{p_3}\right)^{\left(\frac{\kappa-1}{\kappa}\right)_g}\right] \tag{2.12}$$

Analogously, the required compressor power is calculated from the isentropic compressor power and compressor efficiency.

$$P_C = \frac{P_{C,ideal}}{\eta_C} \equiv \frac{\dot{m}_C \Delta h_{sC}}{\eta_C} \tag{2.13}$$

where Δh_{sC} is the increase of the isentropic enthalpy in the compressor.

Using thermodynamic equations for an isentropic process, the required compressor power is calculated from the mass flow rate, inlet temperature, and pressure ratio of the compressor.

$$P_C = \frac{\dot{m}_C c_{p,a} T_1}{\eta_C}\left[\left(\frac{p_2}{p_1}\right)^{\left(\frac{\kappa-1}{\kappa}\right)_a} - 1\right] \tag{2.14}$$

Substituting Eqs. (2.12 and 2.14), one obtains the pressure ratio of compressor π_C in *the first turbocharger equation.*

$$\begin{aligned}\pi_C = \frac{p_2}{p_1} &= \left(1 + \frac{c_{p,g}}{c_{p,a}}\left\langle\frac{\dot{m}_T}{\dot{m}_C}\frac{T_3}{T_1}\eta_{TC}\right\rangle \cdot \left[1 - \left(\frac{p_3}{p_4}\right)^{-\left(\frac{\kappa-1}{\kappa}\right)_g}\right]\right)^{\left(\frac{\kappa}{\kappa-1}\right)_a} \\ &= \left(1 + \frac{c_{p,g}}{c_{p,a}}\left\langle\frac{\dot{m}_T}{\dot{m}_C}\frac{T_3}{T_1}\eta_{TC}\right\rangle \cdot \left[1 - \pi_T^{-\left(\frac{\kappa-1}{\kappa}\right)_g}\right]\right)^{\left(\frac{\kappa}{\kappa-1}\right)_a}\end{aligned} \tag{2.15}$$

within the overall efficiency of the turbocharger η_{TC} is written as

$$\eta_{TC} = \eta_m \eta_T \eta_C \tag{2.16}$$

To achieve the high boost pressure of the charge air p_2, the compressor pressure ratio π_C could be improved according to Eq. (2.15) if

- the overall turbocharger efficiency η_{TC} is high, especially a high mechanical efficiency of the bearing system in low-end torque;
- the exhaust gas temperature T_3 is high due to the large enthalpy. Hence, more turbine power is generated;
- the turbine pressure ratio π_T (turbine expansion ratio) is as high as possible;
- the inlet air temperature T_1 is as low as possible; therefore, the charge-air temperature T_2 is low leading to a high density of the charge air;
- the exhaust gas pressure p_3 is chosen at an optimal pressure in order to compromise between the turbine power and specific fuel consumption;
- the exhaust gas mass flow rate in the turbine should be large.

The first turbocharger Eq. (2.15) shows the behavior of the pressure ratio π_C ($\equiv p_2/p_1$) of the compressor versus the pressure ratio π_T ($\equiv p_3/p_4$) of the turbine. In fact, the dimensionless term δ in the angle brackets < > of Eq. (2.15) does not change so much along the full load curve in the compressor performance map. Therefore, the behavior of the pressure ratios of turbine and compressor can be displayed at various dimensionless parameters δ in Fig. 2.3.

The exhaust gas flow in the turbine can be considered as a compressible flow in a nozzle in which the inlet and outlet pressures are p_3 and p_4, respectively. Based on the flow equation for compressible fluids in the nozzle, *the second turbocharger equation* describes the mass flow rate through the turbine in function of the pressure, temperature at the turbine inlet, and the turbine expansion ratio.

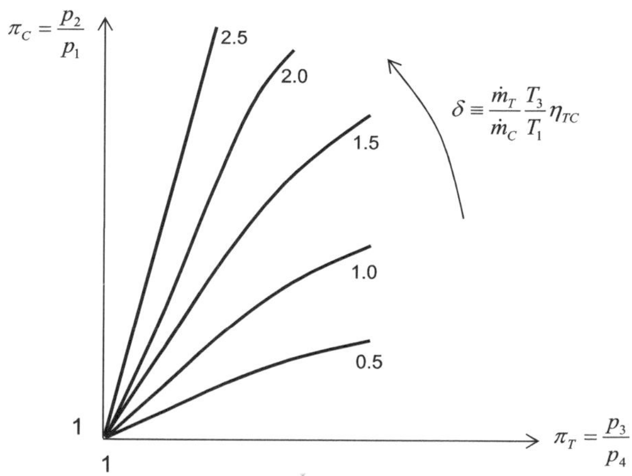

Fig. 2.3 Behavior of the pressure ratios π_C versus π_T

$$\dot{m}_T = \mu A_T p_{3t} \sqrt{\frac{2}{R_g T_{3t}}} \sqrt{\left(\frac{\kappa}{\kappa - 1}\right)_g \left(\left(\frac{p_{3t}}{p_4}\right)^{\frac{-2}{\kappa_g}} - \left(\frac{p_{3t}}{p_4}\right)^{-\left(\frac{\kappa+1}{\kappa}\right)_g}\right)} \quad (2.17)$$

where μ is the flow coefficient due friction and flow contraction at the nozzle outlet, A_T is the throttle cross-sectional area in the turbine wheel.

To eliminate the influences of p_{3t} and T_{3t} on the mass flow rate in the turbine shown in Eq. (2.17), the so-called corrected mass flow rate is defined as

$$\begin{aligned} \dot{m}_{T,cor} &\equiv \frac{\dot{m}_T \sqrt{T_{3t}}}{p_{3t}} = f(\pi_{T,ts}) \\ &= \mu A_T \sqrt{\frac{2}{R_g}} \sqrt{\left(\frac{\kappa}{\kappa - 1}\right)_g \left(\left(\frac{p_{3t}}{p_4}\right)^{\frac{-2}{\kappa_g}} - \left(\frac{p_{3t}}{p_4}\right)^{-\left(\frac{\kappa+1}{\kappa}\right)_g}\right)} \end{aligned} \quad (2.18)$$

Equation (2.18) indicates that the corrected mass flow rate of the turbine is independent of the inlet condition of the exhaust gas of p_{3t} and T_{3t}. It depends only on the turbine expansion ratio $\pi_{T,ts}$.

The performance map of the turbine displays the corrected mass flow rate over the turbine expansion ratio $\pi_{T,ts}$ at various rotor speeds in Fig. 2.4. From a turbine pressure ratio of approximately 3, the mass flow rate has no longer increased, even at higher rotor speeds. In this case, the flow in the turbine becomes a choked flow in which the exhaust gas speed at the throttle area reaches the sonic speed with Mach number $M = 1$. As a result, the exhaust gas mass flow rate through the turbine corresponding to the nominal engine power must be smaller than the mass flow rate

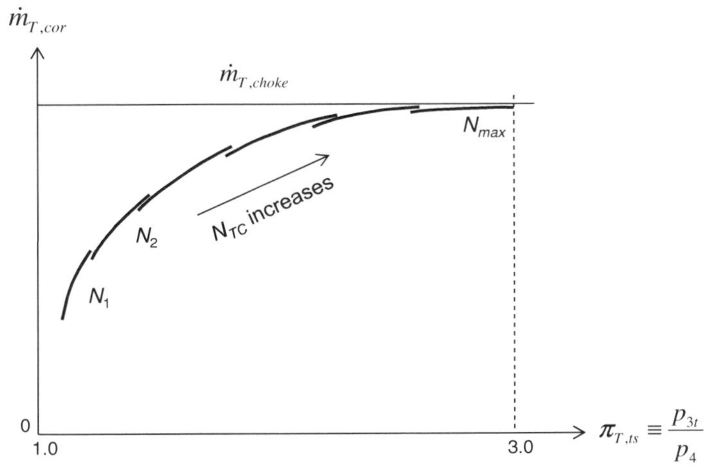

Fig. 2.4 Performance map of the turbine

at choke. Note that the isentropic turbine efficiency at the choked flow is extremely low ($\eta_T < 60\ \%$). This low efficiency is unusable in the automotive turbochargers.

The mechanical efficiency of turbocharger induced by the bearing friction results from Eq. (2.12).

$$\eta_m = \frac{P_C}{P_T} = \frac{\dot{m}_C c_{p,a}(T_2 - T_1)}{\eta_C \eta_T \dot{m}_T c_{p,g} T_3 \left[1 - \left(\frac{p_4}{p_3}\right)^{\left(\frac{\kappa-1}{\kappa}\right)_g}\right]} \tag{2.19}$$

Using Eqs. (2.14 and 2.19), the efficiency product of the mechanical and turbine efficiencies is written as

$$\eta_m \eta_T = \frac{\dot{m}_C c_{p,a} T_1 \left[\left(\frac{p_2}{p_1}\right)^{\left(\frac{\kappa-1}{\kappa}\right)_a} - 1\right]}{\eta_C \dot{m}_T c_{p,g} T_3 \left[1 - \left(\frac{p_3}{p_4}\right)^{-\left(\frac{\kappa-1}{\kappa}\right)_g}\right]} \tag{2.20}$$

The efficiency $\eta_m \eta_T$ described in Eq. (2.20) is a key part in the overall efficiency of the turbocharger. It is determined by measuring the thermodynamic characteristics in Eq. (2.20), such as the mass flow rates in the compressor and turbine, temperatures T_1 and T_3, compressor efficiency η_C, pressures at the compressor inlet and outlet p_1 and p_2, and as well as pressures at the turbine inlet and outlet p_3 and p_4. Figure 2.5 displays the efficiency $\eta_m \eta_T$ versus pressure ratio of turbine $\pi_{T,ts}$ at various rotor speeds.

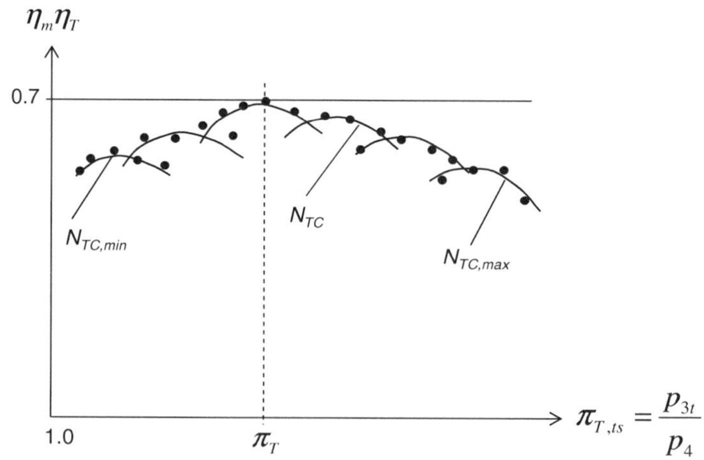

Fig. 2.5 Efficiency $\eta_m \eta_T$ versus $\pi_{T,ts}$ at various rotor speeds N_{TC}

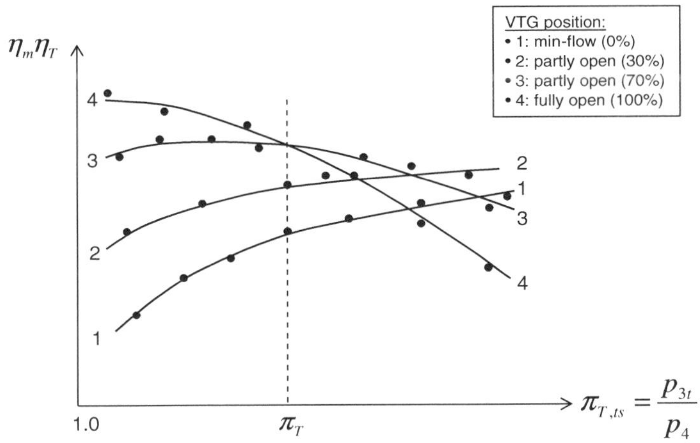

Fig. 2.6 Efficiency $\eta_m \eta_T$ versus $\pi_{T,ts}$ at various VTG positions

At low rotor speeds corresponding to the low-pressure expansion ratios π_T, the mechanical efficiency is quite small due to large bearing friction at the low oil temperature in the bearings. Additionally, the turbine efficiency is still small due to the aerodynamic operating condition at the low rotor speed. Therefore, the resulting efficiency $\eta_m \eta_T$ remains low.

At high rotor speeds, the mechanical efficiency of the bearings increases due to the increased oil temperature and therefore less friction in the bearings. As a result, the efficiency $\eta_m \eta_T$ reaches a maximum value at the turbine pressure ratio at the design point in the performance map.

As the rotor speed further increases to the maximum speed, the turbine efficiency drops due to the aerodynamic operating condition at high rotor speeds; the mechanical efficiency decreases because the bearing friction rises at high rotor speeds. Therefore, the efficiency $\eta_m \eta_T$ decreases with the rotor speed corresponding to the turbine pressure ratio.

However, the efficiency $\eta_m \eta_T$ depends not only on the turbine pressure ratio but also on the position of VTG (variable turbine geometry), as shown in Fig. 2.6.

- Initially, the VTG is open at the min-flow condition for the minimum mass flow rate called 0 % VTG (position 1) at the idle condition of the engine. The efficiency $\eta_m \eta_T$ increases with the turbine pressure ratio because the turbine efficiency becomes larger at high mass flow rates corresponding to a high turbine pressure ratio.
- The VTG begins opening at 30 % VTG (position 2). At increasing turbine pressure ratios, the mass flow rate of exhaust gas rises in the turbine leading to the high turbine efficiency. Hence, the efficiency $\eta_m \eta_T$ increases since the turbine pressure ratio rises to the pressure ratio π_T at the design point, as shown in Fig. 2.5. As a result, the efficiency $\eta_m \eta_T$ is higher than the efficiency at the position 1.

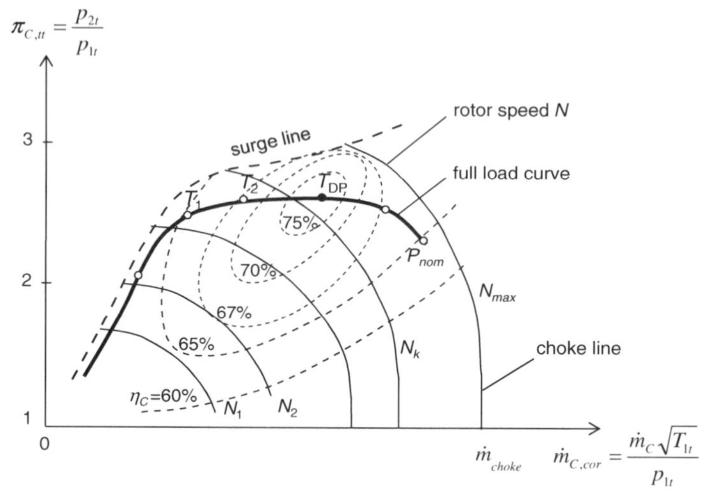

Fig. 2.7 Compressor performance map

- At the position of 70 % VTG (position 3), the turbine efficiency begins decreasing at high mass flow rates corresponding to large turbine pressure ratios.
- Finally, the VTG is fully open at 100 % VTG (position 4). The mass flow rate significantly increases in the turbine at increasing turbine pressure ratio. Therefore, the turbine efficiency reduces with the turbine pressure ratio. As a result, the efficiency $\eta_m \eta_T$ drops with the turbine pressure ratio.

The compressor performance map in Fig. 2.7 shows the compressor ratio π_C versus the corrected mass flow rate.

At low rotor speed in the idle condition of about 30 % of the maximum rotor speed N_{max}, the compressor pressure ratio on the full load curve increases strongly with the mass flow rate to deliver a good transient response at low-end torque. At further increasing the rotor speed N to nearly 70 % of N_{max}, the compressor pressure ratio increases to approximately 2.5 and the engine achieves the maximum torque from T_1 to the design point T_{DP}. Due to the reduced turbine efficiency at the high rotor speeds leading to reducing the overall efficiency of the turbocharger, the compressor pressure ratio decreases at the nominal engine power P_{nom} in the full load curve according to the first turbocharger Eq. (2.15).

2.4 Response Time of Turbochargers

The response time is an important characteristic dealing with turbolag in automotive turbochargers. The turbolag is the delayed time that the turbocharger needs to reach the maximum engine torque from the idle.

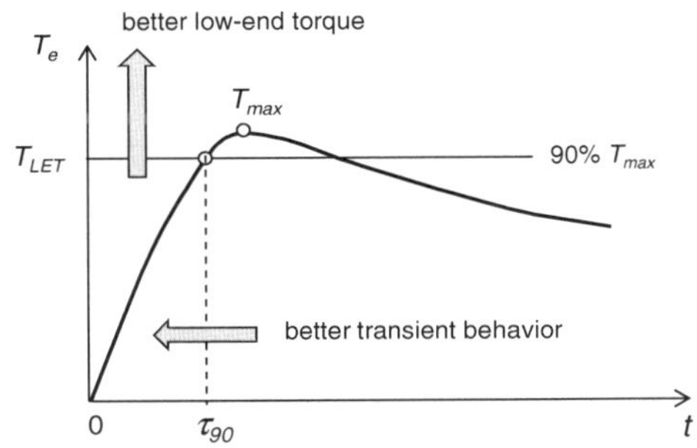

Fig. 2.8 Response time of a turbocharger

To characterize the turbolag of the turbocharger, the response time τ_{90} is defined by the time interval that is required to reach the low-end torque (LET) T_{LET} of 90 % of the maximum engine torque T_{max} in low-end torque LET (see Fig. 2.8). The response time τ_{90} results from the effective turbine power and the polar mass inertia moment of the rotor.

Note that the better the transient behavior of the turbocharger is, the shorter the response time τ_{90} is at a given LET. The better the LET is, the higher the LET T_{LET} is at a given response time.

The angular acceleration of the rotor is calculated from the turbocharger dynamics equation.

$$\eta_m P_T - P_C = I_p \ddot{\theta}\, \Omega \tag{2.21}$$

The mechanical efficiency of the bearing is calculated as [4]

$$\eta_m = 1 - \frac{P_{fB}}{P_T} \tag{2.22}$$

where P_{fB} is the bearing friction power.

The rotor speed N_{TC} of the turbocharger results from Eqs. (2.21 and 2.22) in

$$N_{TC} = \frac{\Omega}{2\pi} = \frac{(\eta_m P_T - P_C)}{2\pi\, I_p \ddot{\theta}} = \frac{(P_T - P_{fB} - P_C)}{2\pi\, I_p \ddot{\theta}} \tag{2.23}$$

Integrating with time the angular acceleration within the response time τ_{90}, one obtains the average angular rotor velocity in LET

$$\Omega_{90} = \int_0^{\tau_{90}} \ddot{\theta}(t)dt \approx \bar{\ddot{\theta}}\tau_{90} \tag{2.24}$$

where $\bar{\ddot{\theta}}$ is the average angular acceleration within the response time.

Substituting Eq. (2.24) into Eq. (2.21) and eliminating the average angular acceleration, the response time of the turbocharger results in

$$\tau_{90} \approx \frac{I_p \Omega_{90}^2}{(\eta_m P_T - P_C)} = \frac{4\pi^2 I_p N_{90}^2}{(\eta_m P_T - P_C)} \tag{2.25}$$

where

I_p is the polar mass inertia moment of the rotor;
P_T is the effective turbine power;
P_C is the effective compressor power;
η_m is the mechanical efficiency of the bearings.

In order to have a good transient response of the turbocharger, it is necessary to keep the response time τ_{90} as small as possible at a given rotor speed N_{90}. According to Eq. (2.25), since the effective turbine power P_T is relatively small in LET, the response time τ_{90} rises leading to a large turbolag of the turbocharger. However, there are some improving measures:

(1) The polar mass inertia moment I_p of the rotor should be as small as possible. The turbine wheel plays a key role in the entire polar mass inertia moment of the rotor due to its heavy mass of Inconel 713C. The polar mass inertia moment of the turbine wheel is proportional to its mass m and squared wheel diameter D^2. Due to $m \sim \rho D^3$, the polar mass inertia moment of the rotor is written as

$$\begin{aligned} I_p &\approx I_{p,TW} \propto mD^2 \\ &\propto \rho D^5 \end{aligned} \tag{2.26}$$

To reduce the inertia moment of the turbine wheel, there are some possibilities as follows:

- The turbine wheel should be as light as possible, such as using light titanium aluminide TiAl6V4 ($\rho = 4.45$ g/cm^3) instead of heavy Inconel 713C ($\rho = 7.91$ g/cm^3);
- The turbine wheel is scalloped at the back face to reduce its mass. However, the turbine efficiency could be reduced by 2–3 % due to the unsuitable aerodynamic flow at the scalloped back face leading to a reduction of the effective turbine power according to Eq. (2.11);
- The inflow diameter D of the turbine wheel should be reduced in order to decrease the polar inertia moment and to increase the turbine efficiency at low rotor speeds leading to the increase of the effective turbine power. However, the small turbine wheel has some negative effects, such as low turbine efficiency at high rotor speeds and small mass flow rate. As a result, the nominal turbine power is reduced.

(2) The mechanical efficiency η_m should be increased using airfoil bearings, magnetic bearings, rolling-element bearings, or rotating floating ring bearings with two oil

films. Generally, the rolling-element bearings generate less friction power, especially in LET. Contrary to the ball bearings, oil-film bearings induce a little more friction power in LET since the oil viscosity in the bearing is high at low rotor speeds.

In fact, the ratio of the bearing friction to turbine power is relatively high in LET due to the small effective turbine power and high bearing friction. However, the discrepancy between the friction coefficients of the rolling element and rotating floating ring bearings is negligibly small at high rotor speeds. Note that the rolling-element bearings cost about 10 times of the oil-film bearings. Therefore, it is recommended to carefully decide which material of the turbine wheel should be used, the suitable inflow diameter of the turbine wheel, either scalloped or un-scalloped turbine wheel. Furthermore, which bearing system should be used in the turbochargers in compromise between the transient response and cost of the bearings?

2.5 Turbocharger Matching

The matching procedure of turbochargers based on the engine characteristics, compressor performance map, turbine performance map, and first turbocharger equation is discussed in the following section (see Fig. 2.9).

The following steps are carried out for the matching procedure of turbochargers [5–7]:

(1) From the engine characteristics (Fig. 2.9, bottom left), the operating field of the turbocharger in the compressor map is determined at the various operating points, such as LET, maximum torque, design point, and nominal power of the engine. At any operating point, the required mass flow rate of the charge air at the given turbocharger speed results from the engine power according to Eqs. (1.1 and 1.2). Additionally, the necessary charge-air density is given by Eq. (1.3). The compression ratio without air intercoolers is calculated from Eq. (1.10). In case of using air intercoolers (see Fig. 2.10), the charge-air temperature T_2 drops at a lower temperature T_{2*} leading to the increase of the charge-air density ρ_{2*}.

The charge-air temperature T_{2*} after the intercooler is calculated from the compressed charge-air temperature T_2, coolant inlet temperature T_c, and intercooler efficiency ε_c (usually between 0.6 and 0.8) as

$$T_{2^*} = (1 - \varepsilon_c)T_2 + \varepsilon_c T_c < T_2 \tag{2.27}$$

Thus, the charge-air temperature in case of using air intercoolers drops with

$$\begin{aligned} \Delta T_{2^*} &\equiv T_{2^*} - T_2 \\ &= \varepsilon_c(T_c - T_2) < 0 \end{aligned} \tag{2.28}$$

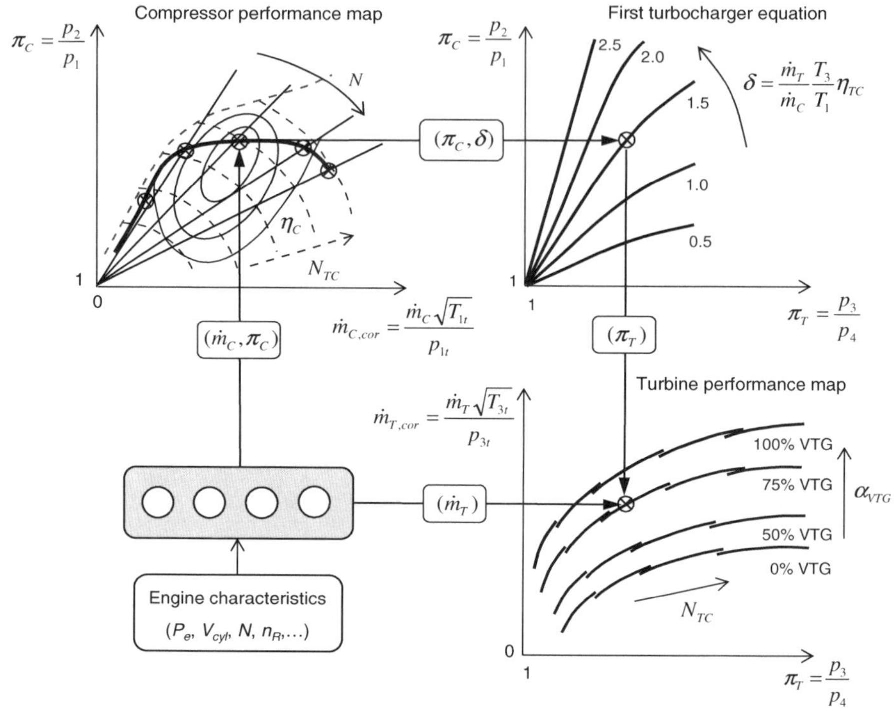

Fig. 2.9 Matching procedure of turbochargers

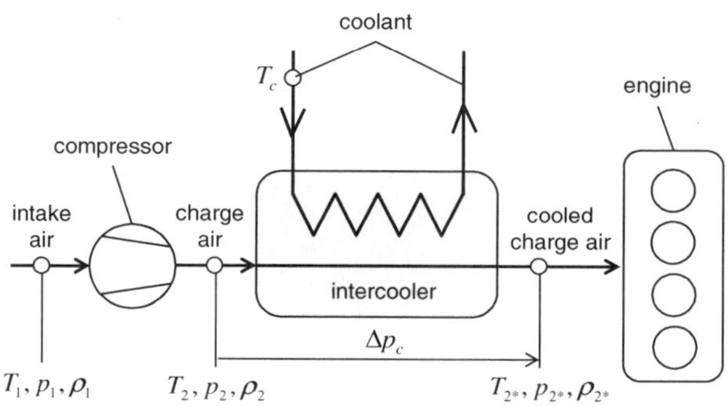

Fig. 2.10 Compressor using a charge-air intercooler

Therefore, the charge-air density rises from ρ_2 to ρ_2 at a small pressure drop in the intercooler. Using the state equation of a perfect gas, the charge-air density is calculated as

$$\rho_{2^*} = \frac{p_{2^*}}{R_a T_{2^*}} = \frac{p_2 - \Delta p_c}{R_a T_{2^*}} > \rho_2 \tag{2.29}$$

where Δp_c is the pressure drop in the air intercooler.

Substituting Eqs. (1.9, 2.27, and 2.29), one obtains the compression ratio of the compressor in case of using the air intercooler

$$\begin{aligned}\pi_C &= \frac{p_2}{p_1} = \frac{p_{2^*} + \Delta p_c}{p_1} = \frac{\rho_{2^*} R_a T_{2^*} + \Delta p_c}{p_1} \\ &= \frac{\rho_{2^*} R_a \left[(1 - \varepsilon_c) T_1 \left(1 + \frac{1}{\eta_C} \left(\pi_C^{\left(\frac{\kappa - 1}{\kappa}\right)_a} - 1 \right) \right) + \varepsilon_c T_c \right] + \Delta p_c}{p_1}\end{aligned} \tag{2.30}$$

where the charge-air density ρ_{2*} is given by the engine requirement and $\kappa_a \approx 1.4$ is the isentropic exponent of the charge air.

The charge-air density ρ_{2*} is determined by iteratively solving Eq. (2.30) under the given boundary conditions for the engine, compressor, and intercooler.

(2) The operating point of the turbocharger is located in the compressor performance map (Fig. 2.9, top left) at the given mass flow rate and compression ratio of the charge air that have been computed in the step 1. The value δ in the diagram of the first turbocharger equation involves the mass flow rates of the charge air and exhaust gas, the temperatures of the exhaust gas and intake air, and the turbocharger efficiency. It results from the operating condition of the engine and the guessed/measured efficiency of the turbocharger.

(3) From the given compression ratio π_C and value δ, the expansion ratio π_T of the turbine corresponding to the operating point is determined in the diagram of the first turbocharger equation (Fig. 2.9, top right).

(4) Both mass flow rate and expansion ratio of the turbine given in the steps 2 and 3 are used to determine the operating point of the turbocharger in the turbine performance map (Fig. 2.9, bottom right).

The operating point gives the corresponding VTG angular position of 75 % VTG. Note that the mass flow rate of the exhaust gas that is identical with the mass flow rate of the VTG turbine, equals the mass flow rates of the charge air and injected fuel in the cylinders based on the air–fuel ratio AFR defined in Eq. (1.2). Therefore, the mass flow rate of the exhaust gas is written as

$$\dot{m}_T = \dot{m}_a + \dot{m}_f = \dot{m}_a \left(\frac{AFR + 1}{AFR} \right) \tag{2.31}$$

Using the turbomachinery theory, the outflow diameter D_2 (exducer diameter) of the compressor wheel and the inflow diameter D_3 (inducer diameter) of the turbine wheel are computed at the given effective powers and mass flow rates of the compressor and turbine, mechanical efficiency, and turbocharger speed. The matching procedure of turbochargers is iterated until the guessed values, such as the efficiencies of the compressor and turbine, efficiency of the air intercooler, etc., are converged. Furthermore, the computed values could be rematched with the measured efficiencies as soon as they are available.

References

1. Aungier, R.H.: Turbine Aerodynamics. ASME Press, New York (2006)
2. Baines, N.C.: Fundamentals of Turbocharging. Concepts ETI, Inc, Wilder (2005)
3. Cumpsty, N.A.: Compressor Aerodynamics. Krieger Publishing Company, Melbourne (2004)
4. Heywood, J.B.: Internal Combustion Engine Fundamentals. McGraw-Hill, New York (1988)
5. Japikse, D.,Baines, N.C.: Introduction to Turbomachinery. Concepts ETI, Inc, Wilder (1994)
6. Japikse, D., et al.: Axial and Radial Turbines. Concepts ETI Inc, Wilder (2003)
7. Whitfield, A., Baines, N.C.: Design of Radial Turbomachines. Pearson Education, Longman Scientific and Technical (1990)

Chapter 3
Vibrations of Turbochargers

3.1 Introduction

Exhaust gas turbochargers used in the automobiles, such as passenger vehicles, commercial vehicles and off-road engines, have some discrepancies to the heavy turbomachines applied to the power plants and chemical industries. The automotive turbochargers are much smaller and operate at high rotor speeds in various operating conditions, such as variable rotor speeds, pressures, temperatures, and mass flow rates. Contrary to the automotive turbochargers, industrial turbomachines are larger, heavier, and mostly operate at stationary conditions. They operate mostly between 3000 and 20,000 rpm since the compressor and turbine wheels are very large compared to the turbochargers. Due to their much smaller wheels, the automotive turbochargers could operate at very high rotor speeds from 100,000 to 400,000 rpm.

The thermomechanical characteristics and lifetime depend on the using material, producing method, and as well as driving cycles. To ensure the durability of materials at various driving cycles, the maximum circumferential velocities of the compressor and turbine wheels are predetermined. In case of automotive turbochargers, the circumferential velocities are limited by 560 m/s for compressor wheels and 530 m/s for turbine wheels in the driving cycles. On the contrary, the maximum circumferential velocities of compressor and turbine wheels in industrial turbomachines are much smaller due to safety for persons and protection of the environment in case of airplanes and nuclear power plants.

The key issue of the discrepancies is the U2W ratio (Unbalance to Weight) of the maximum unbalance force F_U acting upon the rotor to the rotor weight F_W. In the automotive turbochargers, the excited unbalance forces are much larger than their rotor weights due to very high rotor speeds. Hence, the U2W ratio is in the order of approximately 100. Contrary to turbochargers, the U2W ratio of industrial turbomachines is in the order of 1 because of their heavy weights and quite low rotor speeds, as shown in Table 3.1.

H. Nguyen-Schäfer, *Rotordynamics of Automotive Turbochargers*,
Springer Tracts in Mechanical Engineering, DOI 10.1007/978-3-319-17644-4_3

Table 3.1 U2W ratio of turbochargers and turbomachines

Types	Rotor speed N (rpm)	Residual unbalance U (g mm/plane)	Maximum un balance force FU (N)	Weight Fw (N)	Ratio U2W
Automotive turbochargers	150,000–300,000	0.1–1.0	200–500	1–5	>100
Industrial, aero. turbomachines	3 000–20,000	G2.5[a]	550–1 000	1 000–10,000	<1

[a]According to the balancing quality grade G2.5 (DIN-ISO 1940-1)

The U2W ratio shows different behaviors of the rotordynamics between automotive turbochargers and industrial turbomachines. In case of automotive turbochargers using the oil-film bearings, the journal initially lies nearly in the bearing center due to its small weight. At increasing the rotor speeds, it moves from the bearing center outwards to the bearing wall due to the large unbalance force (nearly 100 times higher than its rotor weight at high rotor speeds, U2W $\gg$ 1). After the rotor reaches the resonance, the journal turns backwards to the bearing center due to the self-centering of the rotor. It indicates that the journal of the turbochargers moves with the rotor speed range in the whole bearing clearance.

Contrary to turbochargers, the journal of industrial turbomachines is initially near the bearing wall at low rotor speeds due to their heavy rotor weight. Because the excited unbalance force is much smaller than the rotor weight (U2W < 1), the journal cannot move far away from the equilibrium position. Therefore, the journal orbit is smaller and more stable compared to the journal orbit of turbochargers. As a result, the journal orbit (journal locus) of turbochargers moves with rotor speeds in the whole bearing clearance. The bearing stiffness and damping coefficients vary with the rotor speeds nonlinearly. The larger the journal eccentricity is, the higher the stiffness and damping coefficients of the oil-film bearing are.

The response behavior of the bearing stiffness and damping coefficients versus journal eccentricity ε is displayed in Fig. 3.1. Near the bearing center ($\varepsilon = 0$), the bearing stiffness and damping coefficients have minimum values because the oil film thickness is maximum and nearly equals the radial bearing clearance. As the rotor journal moves near to the bearing wall ($\varepsilon \approx 1$), the oil film is squeezed due to the large eccentricity and high whirl speed of the journal in the bearing. Hence, the bearing stiffness and damping coefficients increase drastically near the bearing wall.

Obviously, the stiffness and damping coefficients of the oil-film bearing have a nonlinear behavior versus the journal eccentricity, as shown in Fig. 3.1. As a result, the induced bearing force acting upon the rotor is nonlinear as the journal locus moves in the entire bearing clearance. That is the case of the automotive turbochargers operating with U2W $\gg$ 1. In the industrial turbomachines, the journal orbit does not move so far from the equilibrium position in the bearing clearance; i.e., the journal locus moves in a small orbit. Therefore, the bearing stiffness and damping coefficients can be linearized at the vicinity of the equilibrium position ε_{eq} of the journal.

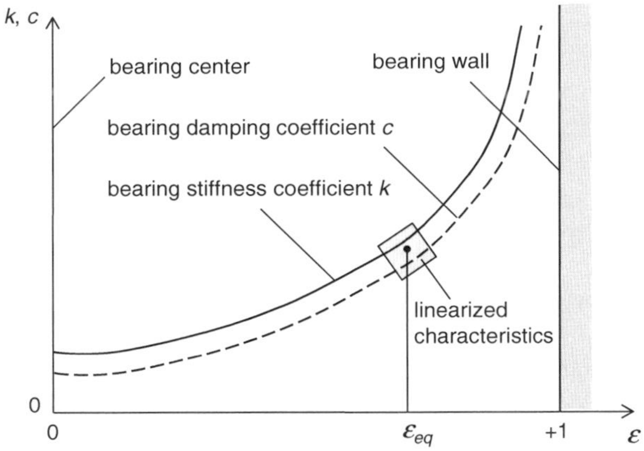

Fig. 3.1 Stiffness and damping coefficients of the oil-film bearings versus journal eccentricity ε

As a result, the response behaviors of large industrial turbomachines using oil-film bearings are usually *quasilinear*; the response behaviors of small automotive turbochargers with oil-film bearings are generally *nonlinear*.

3.2 Vibration Modes of Turbochargers

In the application of automotive turbochargers, the torsional vibration amplitudes of the rotor is relatively small compared to the lateral bending amplitudes caused by the rotor unbalance, especially at extremely high rotor speeds. Therefore, only degrees of freedom (DOF) of the lateral deflection of the rotor are taken into account in the rotordynamic analyses.

The free vibration rotor responses of the linear system consist of many harmonic components with their eigenfrequencies and eigenmodes. On the contrary, the nonlinear vibration responses of turbochargers include not only the synchronous (harmonic) but also sub- and supersynchronous frequency components due to the nonlinear phenomena. Hence, the vibration modes of the nonlinear rotor responses have the shapes resulted from the combination of the modes of the synchronous and nonsynchronous vibrations.

Figure 3.2 shows the setup of a simplified rotor of turbochargers in which the turbine and compressor wheels are mounted and fixed on the rotor shaft. The rotor is supported by two oil-film bearings located between the two wheels.

A residual unbalance U_{TW} remains in the turbine shaft after the low-speed balancing (shop balancing). A residual unbalance U_{CW} exists in the compressor wheel after the high-speed balancing (trim balancing). In case of both unbalance vectors having the same direction, the rotor unbalance is called the “in-phase couple” unbalance and when the directions of the unbalance vectors are opposite to each other, the “out-of-phase couple” unbalance occurs. The out-of-phase couple unbalance normally provides a good rotordynamic behavior because it leads to the

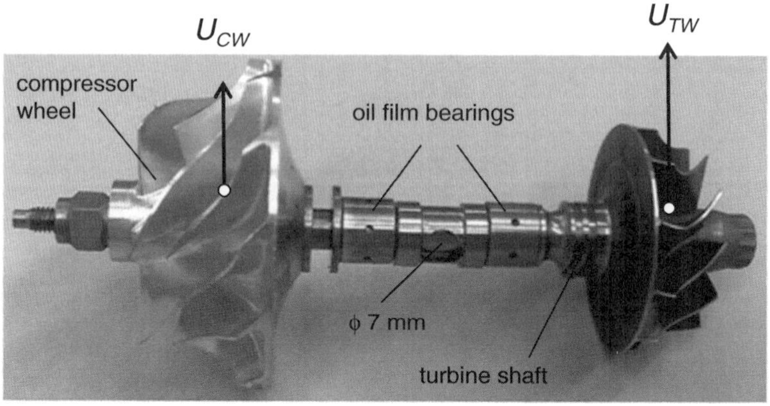

Fig. 3.2 An unbalanced rotor supported by oil-film bearings

smallest residual unbalance vector. In practice, after assembling the compressor wheel on the turbine shaft, the angle of the unbalance vectors is unknown between 0° and 180°.

The unbalance excites the rotor due to the centrifugal forces as the rotor speed Ω increases. The unbalance force equals the product of the unbalance U and squared rotor speed Ω^2, hence $F_{Un} = U\Omega^2$ (see Chap. 8). At the low speeds, the unbalance force remains small and the journal locates nearly at the bearing center due to small weight of the rotor. In this case, the bearing has a small stiffness coefficient, as shown in Fig. 3.1; therefore, the rotor behaves like a rigid rotor.

The vibration modes of the rotor are displayed in Fig. 3.3. At low rotor speeds, the rotor is rigid; the first mode has the cylindrical shape whirling inside the bearing clearance where both compressor and turbine wheels deflect in the same direction with the whirl velocity ω (cylindrical mode). The second mode is the conical mode where the compressor wheel deflects opposite to the turbine wheel deflection (conical mode). Finally, the third mode of a rigid rotor is the bending mode at the very high rotor speeds where both wheels deflect in the same direction again.

At the high rotor speeds, the unbalance force strongly increases with squared rotor speed. Thus, the journal moves outwards to the bearing wall. As a result, the bearing stiffness coefficients become larger, as shown in Fig. 3.1, and the rotor deflects in the lateral direction due to excitation of the unbalance force. In this case, the shaft rotating with the angular speed Ω becomes flexible. The first bending mode has a U shape where both compressor and turbine wheels deflect in the same direction with the whirl velocity ω. The first bending mode induces the first resonance of the rotor relating to the first eigenmode. At further increasing the rotor speed, the rotor traverses the second bending resonance relating to the second eigenmode with an S shape where the compressor wheel deflects opposite to the turbine wheel. Finally, the third bending mode with a W shape relates to the third eigenmode where both wheels deflect in the same direction again.

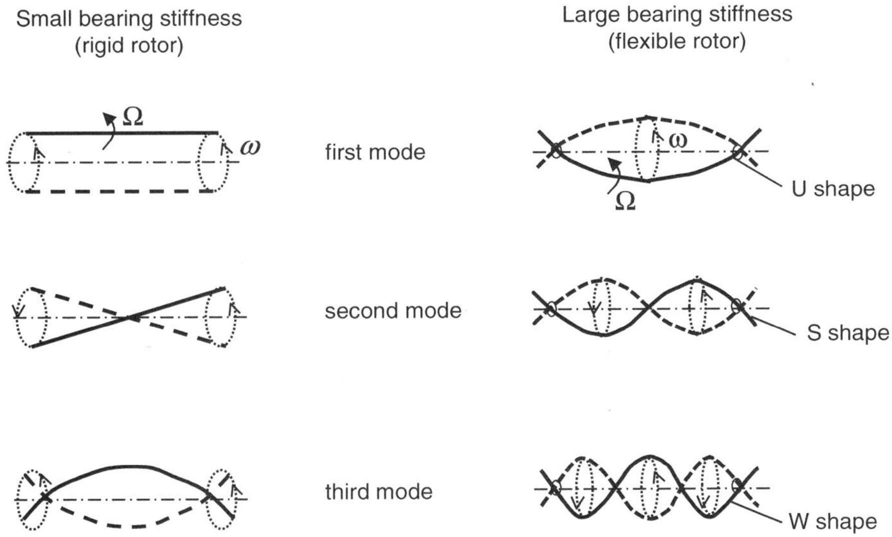

Fig. 3.3 Shapes of the vibration eigenmodes in linear rotordynamics

The vibration modes of the rotor response in nonlinear rotordynamics result from superimposing the modes of the synchronous and nonsynchronous frequency components. Their mode shapes depend on the current stiffness coefficients, damping ratios, and as well as the nonlinear characteristics of the rotor at various rotor speeds.

3.3 Vibration Characteristics of Turbochargers

In this section, the characteristics of vibrations are classified in the frequency and time domains, in which different kinds of vibrations are defined in each domain.

3.3.1 In Frequency Domain

- *Harmonic vibration* has the same frequency of the rotor frequency Ω. The frequency order is called 1X.

$$\omega = \Omega \tag{3.1}$$

- *Subharmonic* has the whirl frequency ω that is smaller than the rotor frequency Ω. The frequency order equals a fractional integer $(1/M) \cdot \text{X}$.

$$\omega = \frac{1}{M}\Omega \tag{3.2}$$

where M is an integer, such as $M = 1, 2, \ldots, N$.

- *Superharmonic* has the whirl frequency ω that is larger than the rotor frequency Ω. The frequency order equals a multiple integer MX.

$$\omega = M\Omega \tag{3.3}$$

where M is an integer, such as $M = 1, 2, \ldots, N$.

- *Subsynchronous vibration* has the frequency that is less than the rotor frequency. The frequency order equals an irrational order, such as 0.35X or 0.47X.
- *Supersynchronous vibration* has the frequency that is larger than the rotor frequency. The frequency order equals a noninteger, such as 2.5X or 3.5X.
- *Forward whirl* is the rotor precession whose velocity has the same direction of the rotor angular velocity.
- *Backward whirl* is the rotor precession whose velocity has the direction opposite to the rotor angular velocity.

3.3.2 In Time Domain

- *Harmonic vibration* has a time function of sine or cosine in a mathematical form

$$x(t) = A \sin(\Omega t + \varphi) \tag{3.4}$$

where

$x(t)$ is the time amplitude of the vibration;
A is the amplitude from the center to peak;
Ω is the rotor angular frequency in rad/s ($\Omega = 2\pi N$);
N is the rotor speed in rps (round per second);
φ is the phase in rad, the rotor angular position at the initial time ($t = 0$).

The excitation unbalance force is a harmonic vibration; its amplitude A equals the unbalance amplitude of $U\Omega^2$.

- *Periodic vibration* includes the sub- and superharmonics. The frequency order of the subharmonics equals a fraction of integer order, such as (1/2)X and (1/3) X. The frequency orders of the superharmonics is a multiple integer order of 2X and 3X. Therefore, the period vibration has the same amplitude after repeating of period T.

$$x(t) = x(t + T) \tag{3.5}$$

- *Quasiperiodic vibration* consists of the sub- and supersynchronous vibrations. The frequency order equals an irrational order, such as 0.37X or 0.45X (subsynchronous), 1.33X or 1.67X (supersynchronous). The quasiperiodic vibration contains at least two incommensurate frequencies ω_1 and ω_2 in which the ratio ω_1/ω_2 is an irrational number. After repeating of period T, the amplitude at time $(t + T)$ is not exactly the same amplitude at the time t. The quasiperiodic vibration often occurs in nonlinear dynamics of the Neimark and Sacker bifurcation. After passing the bifurcation point, the periodic vibration with the frequency ω_1 is bifurcated in another periodic vibration with the frequency ω_2. The resulting quasiperiodic vibration consists of two incommensurate frequencies ω_1 and ω_2. Its quasiperiodic orbit wraps on the surface of a torus and does not intersect the Poincaré map at the same point at every period T (see Chap. 4).

$$x(t) \neq x(t + T) \tag{3.6}$$

- *Chaos vibration* has no rule for its motion in which the unstable vibrations change from the quasiperiodic to chaos motions due to bifurcation.
- *Steady-state operating conditions*
  When the acceleration of the vibration equals zero, the rotor vibration is in a steady-state condition. The steady-state condition is considered when the rotor rotates with a constant angular speed at a certain interval of time.
- *Transient operating conditions*
  When the acceleration of the vibration is taken into account, the vibration is transient because the angular speed changes with time. At run-up or slowdown, the response behavior of the rotor is transient. In the transient condition, the amplitude and critical frequency of the rotor response strongly depend on the rotor acceleration compared to the steady-state condition (see Chap. 5).
- *Free vibration response* results from the rotordynamic system without excitation forces due to unbalance (homogeneous solution of the vibration equation). In a linear system, the free vibration response consists of the eigenvectors and eigenfrequencies of the rotor.
- *Forced vibration response* results from the rotordynamic system with excitation forces of the rotor unbalance (particular solution of the vibration equation). In a linear system, the unbalance forced vibration response is a harmonic vibration with the rotor frequency. On the contrary, the rotor forced vibration response is periodic or quasiperiodic vibrations that include the synchronous and nonsynchronous whirl frequencies in a nonlinear system.

3.4 Linear and Nonlinear Vibrations of Turbochargers

In the following section, the discrepancies between the linear and nonlinear vibrations with synchronous and asynchronous frequencies are discussed. At first, the linear rotordynamic system deals with the unbalance excitation of the rotor. In case of a free vibration without considering the unbalance excitation, the free vibration responses are the sum of the eigenvectors with the corresponding complex eigenvalues of the rotor. They are normally periodic vibrations with the relating eigenfrequencies. The eigenfrequencies of the linear system are the imaginary parts of the complex eigenvalues.

The unbalance response (forced vibration response) traverses the resonances at the first, second, third, or higher critical speeds. The critical speeds are the rotational speeds of the rotor at which the rotor eigenfrequencies equal the rotor frequency. The unbalance responses are harmonic vibrations with the rotor frequency. Thus, their frequency has the first-order 1X and is proportional to the rotor speed.

Campbell diagram (whirl speed map) is only used in linear rotordynamics to find the critical speeds where the rotor eigenfrequencies intersect the synchronous excitation frequency 1X. Each eigenfrequency of the free vibration response corresponds to an eigenmode of the vibration. The eigenmode of the flexible rotor has a U shape for the first, S shape for the second, and W shape for the third eigenfrequency.

In case of linear rotordynamics, the modal analysis is applied to calculate the complex eigenvalues and eigenvectors of the rotor. The eigenfrequencies are the imaginary parts of the complex eigenvalues, in which the positive sign is for the forward whirl and the negative sign, for the backward whirl of the rotor. Similarly, the real parts of the complex eigenvalues show the behavior of the rotor stability that shall be discussed in the next section.

The stiffness and damping coefficients of the rotor are linear only at the small deflections of the well-balanced and damped rotor. As a result, the vibration equation of the rotor is linear. Otherwise, the restoring forces of the rotor are nonlinear at the large rotor deflections; hence, the rotor vibration equation becomes nonlinear. The nonlinear rotordynamic system is caused not only by the nonlinear characteristics of the stiffness and damping coefficients of the oil-film bearings but also by other characteristics, such as oil whirl, rotor misalignment, contact rub between the rotor and bearings, excessive unbalance, and as well as sidebands due to frequency modulations. Therefore, the rotor response of nonlinear rotordynamic systems results from superimposing the unbalance excitation of the rotor on the nonsynchronous frequency components induced by rotordynamic nonlinearity of the system.

The nonsynchronous frequency components contain the sub- and supersynchronous frequency components. The subsynchronous vibrations have frequencies smaller than the rotor frequencies with rational orders of (1/2)X, (1/3)X of the contact rub or irrational orders, such as 0.35X to 0.47X of the oil whirl. On the contrary, the supersynchronous vibrations have the frequencies higher than the rotor frequencies with integer orders of 2X, 3X, 4X or noninteger orders of 2.5X, 3.5X in case of the frequency modulations. Both vibrations consist of the forward and backward whirls that are determined by the sign of their eigenfrequencies. As a result, the rotor locus in a stable condition has mostly the Lissajous curves relating to the synchronous and nonsynchronous vibrations. The nonlinear vibration responses are normally the periodic or quasiperiodic vibrations. In case of the rotor instability they could be changed from the periodic to quasiperiodic and chaos vibrations due to bifurcation.

Due to nonlinearity of the rotor, the eigenfrequencies of the rotor vibration have no longer existed. As a result, the assumption that the vibration response results from the eigenvectors and eigenvalues of the rotor, fails in nonlinear rotordynamics. Therefore, the modal analysis is not valid for analyzing the eigenfrequencies (natural frequencies) of the nonlinear rotordynamic system. In this case, we do not use Campbell diagram to study the rotor responses based on the eigenfrequencies and find the critical frequencies of the rotor. Both eigenfrequencies and therefore critical frequencies are nonexistent in nonlinear rotordynamics. Instead, Waterfall diagram (frequency spectrum diagram) is employed to analyze the whirl frequencies of the nonlinear rotor response. The frequency spectrum analysis in Waterfall diagram will be discussed more in details in Chap. 7.

Some essential remarks of nonlinear rotordynamics of turbochargers are summarized as follows [4–7]:

1. In nonlinear rotordynamics, the vibration responses are not harmonic but periodic, quasiperiodic, and chaotic at the large rotor amplitudes. In fact, they contain not only the synchronous but also nonsynchronous vibrations, such as sub- and supersynchronous vibrations due to the nonlinear phenomena, such as excessive unbalance, misalignment, contact rub in the bearings, oil whirl, and sidebands as well.
2. The nonsynchronous frequencies of the vibration responses are induced by large rotor deflections and other nonlinear characteristics of the rotor and radial bearing forces.
3. The response amplitude of the rotor consists of not only the synchronous vibration amplitude of the unbalance but also the nonsynchronous vibration amplitudes depending on the rotor speed. In general, the amplitudes of the nonsynchronous vibrations are much larger than the synchronous unbalance response amplitude of 1X. Therefore, the bending resonances do not obviously emerge in nonlinear rotordynamics. Instead, the rotor response amplitude is stabilized on the limit cycle (see Chaps. 4 and 7).
4. The bearing stiffness and damping coefficients strongly increase with the rotor deflections leading to the increase of the bearing forces. Due to the force balance between the bearing, unbalance and other forces, the rotor response amplitude increases and reaches the equilibrium position. In fact, the unstable rotor orbit is restabilized on the limit cycle.
5. At an unbalance force acting upon the rotor, different rotor responses could occur in nonlinear rotordynamics due to bifurcation that shall be discussed in

Chap. 4. The responses generally depend on the initial conditions and other parameters, such as rotor speed, bearing clearances, oil inlet temperature and pressure.

6. Campbell diagram (whirl speed map) is used to analyze the eigenfrequencies in linear rotordynamics. However, it has not been applicable in nonlinear rotordynamics any longer. Instead, Waterfall diagram (spectrogram) is applied to study the whirl frequencies and amplitudes of the rotor response versus the rotor speeds in nonlinear rotordynamics.
7. Hurwitz–Routh criterion is generally used to study rotor stability in linear rotordynamics; Hopf bifurcation theory, in nonlinear rotordynamics.

3.5 Measurement of the Rotor Locus

3.5.1 Working Principle of the Eddy-Current Sensor

To measure the locus (orbit) of the certain point on the rotor, two eddy-current sensors are set up perpendicularly to each other near this point. Figure 3.4 shows the contactless measure principle of the eddy-current sensor and explains how it works on the measurement of the rotor locus [1–3].

The working principle of the eddy-current sensor is based on the electromagnetic induction. Applying an alternating voltage U_{AC} with a high frequency to the electric coil of the sensor probe, a high-frequency alternating electromagnetic field is created at the end of the sensor probe according to the induction law. Inserting the sensor probe near the measured point on the rotor shaft at a short distance δ, the

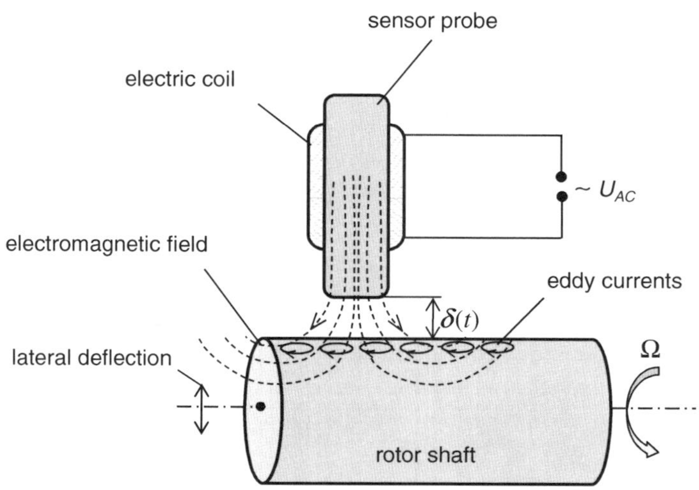

Fig. 3.4 Working principle of the eddy-current sensor

eddy currents are induced in the target material of the rotor shaft. In turn, the eddy currents create an electromagnetic field that reacts against the initial electromagnetic field of the sensor probe leading to a change of the electric coil impedance with time. The impedance varies with the current distance δ between the sensor probe and rotor shaft. Therefore, the time-dependent distance $\delta(t)$ is determined by measuring the change of the electric coil impedance with time. To make sure that the measuring principle works, the target surface on the rotor shaft must be at least three times larger than the cross-sectional area of the sensor probe and the distance δ between the sensor probe and the rotor shaft is about 1–2 mm.

3.5.2 Measurement of the Rotor Locus

Automotive turbochargers are applied to high temperatures of the exhaust gases, such as 820–850 °C in diesel engines and 950–1050 °C in gasoline engines. Therefore, it is not easy to measure the rotor locus at the turbine side due to very high temperatures of the exhaust gas. Additionally, there is quite difficult to install the sensors in the bearing casing to measure the orbit of the journal in the radial bearings. As a result, one sets up the two eddy-current sensors perpendicularly to each other at the compressor housing to measure the orbit of the rotor shaft at the compressor inlet, as displayed in Figs. 3.5 and 3.6.

The measured orbit shows only the displacement of the shaft at this position but not the actual displacements of the journal inside the bearings or the rotor deflection between both radial bearings. However, analyzing the spectrum of the rotor frequencies and amplitudes at various rotor speeds in Waterfall diagram helps us to know the rotor response behavior.

Additionally, the acceleration in the bearing casing near the radial bearings is also measured using an acceleration sensor. Its acceleration spectrum in Waterfall

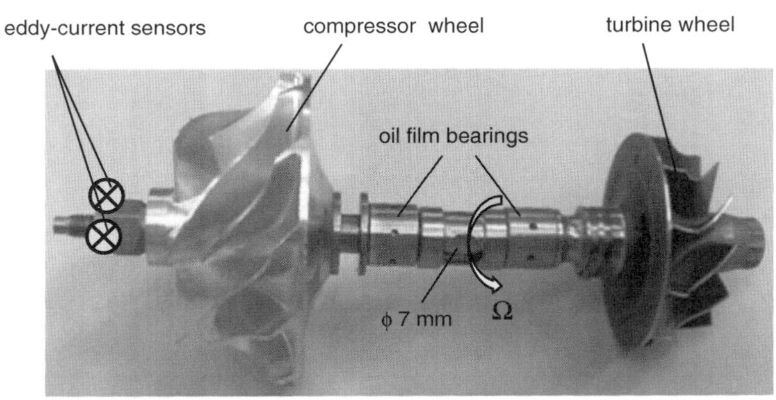

Fig. 3.5 Position of the eddy-current sensors in the rotor

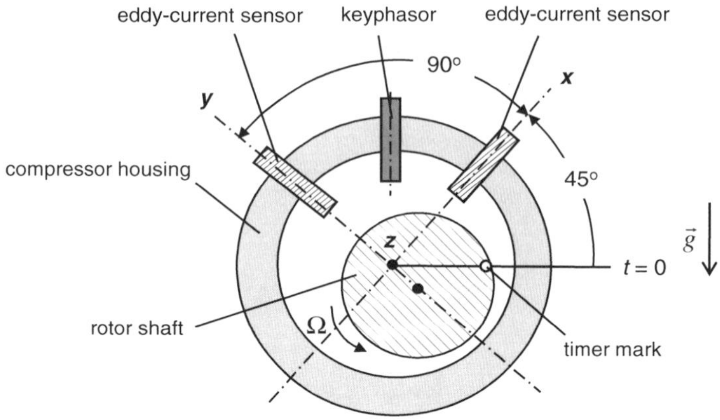

Fig. 3.6 Setup of the eddy-current sensors on the compressor housing

diagram determines whether the contact rub between the rotor and bearings occurs. If the contact rub happens in the bearings, the large amplitudes of subsynchronous vibrations of (1/2)X or (1/3)X, and sideband frequencies will appear in Waterfall diagram, especially at the high rotor speeds. In addition, the computation of nonlinear rotordynamics provides all characteristics of the rotor response, like the rotor deflections, whirl frequencies, and vibration modes at any point of the rotor versus rotor speeds at the run-up and slowdown simulation.

The eddy-current sensors are mounted perpendicularly to each other in the axes x and y on the compressor housing, as shown in Fig. 3.6. The distance of the sensors to the target is approximately 1 mm. The Bently keyphasor [1] located between the eddy-current sensors is used to indicate the rotor speed in the orbit. Each sensor measures the displacement of the rotor in both directions x and y. The rotor orbit has the form of Lissajous curve in the phase diagram x-y.

The orbit amplitude of the measured point on the rotor results from its amplitude components in the directions x and y.

$$r(t) = \sqrt{x(t)^2 + y(t)^2} \tag{3.7}$$

where $x(t)$, $y(t)$ are the measured vibration signals in the directions x and y, respectively.

The vibrations of the rotor response are measured by the eddy-current sensors in the directions x and y, as shown in Fig. 3.6. The orbit amplitude of the rotor is calculated using Eq. (3.7) and displayed in the phase plane x-y (cf. Figs. 3.8, 3.11, 3.14, 3.17, and 3.20).

3.5.3 Studying Cases of the Rotor Orbit

The vibration signals in the directions x and y are measured by the eddy-current sensors, as shown in Fig. 3.6. By means of the FFT spectra analyzer, the corresponding frequencies are displayed in the spectrum diagram, and the orbit is computed and plotted in the phase diagram x-y.

In nonlinear rotordynamics, the harmonic vibration of the unbalance excitation is superimposed on the sub- and supersynchronous frequency components of the rotor resulting in the rotor vibration response. The vibration response of the rotor contains vibration components in the directions x and y and can be written in the real and imaginary terms in the phase plane x-y ([6, 7]).

$$\begin{aligned} r(t) &= x(t) + jy(t) \\ &= \left(r_{fw}e^{+j\Omega t} + r_{bw}e^{-j\Omega t}\right) + \sum_{k=1}^{K}\left(r_{k,fw}e^{+j\omega_k t} + r_{k,bw}e^{-j\omega_k t}\right) \end{aligned} \tag{3.8}$$

where

- the first term of the right-hand side is the forward and backward unbalance harmonics with the rotor frequency Ω;
- the second term is the asynchronous vibrations at the whirl frequencies $\omega_k = n_k\Omega$;
- K is the asynchronous frequency components of the nonlinear rotor.

Depending on the sub- or supersynchronous vibrations, the whirl frequency orders n_k could be integer, fractional, and as well as irrational numbers ($n_k < 1$ or $n_k > 1$). The positive and negative signs of the whirl frequencies indicate forward and backward whirls, respectively.

3.6 Study of Case Histories

In the following section, some rotor orbits of the turbochargers are computed in which the rotor vibration response results from the superimposition of the unbalance vibration (1X) on the nonsynchronous frequency components that are induced by the nonlinearity characteristics of the rotor, such as the nonlinear bearing forces, excessive unbalance, rotor misalignment, and contact rub in the bearings.

In order to study the case histories, the frequencies and amplitudes of the vibration components are predetermined and displayed in the frequency spectrum diagram [3, 4, 8]. The time vibrations in the directions x and y, and the rotor response are plotted in the time domain. Finally, the orbit of the rotor response is displayed in the phase plane x-y. These study cases aim to make the readers familiar with the nonlinear vibration responses that include various synchronous and asynchronous frequency components in nonlinear rotordynamics of automotive turbochargers.

Case History #1:
An unbalanced rotor

Anisotropic (orthotropic) radial bearings are defined when the bearing stiffness coefficients K_x and K_y in the directions x and y are different from each other. In other case, they are called isotropic bearings. Due to the different stiffness coefficients, the rotor deflection in the direction x is obviously different from the rotor deflection in the direction y. As a result, the orbit shape of the rotor is elliptic where the maximal rotor deflection is in the direction with the lower bearing stiffness coefficient.

The rotor vibration response results from superimposing the forward and backward whirls of the unbalanced rotor.

- Component in the direction x:

$$x(t) = r\cos\Omega t + r_1\cos(-\Omega t + \varphi_1)$$

- Component in the direction y:

$$y(t) = r\sin\Omega t + r_1\sin(-\Omega t + \varphi_1)$$

where

r is the forward vibration amplitude (peak amplitude);
r_1 is the backward vibration amplitude (peak); given $r_1 = 0.5r$;
Ω is the rotor angular speed;
t is time;
φ_1 is the vibration phase (= $\pi/4$).

The frequency spectrum of the vibration components are displayed in Fig. 3.7. The orbit of the rotor response results from superimposing the vibration components, as displayed in Fig. 3.8. In case of $K_x < K_y$, the orbit is elliptical with the maximal deflection in the direction x.

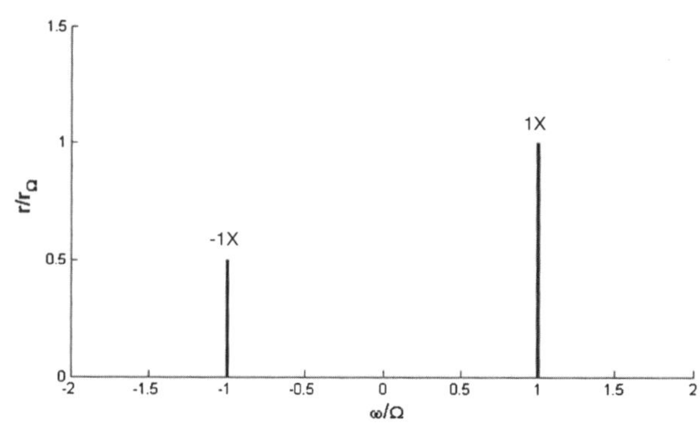

Fig. 3.7 Frequency spectrum of the unbalance vibration

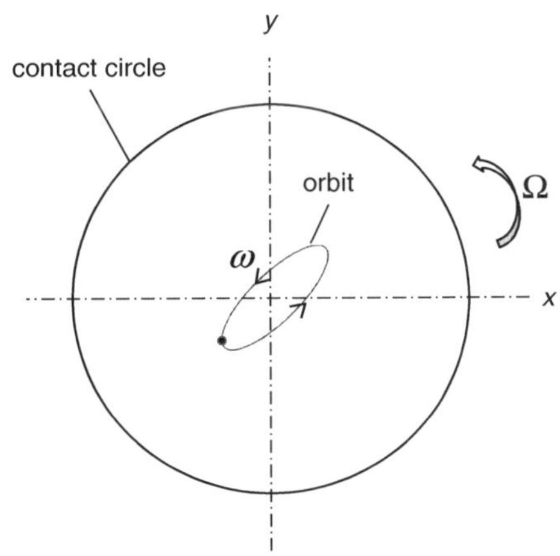

Fig. 3.8 Orbit of the response of an unbalanced rotor

The whirl precession with the frequency $\omega = \Omega$ has the same rotating direction of the rotor velocity. In this case, the rotor orbit is a forward whirl because the forward amplitude is larger than the backward amplitude. Note that the whirling direction of the rotor follows the whirling direction of the vibration component with the larger amplitude.

The rotor locus moves forwards from the equilibrium position (nearly in the middle of the contact circle) in an elliptical orbit. The contact circle is determined by manually moving and rotating the end of the compressor wheel at the screw nut with a slight contact occurring somewhere between the rotor and bearings. In reality, the contact circle becomes larger in the operating condition of the turbocharger. In the turbocharger operation, the inner bearing clearance increases due to the thermal expansion of the bearing ring in radial direction; and the deflection of the flexible rotor end at the compressor wheel is higher compared to the rigid rotor at the calibration. As a result, both effects are to blame for the larger contact circle during the operation.

Case History #2:
An unbalanced rotor with inner oil whirl

An unbalanced rotor operating under influence of the inner oil whirl occurred in the radial bearings is studied. Oil whirl is subsynchronous self-excited vibration with an irrational frequency order between 0.35X and 0.47X (half-frequency whirl).

The effect of the inner oil whirl is induced by the whirling of lubricating oil in the bearing clearance that causes the rotor whirling at the irrational frequency. Therefore, the oil whirl mostly occurs in the bearing clearance without the precondition of the rotor unbalance. Note that the unbalanced rotor causes the unbalance response on the rotor; the whirling of oil in the bearing clearance induces the oil whirl. Therefore, both effects have nothing to do with each other.

In fact, the oil whirl does not happen as long as the damping in the bearing is large enough to keep the energy balance between the whirling kinetic energy of oil induced by the destabilizing force and the dissipated damping energy in the bearing clearance. On the one hand, the damping coefficient of bearing decreases with the oil temperature in the bearing. On the other hand, the cross-coupled stiffness coefficient of the bearing that causes the destabilizing force of the rotor increases with the rotor speed. Therefore, the whirling kinetic energy of the rotor exceeds the dissipated damping energy in the oil film as the rotor speed increases leading to oil whirl. The oil whirl excites itself from a stable to unstable rotor response.

The rotor vibration response results from the superimposition of the unbalance excitation and subsynchronous frequency component of the oil whirl (see Fig. 3.9).

- Component in the direction x:

$$x(t) = r\cos\Omega t + r_1\cos(\omega_1 t + \varphi_1)$$

- Component in the direction y:

$$y(t) = r\sin\Omega t + r_1\sin(\omega_1 t + \varphi_1)$$

where

r is the harmonic vibration amplitude (peak);
r_1 is the oil whirl vibration amplitude (peak); given $r_1 = r$;
Ω is the rotor frequency;
ω_1 is the oil whirl frequency with an irrational frequency order 0.47X;
φ_1 is the vibration phase.

Figure 3.10 displays the superimposed vibrations in the directions x and y and the vibration response of the rotor. The phase difference between the vibrations in the directions x and y is 90° because the eddy-current sensors are perpendicular to each other (see Fig. 3.6b). At the rotor speed of 200,000 rpm, the period T_Ω is about 0.3 ms; the period of the oil whirl vibration, $T_{\omega 1}$ nearly 0.64 ms.

The rotor orbit resulted from the periodic vibrations of the unbalance and self-excited oil whirl in the directions x and y are calculated at the rotor speed of 200,000 rpm and displayed in Fig. 3.11. The rotor precession is a forward whirl that has the same direction of the rotor speed. Due to the irrational frequency order of

Fig. 3.9 Frequency spectrum of the vibration components

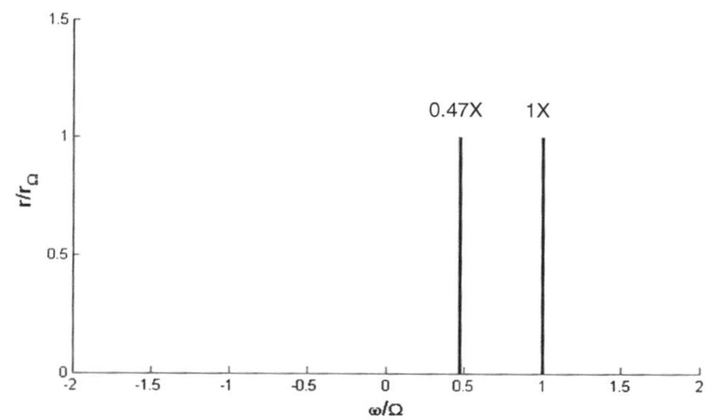

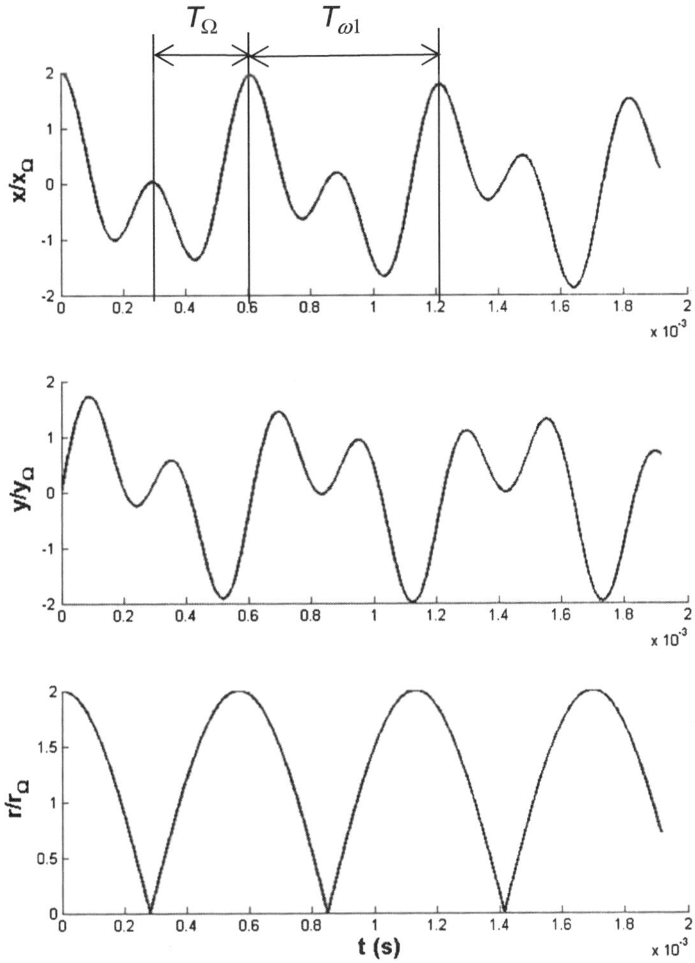

Fig. 3.10 Vibrations in x and y directions and the rotor response

the oil whirl (0.47X), the rotor response is quasiperiodic, in which the orbits are not the same after repeat of period T. The timer mark taken by the keyphasor indicates the number of *revolution* of the rotor in an orbit cycle, which is called the *convolution*.

Having considered a single orbit of the rotor response, there are 2 timer marks on it; i.e., the orbit completes one convolution in every two revolutions of the rotor due to 2 timer marks. In other words, the oil whirl is a subsynchronous vibration and has a frequency of nearly one-half of the rotor frequency ($\omega = 0.47\Omega$). Generally, in the subsynchronous frequency components with a frequency order of $(1/N)$X, the orbit of the rotor response completes one convolution in every N revolutions with N timer marks. In the supersynchronous components with a frequency order of NX,

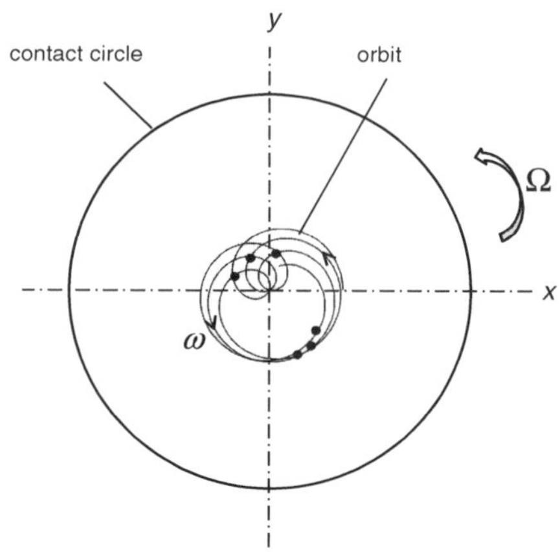

Fig. 3.11 Orbit of the rotor response (oil whirl)

the rotor orbit completes N convolutions in one revolution of the rotor with 1 timer mark.

The asynchronous frequency components are summarized as

- Subsynchronous frequency order $(1/N)$X means N revolutions per convolution;
- Supersynchronous frequency order NX means N convolutions per revolution.

If the whirl orbit has N inner loops, it corresponds to a subsynchronous forward whirl with a frequency order of $1/(N + 1)$X. As a rule of thumb, the frequency order is the inversion of the number of loops $(N + 1)$. In case of the oil whirl with a forward whirl frequency order of (1/2)X, it results in one inner loop in the orbit, as shown in Fig. 3.11. On the contrary, the backward whirls have N outer loops; the whirl frequency order is $1/(N - 1)$X.

Case History #3:
An unbalanced rotor with inner and outer oil whirls

To reduce the bearing friction, rotating floating ring bearings with two oil films have been usually applied to automotive turbochargers. The vibration response of the rotor results from superimposing the harmonic unbalance excitation on the subsynchronous components, the inner and outer oil whirl vibrations. The self-excited frequency components have irrational frequency orders from 0.25X to 0.70X for the inner oil whirl, from 0.1X to 0.3X for the outer oil whirl. The whirl frequency orders depend on the rotor speed and as well as the oil temperature in the bearing.

The rotor vibration response results from superimposing the unbalance vibration on the subsynchronous frequency components of the oil inner and outer whirls.

- Component in the direction x:

$$x(t) = r \cos \Omega t + r_1 \cos(\omega_1 t + \varphi_1) + r_2 \cos(\omega_2 t + \varphi_2)$$

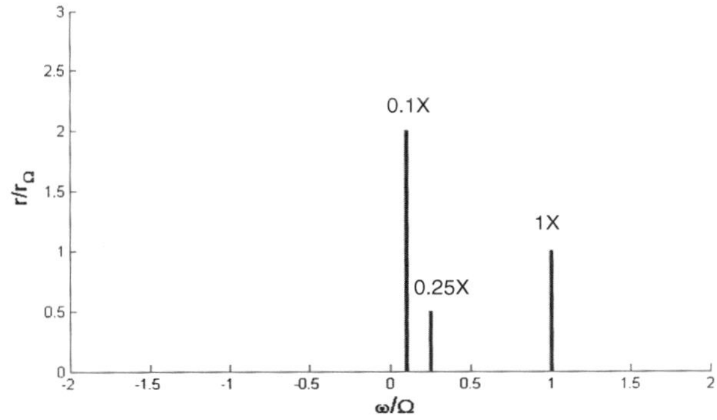

Fig. 3.12 Frequency spectrum of the vibration components

- Component in the direction y:

$$y(t) = r\sin\Omega t + r_1 \sin(\omega_1 t + \varphi_1) + r_2 \sin(\omega_2 t + \varphi_2)$$

where

r	is the harmonic vibration amplitude (peak);
r_1	is the vibration amplitude (peak) of the inner oil whirl; given $r_1 = 0.5r$;
r_2	is the vibration amplitude (peak) of the outer oil whirl; given $r_2 = 2r$;
Ω	is the rotor frequency;
ω_1	is the inner oil whirl frequency with an irrational frequency order 0.25X;
ω_2	is the outer oil whirl frequency with an irrational frequency order 0.1X;
φ_1, φ_2	are the vibration phases of the vibration components.

The frequency spectrum of the vibration components is displayed in Fig. 3.12 where the frequency components of the inner and outer oil whirls have frequency orders of 0.25X and 0.1X, respectively.

The amplitude of the rotor response consisting of the unbalance, inner and outer oil whirls at the rotor speed of 200,000 rpm is calculated according to Eqs. (3.7 and 3.8), as shown in Fig. 3.13. The vibration response is a periodic or quasiperiodic vibration because their subsynchronous whirls have irrational frequency orders. In case of the quasiperiodic response, the amplitudes after repeat of period T are not always the same amplitude. Its response orbit shown in Fig. 3.14 does not have the same orbit at times $(t + nT)$. In Fig. 3.13 the period of the rotor T_Ω is found at 0.3 ms at 200,000 rpm; the period of the inner oil whirl $T_{\omega 1}$ nearly 1.2 ms; the period of the outer oil whirl $T_{\omega 2}$, approximately 3 ms.

At high rotor speeds, the temperature of the inner oil film becomes higher than the temperature of the outer oil film due to the smaller inner bearing clearance. As a result, the ring speed of the bearing decreases with the rotor speed and the frequency of the inner oil whirl reduces to nearly 25 % of the rotor frequency (0.25X) at high rotor speeds. The frequency of the outer oil whirl is lower than the inner oil whirl. It is assumed approximately 10 % of the rotor frequency (0.1X).

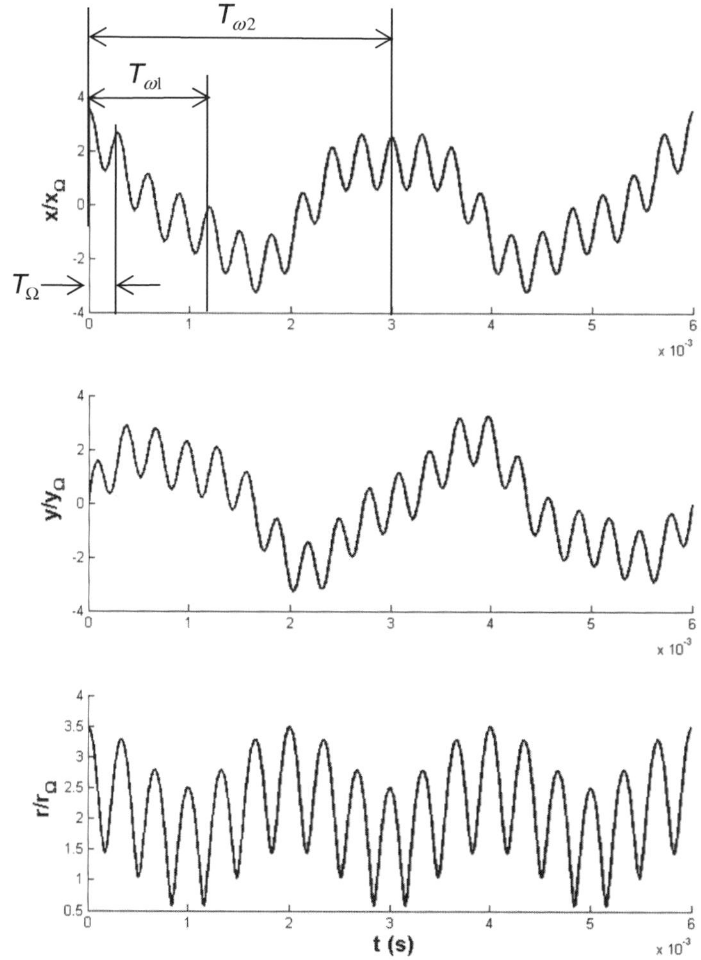

Fig. 3.13 Vibrations in the *x* and *y* directions and the rotor response

The rotor orbit resulted from the periodic vibrations of the unbalance and self-excited inner and outer oil whirls in the directions *x* and *y* are calculated and displayed in Fig. 3.14. The inner loops of the rotor orbit indicate that the rotor precession is a forward whirl and has the same direction of the rotor velocity. The envelope of the rotor orbit is called the limit cycle, on which the rotor locus stabilizes. Note that the limit cycle is different to the contact circle of the rotor (see Fig. 3.14).

Case History #4:
A rotor with heavy rub in the bearings

In a heavy rub at normal tight condition in the bearings, the unbalance force excites the subsynchronous frequency components of the nonlinear rotor induced by the heavy rub resulting to the rotor response. At the heavy rub in the bearings, the rub

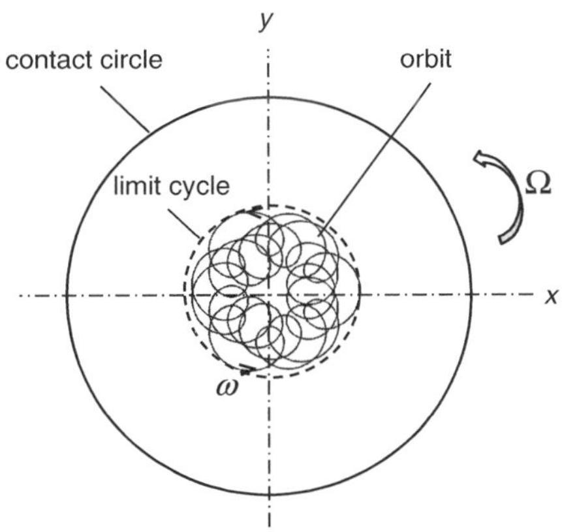

Fig. 3.14 Orbit of the rotor response with the inner and outer oil whirls

subsynchronous frequency components have fractional frequency orders of (1/2)X, (1/3)X, or (1/4)X, and high harmonic orders of 2X, 3X.

The rotor response results from superimposing the unbalance vibration on the subsynchronous frequency component of the heavy rub.

- Component in the direction x:

$$x(t) = r\cos\Omega t + r_{bw}\cos(-\Omega t) + r_1\cos(\omega_1 t + \varphi_1)$$

- Component in the direction y:

$$y(t) = r\sin\Omega t + r_{bw}\sin(-\Omega t) + r_1\sin(\omega_1 t + \varphi_1)$$

where

r is the harmonic vibration amplitude (peak);
r_{bw} is the backward harmonic vibration amplitude (peak); given $r_{bw} = 0.5r$;
r_1 is the vibration amplitude (peak); given $r_1 = 1.5r$;
Ω is the rotor frequency;
ω_1 is the rub-related frequency with a fractional frequency order (1/2)X;
φ_1 is the vibration phase.

The frequency spectrum of the vibrations is displayed in Fig. 3.15 where the rub frequency component has a fractional frequency order of (1/2)X.

The amplitude of the rotor response consisting of the unbalance and subsynchronous whirl of heavy rub at the rotor speed of 200,000 rpm is calculated according to Eqs. (3.7 and 3.8), as shown in Fig. 3.16. The rotor response is periodic

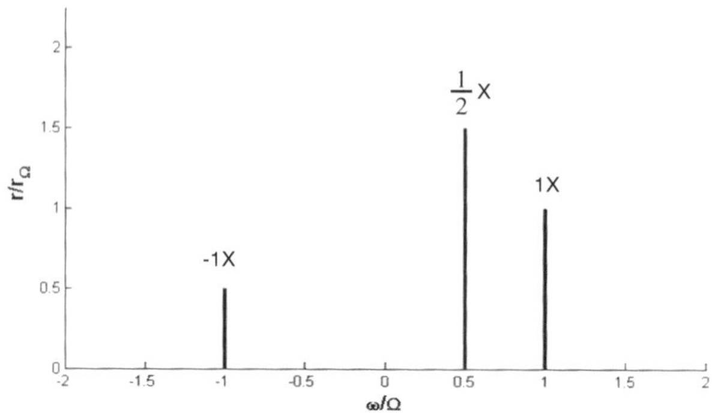

Fig. 3.15 Frequency spectrum of the vibration components

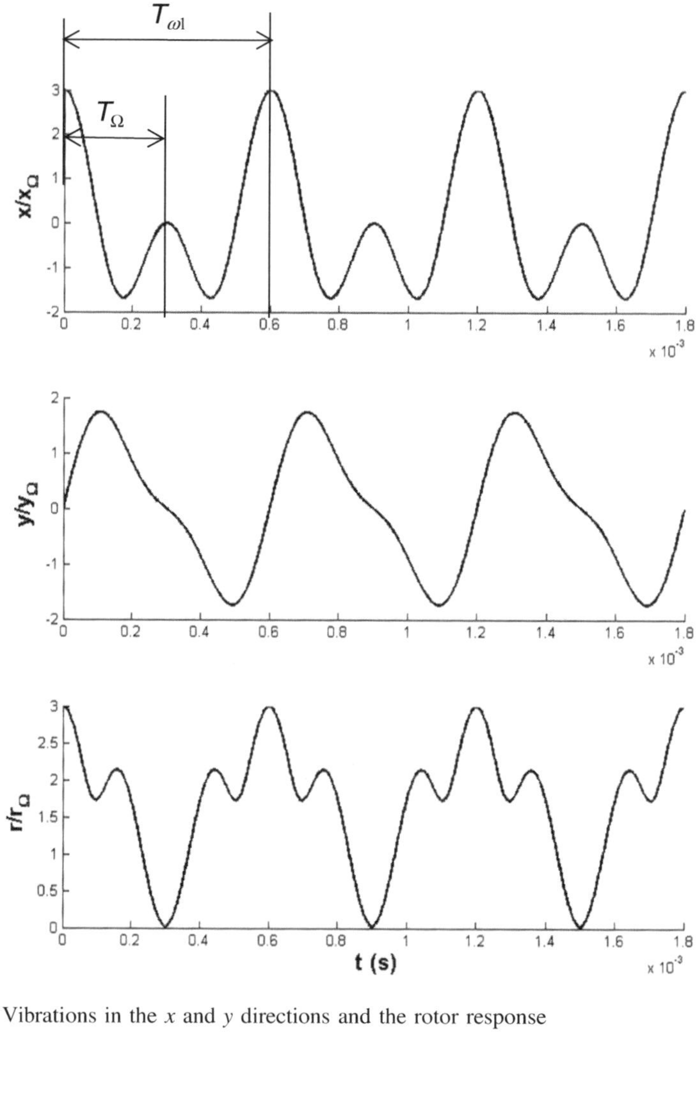

Fig. 3.16 Vibrations in the x and y directions and the rotor response

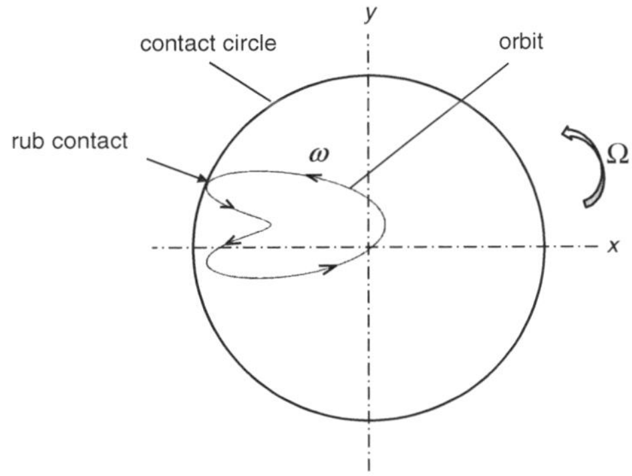

Fig. 3.17 Orbit of the rotor response with a heavy rub

because the subsynchronous whirl has a fractional frequency order of (1/2)X. In Fig. 3.16, the period of the rotor T_Ω is found at 0.3 ms at 200,000 rpm; the period of the contact rub whirl $T_{\omega 1}$ nearly 0.6 ms.

The rotor orbit resulted from the periodic vibrations of the unbalance harmonics and rub subsynchronous whirl in the directions x and y are calculated at the rotor speed of 200,000 rpm and plotted in Fig. 3.17. The rotor response is a forward whirl with a fractional frequency component of (1/2)X. It indicates that the rub contact of the rotor to the bearing occurs once in every two revolutions of the rotor. After contacting the bearing wall, the rotor bounces back; the rub contact takes place again after two revolutions of the rotor.

Case History #5:
A rotor with high radial loads due to misalignment

The rotor misalignment induces high radial loads on the bearings because the journal eccentricity increases. As a result, the oil film in the bearing clearance is squeezed so that the rotordynamic behavior of the system becomes strongly non-linear at the large journal eccentricity. In fact, the unbalance forces excite the supersynchronous frequency components at high radial bearing loads resulting to the rotor response. The supersynchronous frequency components normally have integer frequency orders of 2X and 3X at high radial loads due to misalignment.

The rotor response results from superimposing the unbalance vibration on the supersynchronous frequency component of the rotor misalignment.

- Component in the direction x:

$$x(t) = r\cos\Omega t + r_{1,fw}\cos(\omega_1 t + \varphi_1) + r_{1,bw}\cos(-\omega_1 t + \varphi_1)$$

- Component in the direction y:

$$y(t) = r\sin\Omega t + r_{1,fw}\sin(\omega_1 t + \varphi_1) + r_{1,bw}\sin(-\omega_1 t + \varphi_1)$$

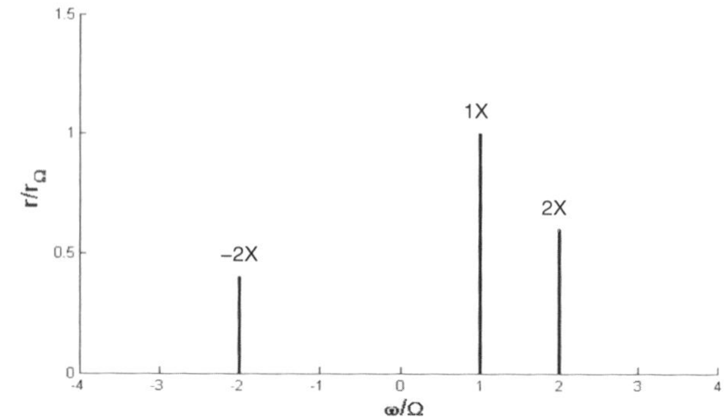

Fig. 3.18 Frequency spectrum of the vibration components

where

r	is the harmonic vibration amplitude (peak);
$r_{1,fw}$	is the forward vibration amplitude 2X (peak); given $r_{1,fw} = 0.6r$;
$r_{1,bw}$	is the backward vibration amplitude −2X (peak); given $r_{1,bw} = 0.4r$;
Ω	is the rotor frequency;
ω_1	is the whirl frequency 2X;
φ_1	is the vibration phase.

The frequency spectrum of the vibrations is displayed in Fig. 3.18, in which the supersynchronous frequency components have an integer frequency order of 2X of the misalignment or extremely large radial load; the unbalance excitation frequency has an order of 1X.

The amplitude of the rotor response consisting of the unbalance harmonics and misalignment supersynchronous whirl at the rotor speed of 200,000 rpm is calculated according to Eqs. (3.7 and 3.8) and displayed in Fig. 3.19. The vibration response is periodic because the supersynchronous whirl has an integer frequency order of 2X.

In Fig. 3.19, the period of the rotor T_Ω is found at 0.3 ms at 200,000 rpm; the period of the misalignment frequency whirl $T_{\omega 1}$, nearly 0.15 ms. The response orbit resulted from the superimposition of the unbalance harmonics and misalignment supersynchronous whirls in the directions x and y are calculated at the rotor speed of 200,000 rpm and plotted in Fig. 3.20.

At increasing the radial bearing load due to the misalignment, the rotor orbit changes from the "banana" shape (see Fig. 3.20) to the "lying-eight" shape (see Fig. 3.21). The banana shape is a forward whirl and bounces back as it comes close to the bearing wall. In case of very high radial loads on the bearings, the orbit of the

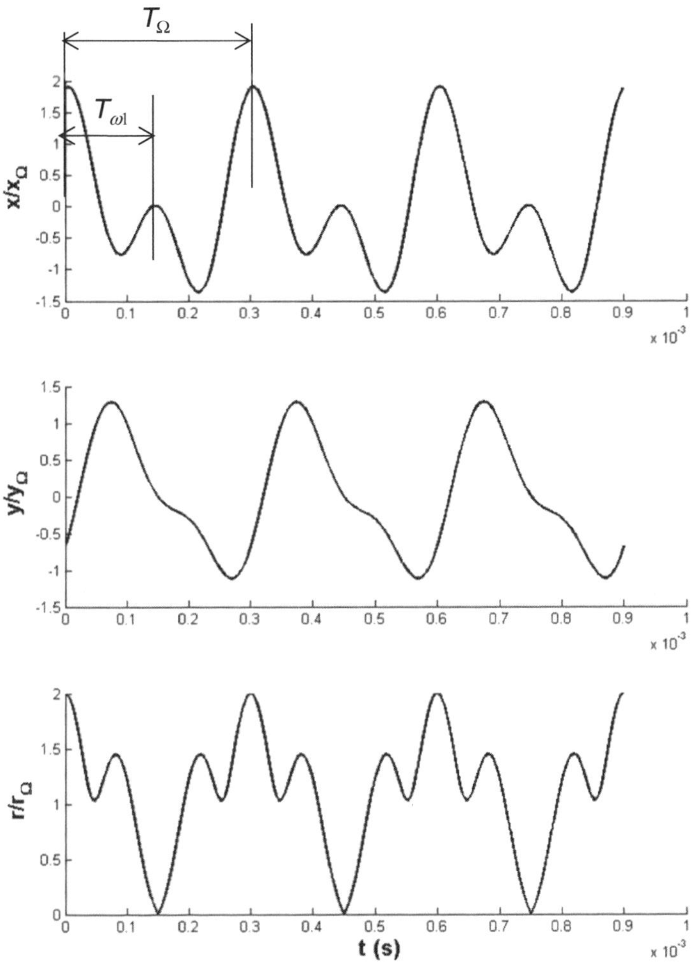

Fig. 3.19 Vibrations in the x and y directions and the rotor response

rotor response turns into the lying-eight shape. Contrary to the banana shape, the lying-eight shape has two loops of forward and backward whirls. The journal bounces back in the backward loop near the bearing wall. Far away from it, the radial load is relieved. As a result, the rotor turns forwards and has the same direction of the rotor velocity (see Fig. 3.21).

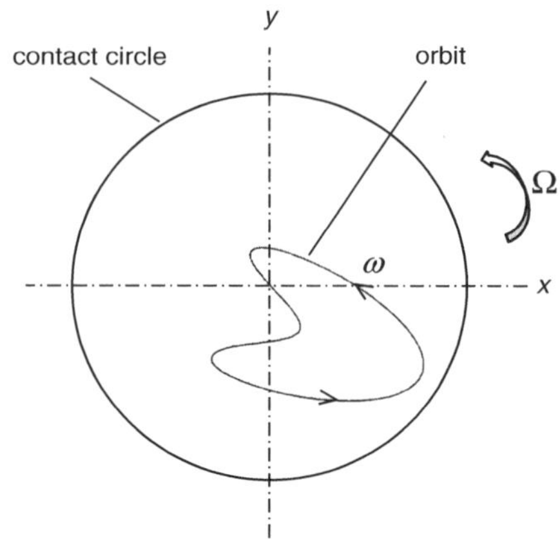

Fig. 3.20 Orbit of the rotor response with a banana shape

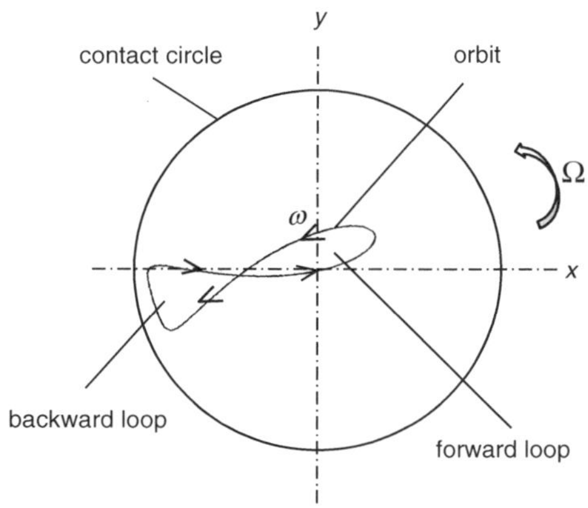

Fig. 3.21 Orbit of the rotor response with a lying-eight shape

The rotor misalignment with an extremely high radial load or the excessively unbalanced rotor could cause the contact rub between the journal and bearing. The frequencies have multiple harmonic frequency orders of 2X, 3X, or higher, and as well as fractional frequency orders of (1/2)X, (1/3)X, as discussed in the case history #4.

References

1. Bently, D.E., Hatch, C.: Fundamentals of Rotating Machinery Diagnostics. Bently Pressurized Bearing Press, Minden (2002)
2. Ehrich, F.: Handbook of Rotordynamics. Krieger Publishing Company, Malabar (2004)
3. Gasch, R., Nordmann, R., Pfützner, H.: Rotordynamik, 2. Auflage. Springer, New York (2006)
4. Kraemer, E.: Rotordynamics of Rotors and Foundations. Springer, New York (1993)

5. Newland, D.E.: Mechanical Vibration Analysis and Computation. Dover Publications Inc, New York (2006)
6. Schmidt, G., Tondl, A.: Nonlinear Vibrations. Cambridge University Press, Cambridge (2009)
7. Tondl, A.: Some Problems of Rotor Dynamics. Chapman & Hall, London (1965)
8. Vance, J.: Rotordynamics of Turbomachinery. Wiley, New York (1988)

Chapter 4
Stability Analysis of Rotordynamic Behaviors

4.1 Introduction

Two important aspects in rotordynamics of the automotive turbochargers are resonance and instability of the rotor. Both have a common negative effect that causes damage of the turbochargers during operation. However, there is a big difference between the resonance and instability if we take a close look on them.

At the resonance, the rotor reaches maximum deflection at the critical frequencies that depend on the rotor characteristics, such as mass, mass inertia moments, stiffness coefficient of the rotor. The excitation unbalance force is proportional to squared rotor speed. The unbalance force is balanced by the inertia force, bearing force, stiffness, and damping forces acting upon the rotor at any time. The more the rotor deflects, the larger the stiffness force acts upon the rotor. In case of lacking or small damping effect, the deflection of the rotor extremely increases at the resonance so that the normal stress of the shaft exceeds the ultimate tensile stress, or the contact between the journal and bearings occurs at the breakdown of the oil film. Therefore, the rotor is damaged at the critical frequency. Due to self-centering effect of the rotor, the maximum deflection amplitude is reduced at once when the rotor speed passes the critical frequency.

Contrary to the resonance, the rotor amplitude in an unstable condition extremely increases with time and is independent of the rotor speed. As soon as the rotor speed exceeds the onset of instability (threshold of instability), the rotor becomes unstable at which the rotor amplitude increases with time without limit. Its amplitude is rising further even at changing the rotor speed. This phenomenon is very different to the case of resonance. The rotor instability occurs when the oil whirl takes place in the oil-film bearings. In this case, the dissipative damping force is small compared to the destabilizing force induced by the cross-coupled stiffness coefficient of the bearing leading to the oil whirl development.

H. Nguyen-Schäfer, *Rotordynamics of Automotive Turbochargers*,
Springer Tracts in Mechanical Engineering, DOI 10.1007/978-3-319-17644-4_4

The oil whirl is a kind of the self-excitation instability where the rotor amplitude increases without precondition of the unbalance. When the oil whirl frequency reaches the bending critical frequency, mostly the first critical frequency, the *oil whirl* turns to *oil whip*. At the oil whip, the rotor becomes a real instability where the rotor deflection continuously increases faster and the journal eccentricity exceeds the bearing clearance leading to damage of the radial bearings and turbochargers as well.

The resonance phenomenon is relatively easy to deal with; thus, it will be discussed later in Chap. 5. On the contrary, the rotor instability due to oil whirl and oil whip is more dangerous for the turbochargers and much more difficult to deal with, especially in nonlinear rotordynamics. Therefore, it is discussed first in the following section. In order to avoid damages of the turbochargers due to rotor instability, the rotor characteristics of the compressor, turbine wheels, shaft, sealing, and fluid film bearings must be analyzed at the beginning of the rotordynamic design.

There are two methods to analyze the stability of the turbochargers. First, eigenvalues analysis with Routh–Hurwitz criterion is for linear rotordynamic problems. Second, bifurcation method with Hopf theory is for nonlinear rotordynamic problems.

4.2 Stability Analysis of Linear Rotordynamics

4.2.1 Eigenvalues of the Free Vibration Response

The eigenvalues analysis is based on the free vibration response of the linear rotordynamic system. The free vibration response results from superimposing the harmonic components of the rotor without excitation of the rotor unbalance. In fact, the free vibration response is exactly the homogeneous solutions of the vibration equations of the rotor. Each harmonic component has a complex eigenvalue in which the imaginary part is the eigenfrequency of the rotor (natural frequency), and the real part indicates the stability characteristic of the rotor [2, 3].

The free vibration response of the rotor is written as

$$r(t) = \sum_{i=1}^{N} r_i e^{\lambda_i t} \tag{4.1}$$

where

r_i is the rotor amplitude corresponding to the eigenvalue i;
λ_i is the complex eigenvalue of the eigenmode i;
N is the number of degrees of freedom (DOF) of the rotor.

The complex eigenvalue consists of the real and imaginary parts

$$\lambda = \alpha \pm j\omega_d \tag{4.2}$$

within

α is the real part of the complex eigenvalue and called the growth/decay rate;

ω_d is the imaginary part of the complex eigenvalue (eigenfrequency or damped natural frequency).

Substituting λ into Eq. (4.1), one obtains the rotor free response applying Euler's theorem.

$$\begin{aligned} r(t) &\equiv x(t) + jy(t) \\ &= \sum_{i=1}^{N} r_{i,fw} e^{\alpha_i t} e^{j\omega_i t} + \sum_{i=1}^{N} r_{i,bw} e^{\alpha_i t} e^{-j\omega_i t} \\ &= \sum_{i=1}^{N} r_{i,fw} e^{\alpha_i t} [\cos(\omega_i t) + j\sin(\omega_i t)] + \sum_{i=1}^{N} r_{i,bw} e^{\alpha_i t} [\cos(-\omega_i t) + j\sin(-\omega_i t)] \end{aligned} \tag{4.3}$$

The parts in the square brackets [ ] in Eq. (4.3) are the harmonic vibrations with the eigenfrequencies of the forward and backward frequency components. Hence, their maximal amplitudes are limited at 1 at any time t. On the contrary, the terms in front of the square brackets [ ] containing an exponential function $\exp(\alpha_i t)$ change with time. Their amplitudes of the rotor response depend on the growth/decay rate α. If α is a growth rate ($\alpha > 0$), the rotor amplitude rises with time t. Hence, the rotor response is unstable at any growth rate leading to the rotor instability. However, in case of a decay rate ($\alpha < 0$), the rotor amplitude dies down with time t leading to the rotor stability. As a result, α is the key parameter for the stability analysis of the rotor in linear rotordynamics (cf. Fig. 4.1).

Cases of the growth/decay rate α for the rotor response behavior:

- $\alpha < 0$: decay rate → rotor stability;
- $\alpha = 0$: threshold of instability (onset of instability);
- $\alpha > 0$: growth rate → rotor instability.

The stability behaviors depending on the growth/decay rates are displayed in Fig. 4.1. At a decay rate ($\alpha < 0$), the vibration amplitude with the frequency ω diminishes in a short time. The envelope of the vibration is the exponential function $r_0 \exp(\alpha t)$, which begins from the initial position r_0 at time $t = 0$ and decreases with time to zero. The corresponding forward orbit decays from the initial position r_0 to zero as $t \to \infty$. In this case, the free vibration is stable or unconditionally stable (see Fig. 4.1a).

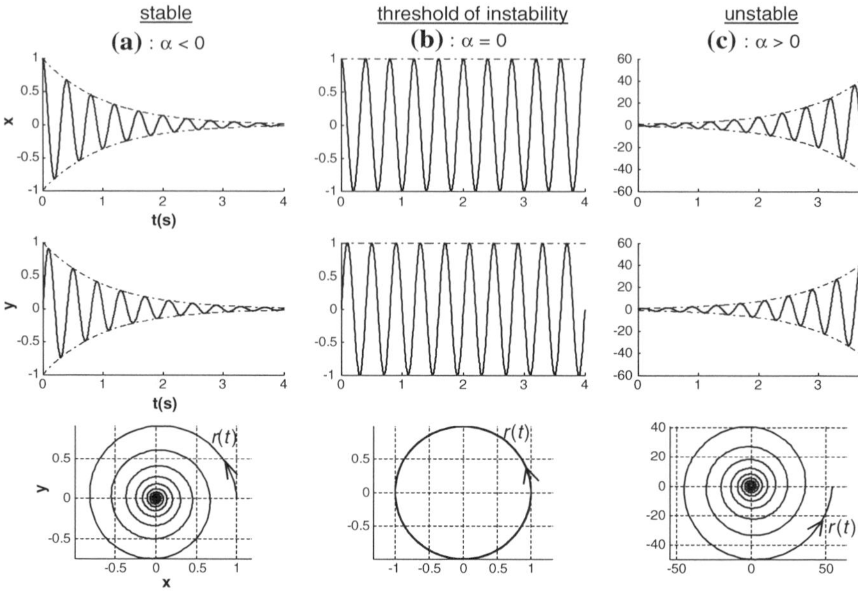

Fig. 4.1 Cases of the growth/decay rates α in the phase plane *x*-*y*

On the contrary, at a growth rate ($\alpha > 0$), the vibration begins at the initial position and increases with time exponentially; therefore, the vibration is unstable (see Fig. 4.1c).

At the rate of $\alpha = 0$, the vibration amplitude is unchanged with time, neither decreasing nor increasing with time because the amplitude of the function $r_0 \exp(\alpha t)$ always remains constant at r_0 at any time t (see Fig. 4.1b). The rotor vibration moves in a circular orbit that is called the limit cycle that occurs in the Hopf bifurcation. The vibration is at the threshold of instability (onset of instability) [1].

To study the rotor stability, the complex eigenvalues λ are analyzed in the λ plane, as shown in Fig. 4.2. The complex eigenvalue and its conjugated eigenvalue have the common real part and opposite imaginary part with positive and negative signs that correspond to forward and backward damped natural frequencies. If both these eigenvalues lie in the left half-plane ($\alpha < 0$), the rotor behavior is stable; on the right half-plane ($\alpha > 0$), it is unstable. In case of the threshold of instability ($\alpha = 0$), the rotor vibration moves with the damped natural frequency ω_d at the constant amplitude, neither decreasing nor increasing with time.

In order to maintain the rotor in the stability, all eigenvalues of the rotor must lie on the left half-plane to ensure all decay rates are always negative ($\alpha < 0$).

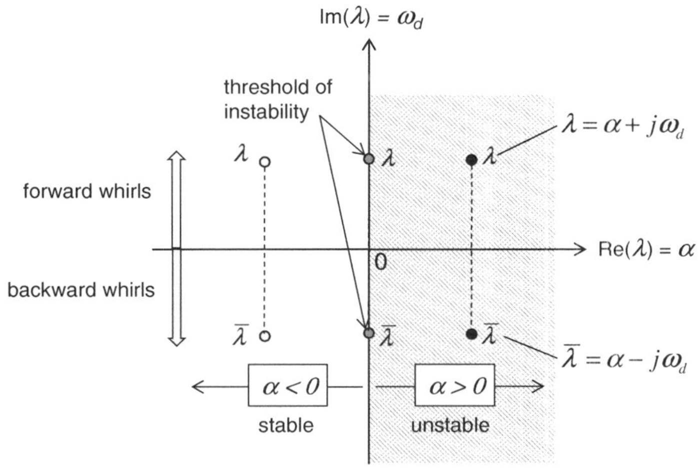

Fig. 4.2 Stability analysis in the λ plane

In case of the decay rate, there are two possibilities of stability:

- In the upper left half-plane ($\alpha < 0$, $\omega_d > 0$), the vibration response is a stable forward whirl and decays with time from the initial position to zero;
- In the lower left half-plane ($\alpha < 0$, $\omega_d < 0$), the vibration response is a stable backward whirl and decays with time from the initial position to zero, but the whirl direction is opposite to the upper left half-plane.

4.2.2 A Study Case of Calculating the Eigenvalues

In the following section, a very simple model of the Jeffcott rotor is chosen to demonstrate the principle of calculating the eigenvalues of the rotor. The key issue is preferred to understand the physical meaning of the problem rather than to show off the complicated mathematical calculations. Therefore, a simple Jeffcott rotor is used for the study case in which the gyroscopic effect of the wheel is not taken into account.

Figure 4.3 shows the simulated Jeffcott rotor that consists of the wheel, rotor shaft, and two orthotropic bearings. The wheel with mass m is located in the middle of the shaft that is supported by the orthotropic bearings ($K_1 \neq K_2$) at the ends of the shaft. The rotor is excited by the unbalance force F_{Un}, which is proportional to squared rotor speed.

The bearing stiffness coefficients are K_1 and K_2; bearing damping coefficients, c_1 and c_2 in the directions x_1 and x_2, respectively. The shaft has an own stiffness coefficient K_S. The impedance method is used to calculate the equivalent stiffness coefficient acting upon the rotor. The equivalent stiffness coefficient of the rotor results from the shaft and bearing stiffness coefficients in each direction (Fig. 4.4).

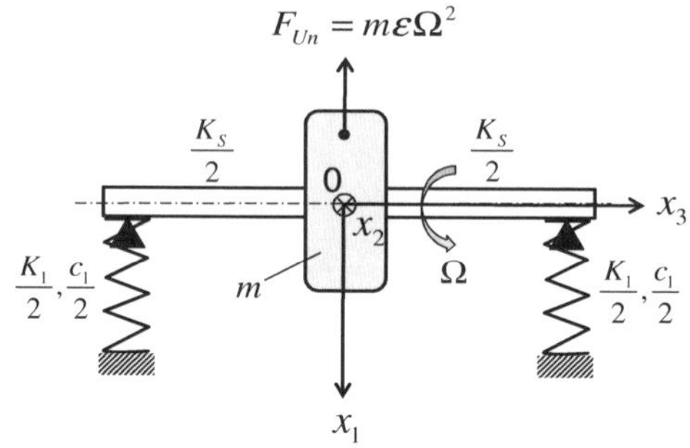

Fig. 4.3 Simulation model of the Jeffcott rotor

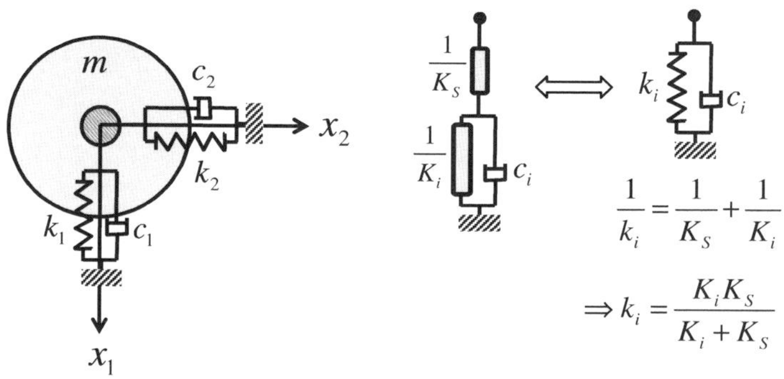

Fig. 4.4 Equivalent stiffness coefficients k_i of the rotor

The equivalent impedances acting upon the rotor $1/k_1$ and $1/k_2$ in the directions x_1 and x_2 are calculated as

$$\frac{1}{k_i} = \frac{1}{K_S} + \frac{1}{K_i}; \; i = 1, 2 \tag{4.4}$$

Therefore, the equivalent stiffness coefficient of the rotor results as

$$k_i = \frac{K_i K_S}{K_i + K_S}; \; i = 1, 2 \tag{4.5}$$

Applying Newton's second law to the rotor, the vibration equations without the gyroscopic effect of the 2-DOF rotor are written as

$$\begin{cases} m\ddot{x}_1 + c_1\dot{x}_1 + k_1x_1 = m\varepsilon\Omega^2 \cos\Omega t \\ m\ddot{x}_2 + c_2\dot{x}_2 + k_2x_2 = m\varepsilon\Omega^2 \sin\Omega t \end{cases} \tag{4.6}$$

The homogeneous equations are taken into account in studying the rotor stability.

$$\begin{cases} m\ddot{x}_1 + c_1\dot{x}_1 + k_1x_1 = 0 \\ m\ddot{x}_2 + c_2\dot{x}_2 + k_2x_2 = 0 \end{cases} \quad (4.7)$$

Equation (4.7) can be expressed in the matrix formulation:

$$\begin{bmatrix} m & 0 \\ 0 & m \end{bmatrix} \cdot \begin{bmatrix} \ddot{x}_1 \\ \ddot{x}_2 \end{bmatrix} + \begin{bmatrix} c_1 & 0 \\ 0 & c_2 \end{bmatrix} \cdot \begin{bmatrix} \dot{x}_1 \\ \dot{x}_2 \end{bmatrix} + \begin{bmatrix} k_1 & 0 \\ 0 & k_2 \end{bmatrix} \cdot \begin{bmatrix} x_1 \\ x_2 \end{bmatrix} = \begin{bmatrix} 0 \\ 0 \end{bmatrix} \quad (4.8)$$

The free vibration solutions of Eq. (4.8) are assumed as

$$\begin{bmatrix} x_1 \\ x_2 \end{bmatrix} = \begin{bmatrix} X_1e^{\lambda t} \\ X_2e^{\lambda t} \end{bmatrix} \quad (4.9)$$

Where

X_1 and X_2 are the amplitudes of x_1 and x_2, respectively;

λ is the complex eigenvalue; $\lambda = \alpha \pm j\omega_d$.

Inserting Eq. (4.9) and its first- and second-order derivatives into Eq. (4.8), one obtains

$$\begin{bmatrix} m\lambda^2 + c_1\lambda + k_1 & 0 \\ 0 & m\lambda^2 + c_2\lambda + k_2 \end{bmatrix} \cdot \begin{bmatrix} X_1e^{\lambda t} \\ X_2e^{\lambda t} \end{bmatrix} = \begin{bmatrix} 0 \\ 0 \end{bmatrix} \quad (4.10)$$

For nontrivial solutions x_1 and x_2, the determinant of the matrix must equal zero.

$$D(\lambda) = \begin{vmatrix} m\lambda^2 + c_1\lambda + k_1 & 0 \\ 0 & m\lambda^2 + c_2\lambda + k_2 \end{vmatrix} = 0 \quad (4.11)$$

Thus, one obtains the eigenvalue equations

$$D(\lambda) = (m\lambda^2 + c_1\lambda + k_1) \cdot (m\lambda^2 + c_2\lambda + k_2) = 0 \quad (4.12)$$

Equation (4.12) is called the characteristic equation of the rotor with fourth order of λ. It results from twice degrees of freedom (2 × DOF) of the rotor due to forward and backward whirls. In this case, the number of DOF (x_1 and x_2) equals 2; hence, the characteristic equation is in the fourth order of λ. Note that DOF of the rotor is the number of the independent spatial coordinates that describe the rotor vibration behavior at any time. Therefore, it equals the number of the ordinary differential equations (ODE) of the rotor vibration; hence, DOF = 2, as shown in Eq. (4.6).

Roots of the characteristic equation $D(\lambda)$ are the eigenvalues of the system that are generally complex numbers. Equation (4.12) can be written in two homogeneous equations

$$m\lambda^2 + c_i\lambda + k_i = 0;\ i = 1, 2 \tag{4.13}$$

Equation (4.13) can be transformed with dropping the index i into

$$\lambda^2 + 2\omega_n\zeta\lambda + \omega_n^2 = 0 \tag{4.14}$$

where ζ is the damping ratio; ω_n is the undamped natural frequency.

The undamped natural frequency is defined as

$$\omega_n \equiv \sqrt{\frac{k}{m}} \tag{4.15}$$

The dimensionless damping ratio is defined as

$$\zeta \equiv \frac{c}{2m\omega_n} = \frac{c}{2\sqrt{km}} \tag{4.16}$$

The eigenvalues are obtained solving Eq. (4.14)

$$\begin{aligned}\lambda &= -\omega_n\zeta \pm j\omega_n\sqrt{1-\zeta^2} \\ &= -\frac{c}{2m} \pm j\omega_n\sqrt{1-\zeta^2} \\ &\equiv \alpha \pm j\omega_d\end{aligned} \tag{4.17}$$

Resulting from Eq. (4.17), there are four eigenvalues λ_i of the characteristic equation:

$$\lambda_i = \alpha_i \pm j\omega_{d,i};\quad i = 1, 2 \tag{4.18}$$

within

$$\alpha_i = -\frac{c_i}{2m} = -\omega_{n,i}\zeta_i \text{ is the growth/decay rate } (1/\text{s}) \tag{4.19}$$

$$\omega_{d,i} = \omega_{n,i}\sqrt{1-\zeta_i^2} \text{ is the damped natural frequency } (1/\text{s}) \tag{4.20}$$

To keep the rotor stable in any operating condition, the growth/decay rate must be always negative. According to Eq. (4.19), the rotor is stable when the damping coefficient c is positive. Otherwise, the rotor is unstable at the negative damping coefficient, e.g., in case of suction in the damping absorber.

The damped natural frequency ω_d is a little smaller than the undamped frequency ω_n that depends on the damping ratio according to Eq. (4.20).

There are three cases of the dimensionless damping ratio ζ at $\alpha < 0$:

- $\zeta < 1$: the vibration response is underdamped or subcritically damped with the underdamped frequency $\omega_d < \omega_n$. The rotor orbit is stable and its amplitude decreases with time. In this case, the growth/decay rate results in

$$\alpha = -\omega_n \zeta \tag{4.21}$$

- $\zeta = 1$: the vibration response is critically damped, $\omega_d = 0$. There is no vibration, and its amplitude decreases much more quickly to zero without oscillation. In this case, the growth/decay rate is opposite to the undamped natural frequency.

$$\alpha = -\omega_n \tag{4.22}$$

- $\zeta > 1$: the vibration response is overdamped or supercritically damped, $\omega_d = 0$. There is no vibration, and its amplitude decreases quickly to zero without oscillation. In this case, the growth/decay rate is written as

$$\alpha = -\omega_n(\zeta \pm \sqrt{\zeta^2 - 1}) \tag{4.23}$$

The decay rate with the plus sign in Eq. (4.23) is much larger than the decay rate with the minus sign. Therefore, its response amplitude decays much faster than the amplitude with the minus sign. In fact, the decay rate with the minus sign dominates the response amplitude in the long-term motion.

The influence of the damping ratio ζ on the decay rate $\alpha < 0$ is displayed in Fig. 4.5. First, at the undamped condition ($\zeta < 1$), the growth/decay rate linearly decreases with ζ. Second, at the critically damped condition ($\zeta = 1$), the rate α equals to $-\omega_n$ according to Eq. (4.22). Third, at the overdamped condition ($\zeta > 1$), the decay rate with minus sign decreases from ω_n to zero (long-term motion $\zeta^- > 1$), and the decay rate with the plus sign increases from ω_n to ∞ (short-term motion $\zeta^+ > 1$) as the damping ratio ζ increases to ∞.

The behaviors of the vibration response at various damping ratios are shown in Fig. 4.6. The underdamped vibration response ($\zeta < 1$) with a damped natural frequency ω_d decays much more slowly than the critically and overdamped vibration responses because of its small decay rate α, as shown in Fig. 4.5.

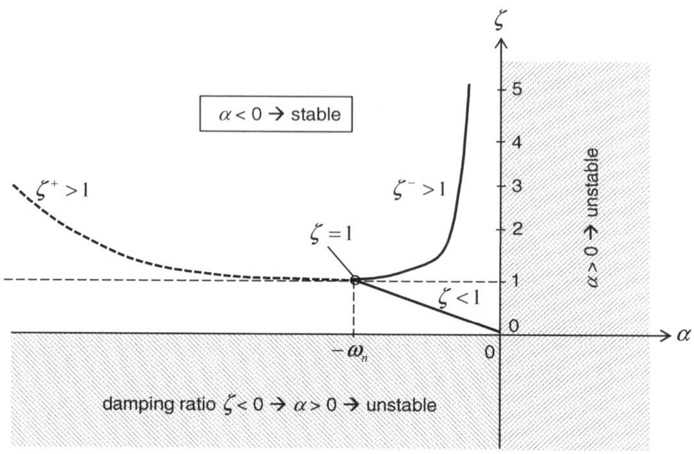

Fig. 4.5 Growth/decay rate α at various damping ratios ζ

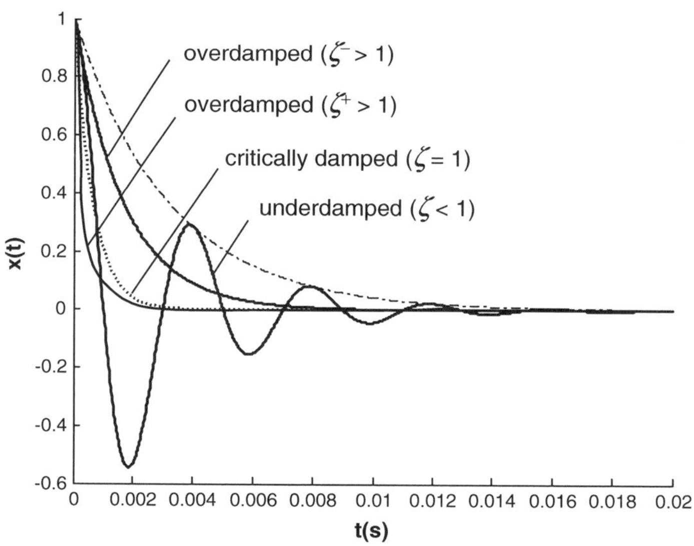

Fig. 4.6 Behaviors of the response amplitudes at various damping ratios ζ

4.2.3 Stability Analysis with Routh–Hurwitz Criterion

In the earlier section, a simple model of the Jeffcott rotor without the gyroscopic effect has been studied. In fact, the computed model of turbochargers is much more complex than the Jeffcott rotor model. In the computed turbocharger model, the gyroscopic effect of the compressor and turbine wheels are taken into account, orthotropic oil-film bearings must be also considered. In this case, the numbers of degree of freedom N are very large, such as 15, 20, or more; the eigenvalues result in two times of DOF (=2 N). Therefore, it is very difficult to solve the $2N$-order characteristic equation of λ analytically.

The vibration equations of the rotordynamic model of turbochargers are described in the matrix form

$$\mathbf{M}\ddot{\mathbf{x}} + \mathbf{C}\dot{\mathbf{x}} + \mathbf{K}\mathbf{x} = \mathbf{f}(t) \tag{4.24}$$

where

$\mathbf{M} = (N \times N)$	mass matrix containing the mass and inertia moments of the rotor;
$\mathbf{C} = (N \times N)$	damping coefficient and gyroscopic matrix containing the diagonal and cross-coupled damping coefficients;
$\mathbf{K} = (N \times N)$	stiffness matrix containing the diagonal and cross-coupled stiffness coefficients;
$\mathbf{x}(t) = (N \times 1)$	vibration response vector;
$\mathbf{f}(t) = (N \times 1)$	external forces vector.

In order to reduce the second to first-order equation set, a $(2N \times 1)$ new vector $\mathbf{z}$ of the vibration response is defined as

$$\mathbf{z} \equiv \begin{bmatrix} \mathbf{x} \\ \mathbf{y} \end{bmatrix} = \begin{bmatrix} \mathbf{x} \\ \dot{\mathbf{x}} \end{bmatrix} \Rightarrow \dot{\mathbf{z}} = \begin{bmatrix} \dot{\mathbf{x}} \\ \ddot{\mathbf{x}} \end{bmatrix} = \begin{bmatrix} \mathbf{y} \\ \dot{\mathbf{y}} \end{bmatrix} \tag{4.25}$$

where $\mathbf{y} = \dot{\mathbf{x}};\ \dot{\mathbf{y}} = \ddot{\mathbf{x}}$.

Substituting the vector $\mathbf{z}$ into Eq. (4.24), one obtains $2N$ first-order vibration equations of the rotor.

$$\begin{aligned} \begin{bmatrix} \dot{\mathbf{x}} \\ \ddot{\mathbf{x}} \end{bmatrix} &= \begin{bmatrix} \mathbf{0} & \mathbf{I} \\ -\mathbf{M}^{-1}\mathbf{K} & -\mathbf{M}^{-1}\mathbf{C} \end{bmatrix} \cdot \begin{bmatrix} \mathbf{x} \\ \dot{\mathbf{x}} \end{bmatrix} + \begin{bmatrix} \mathbf{0} \\ \mathbf{M}^{-1}\mathbf{f}(t) \end{bmatrix} \\ \Leftrightarrow \dot{\mathbf{z}} &= \begin{bmatrix} \mathbf{0} & \mathbf{I} \\ -\mathbf{M}^{-1}\mathbf{K} & -\mathbf{M}^{-1}\mathbf{C} \end{bmatrix} \cdot \mathbf{z} + \begin{bmatrix} \mathbf{0} \\ \mathbf{M}^{-1}\mathbf{f}(t) \end{bmatrix} \\ &= \mathbf{A}\mathbf{z} + \mathbf{b}(t) \end{aligned} \tag{4.26}$$

To analyze the dynamic stability, only the homogeneous equations of Eq. (4.26) are considered where the excitation force $\mathbf{f}(t)$ equals zero; hence, $\mathbf{b}(t) = \mathbf{0}$.

$$\dot{\mathbf{z}} = \mathbf{A}\mathbf{z} \tag{4.27}$$

where

$\mathbf{z} = (2N \times 1)$	vector, the vibration response vector;
$\mathbf{A} = (2N \times 2N)$	matrix of the rotor characteristic;
N	degrees of freedom of the rotor (DOF).

The free vibration response vector is assumed as

$$\mathbf{z} = \hat{\mathbf{z}} e^{\lambda t} \tag{4.28}$$

within $\hat{\mathbf{z}}$ is the $(2N \times 1)$ eigenvector.

Substituting the response vector $\mathbf{z}$ and its first-order derivative into Eq. (4.27), one obtains

$$(\mathbf{A} - \lambda \mathbf{I})\hat{\mathbf{z}} e^{\lambda t} = \mathbf{0} \tag{4.29}$$

The characteristic equation of the eigenvalues results from Eq. (4.29) in

$$\det(\mathbf{A} - \lambda \mathbf{I}) = 0 \tag{4.30}$$

Because $\mathbf{A}$ has the order of a $(2N \times 2N)$ matrix, the characteristic equation is a polynomial equation of $2N$ order.

$$D(\lambda) = a_n \lambda^n + a_{n-1} \lambda^{n-1} + \cdots + a_2 \lambda^2 + a_1 \lambda + a_0 = 0 \tag{4.31}$$

where

λ are the complex eigenvalues with $\lambda = \alpha + j\omega_d$;

a_i is the equation coefficients, $i = 0, 1, 2, \ldots, n$;

n is the polynomial exponent that equals twice DOF ($n = 2N$).

Using Gauss theorem, Eq. (4.31) can be written in the polynomial with the n order of λ where λ_i are the roots of $D(\lambda)$.

$$D(\lambda) = a_n(\lambda - \lambda_1)(\lambda - \lambda_2)\ldots(\lambda - \lambda_{n-1})(\lambda - \lambda_n) \tag{4.32}$$

Finding the roots λ_i of the polynomial $D(\lambda)$ in Eq. (4.31) is quite difficult, especially at a large number of DOF. In fact, it need not to know what the exact values of the roots λ_i are, but only the sign of the real parts α of the eigenvalues are necessary for the stability analysis. In this case, the Routh–Hurwitz stability criterion is applied to analyze the eigenvalues in linear rotordynamics.

According to the Routh–Hurwitz criterion, the vibration response is stable ($\alpha < 0$) when the following two conditions are satisfied:

(1) All coefficients a_i ($i = 0, 1, 2, \ldots, n$) in the characteristic Eq. (4.31) must be nonzero and have the same positive or negative signs (necessary condition).
(2) Hurwitz determinant D_{n-1}, and its diagonal determinants D_i for $i = 1, 2, \ldots, n - 2$ must be positive (sufficient condition).

The Hurwitz determinant D_{n-1} and diagonal determinants D_i for the characteristic equation $D(\lambda)$ of n order ($n = 2$ DOF) can be written as

$$D_1 = |a_{n-1}| > 0; \tag{4.33a}$$

$$D_2 = \begin{vmatrix} a_{n-1} & a_{n-3} \\ a_n & a_{n-2} \end{vmatrix} > 0; \tag{4.33b}$$

$$D_3 = \begin{vmatrix} a_{n-1} & a_{n-3} & a_{n-5} \\ a_n & a_{n-2} & a_{n-4} \\ 0 & a_{n-1} & a_{n-3} \end{vmatrix} > 0\ldots \tag{4.33c}$$

$$D_{n-2} = \begin{vmatrix} a_{n-1} & a_{n-3} & a_{n-5} & \cdots & . & 0 \\ a_n & a_{n-2} & a_{n-4} & \cdots & . & 0 \\ 0 & a_{n-1} & a_{n-3} & \cdots & . & 0 \\ 0 & a_n & a_{n-2} & \cdots & . & a_0 \\ 0 & 0 & . & \cdots & a_3 & a_1 \\ 0 & 0 & . & \cdots & a_4 & a_2 \end{vmatrix} > 0 \tag{4.33d}$$

$$D_{n-1} = \begin{vmatrix} a_{n-1} & a_{n-3} & a_{n-5} & . & \cdots & . & 0 \\ a_n & a_{n-2} & a_{n-4} & . & \cdots & . & 0 \\ 0 & a_{n-1} & a_{n-3} & a_{n-5} & \cdots & . & . \\ 0 & a_n & a_{n-2} & a_{n-4} & \cdots & . & . \\ 0 & 0 & . & . & \cdots & a_1 & 0 \\ 0 & 0 & . & . & \cdots & a_2 & a_0 \\ 0 & 0 & . & . & a_5 & a_3 & a_1 \end{vmatrix} > 0 \tag{4.33e}$$

$D_n = a_0 D_{n-1} > 0$ is not necessary because a_0 and D_{n-1} are positive. (4.33f)

where the coefficients a_{n-i} equal zero at all negative indices $(n - i)$.

The Routh–Hurwitz criterion is applied to the Jeffcott rotor, as shown in Fig. 4.3 to analyze rotor stability of the vibration equations with the fourth order of λ. The characteristic equation of the rotor vibration is written as

$$\begin{aligned} &(m\lambda^2 + c_1\lambda + k_1)\cdot(m\lambda^2 + c_2\lambda + k_2) \\ &= m^2\lambda^4 + m(c_1 + c_2)\lambda^3 + (mk_1 + mk_2 + c_1c_2)\lambda^2 \\ &+ (c_1k_2 + c_2k_1)\lambda + k_1k_2 = 0 \end{aligned} \tag{4.34}$$

Checking stability conditions of Eq. (4.34):

- the first condition is satisfied because all coefficients are positive and nonzero;
- the second condition of D_1, D_2, and D_3 has to be checked whether they are positive.

The Hurwitz determinant and diagonal determinants are calculated as

$$D_1 = |m(c_1 + c_2)| = m(c_1 + c_2)$$

$$D_2 = \begin{vmatrix} m(c_1 + c_2) & (c_1k_2 + c_2k_1) \\ m^2 & (mk_1 + mk_2 + c_1c_2) \end{vmatrix} = m^2(c_1k_1 + c_2k_2) + mc_1c_2(c_1 + c_2)$$

$$D_3 = \begin{vmatrix} m(c_1 + c_2) & (c_1k_2 + c_2k_1) & 0 \\ m^2 & (mk_1 + mk_2 + c_1c_2) & k_1k_2 \\ 0 & m(c_1 + c_2) & (c_1k_2 + c_2k_1) \end{vmatrix}$$

After a few calculating steps, the determinant D_3 results in

$$D_3 = mc_1c_2\Big(m(k_1 - k_2)^2 + (c_1 + c_2) \cdot (c_1k_2 + c_2k_1)\Big)$$

All determinants D_1, D_2, and D_3 are positive due to positive damping coefficients c_1 and c_2. Hence, the vibration response is stable. Otherwise, the rotor response is unstable when the damping coefficients of the rotor are negative at a suction damper.

4.3 Stability Analysis of Nonlinear Rotordynamics

The eigenvalues have not existed any longer in nonlinear rotordynamics; therefore, the eigenvalues analysis is not suitable in this case. Instead, the bifurcation theory is normally used to analyze the stability of nonlinear rotordynamics.

The nonlinear characteristics of the oil-film bearings are to blame for nonlinear rotordynamics of the turbochargers. The nonlinear bearing forces acting upon the rotor contain two parts, the rotation and damping forces. According to the bearing dynamics, the bearing rotation forces are nonlinear with the journal eccentricity and angular position of the journal locus. The bearing damping forces are nonlinear not only with the journal eccentricity and its angular position but with the eccentricity velocity and whirl velocity of the journal in the bearing clearance. As a result, the vibration equations are generally nonlinear in the rotordynamics of automotive turbochargers.

4.3.1 Vibration Equations in the Autonomous Systems

Using Newton's second law for the rotor of N degrees of freedom (DOF), one obtains the nonlinear vibration equations of the rotor in the matrix formulation.

$$\begin{bmatrix} \dot{\mathbf{x}} \\ \ddot{\mathbf{x}} \end{bmatrix} = \begin{bmatrix} \mathbf{0} & \mathbf{I} \\ -\mathbf{M}^{-1}\mathbf{K}_S & -\mathbf{M}^{-1}\mathbf{C} \end{bmatrix} \cdot \begin{bmatrix} \mathbf{x} \\ \dot{\mathbf{x}} \end{bmatrix} + \begin{bmatrix} \mathbf{0} \\ \mathbf{M}^{-1}\mathbf{F}_{\mathbf{Un}}(t) + \mathbf{M}^{-1}\mathbf{F}_{\mathbf{B}}(\mathbf{x}, \dot{\mathbf{x}}) \end{bmatrix} \quad (4.35)$$

where

$\mathbf{M} = (N \times N)$	diagonal matrix containing mass and mass inertia moments;
$\mathbf{K_S} = (N \times N)$	shaft stiffness coefficient matrix;
$\mathbf{C} = (N \times N)$	rotor damping coefficient and gyroscopic matrix;
$\mathbf{I} = (N \times N)$	unit matrix;
$\mathbf{x} = (N \times 1)$	DOF vector including all degrees of freedom of the system;
$\mathbf{F_{Un}} = (N \times 1)$	unbalance force vector;
$\mathbf{F_B} = (N \times 1)$	bearing force vector consisting of the rotation and damping forces;
N	degrees of freedom (DOF).

The $(2N \times 1)$ vector $\mathbf{z}$ is defined as

$$\mathbf{z} \equiv \begin{bmatrix} \mathbf{x} \\ \mathbf{y} \end{bmatrix} = \begin{bmatrix} \mathbf{x} \\ \dot{\mathbf{x}} \end{bmatrix}$$

within

$$\mathbf{x} = \begin{pmatrix} x_1 \\ x_2 \\ \cdot \\ x_N \end{pmatrix}; \mathbf{y} = \dot{\mathbf{x}} = \begin{pmatrix} \dot{x}_1 \\ \dot{x}_2 \\ \cdot \\ \dot{x}_N \end{pmatrix} \equiv \begin{pmatrix} y_1 \\ y_2 \\ \cdot \\ y_N \end{pmatrix}$$

Substituting the vector $\mathbf{z}$ into Eq. (4.35), the vibration equations become

$$\begin{aligned} \dot{\mathbf{z}} &= \begin{bmatrix} \mathbf{0} & \mathbf{I} \\ -\mathbf{M}^{-1}\mathbf{K_S} & -\mathbf{M}^{-1}\mathbf{C} \end{bmatrix} \cdot \mathbf{z} + \begin{bmatrix} \mathbf{0} \\ \mathbf{M}^{-1}\mathbf{F_{Un}}(t) + \mathbf{M}^{-1}\mathbf{F_B}(\mathbf{z}, \Omega) \end{bmatrix} \\ &= \mathbf{A}\mathbf{z} + \mathbf{b}(\mathbf{z}, \Omega, t) \end{aligned} \tag{4.36}$$

where

$$\mathbf{A} = \begin{bmatrix} \mathbf{0} & \mathbf{I} \\ -\mathbf{M}^{-1}\mathbf{K_S} & -\mathbf{M}^{-1}\mathbf{C} \end{bmatrix}$$

is the $(2 \times 2N)$ matrix of the rotor characteristic;

$$\mathbf{b} = \begin{bmatrix} \mathbf{0} \\ \mathbf{M}^{-1}\mathbf{F_{Un}}(t) + \mathbf{M}^{-1}\mathbf{F_B}(\mathbf{z}, \Omega) \end{bmatrix}$$

is the $(2N \times 1)$ vector of the unbalance and bearing forces.

The system is called autonomous when it is not explicitly dependent on time t. We transform the nonautonomous nonlinear equations into the autonomous nonlinear equations where the time t is considered as a parameter θ during iteratively solving the equations. Hence, the vibration equations in Eq. (4.36) can be considered as the autonomous equations.

$$\dot{\mathbf{u}} = \mathbf{A}^{*}\mathbf{u} + \mathbf{b}^{*}(\mathbf{u}, \Omega, \theta) \tag{4.37}$$

where

u is the new vector of $(2N + 1 \times 1)$ consisting of the vector **z** and time parameter t;

b* is the external force vector that is not explicitly dependent on time t. In this case, t is considered as a time parameter θ.

Equation (4.37) is the $2N + 1$ first-order differential equations with an additional equation ($\dot{\theta} = 1$) in which the rotor speed Ω is chosen as a bifurcation parameter for the stability analysis.

To study the interactive bifurcation analysis of dynamic systems, the software package MATCONT [4] based on the MATLAB code is recommended.

4.3.2 Stability Analysis with Bifurcation Theory

The bifurcation occurs at bifurcation points where the vibrations change from the steady state at the equilibrium into an unsteady state. The bifurcation point is defined as the bifurcation parameter μ reaches the critical threshold value μ_0. At the bifurcation point, the dynamic behavior of the system is transferred from the steady state to periodic or quasiperiodic vibrations, from the stable to unstable dynamic behaviors, and from the regular to irregular and as well as chaotic vibrations. Hopf bifurcation is a special kind of the bifurcation that connects the equilibrium to the periodic oscillation at the bifurcation point [7–10].

The loss of stability at Hopf bifurcation takes place as the eigenvalue λ and its conjugated eigenvalue cross the imaginary axis ($\mu_0 = 0$) at the points $\lambda = \pm j\omega_0$. In this case, the bifurcation parameter μ equals the critical value μ_0 that depends on the characteristics of the system, such as the rotor unbalance, bearing clearances and geometries, and oil operating condition (oil temperature and pressure). At the bifurcation point, the dynamic behavior changes from the equilibrium into periodic oscillations with the initial frequency ω_0. The rotor locus is limited on the limit cycle (Fig. 4.7).

4.3.3 Characteristics of Hopf Bifurcation Theory

(a) Equilibrium Points

Equilibrium points are defined by the state vectors $\mathbf{u}^{*}$ where the system is in a steady-state condition ($\dot{\mathbf{u}} = \mathbf{0}$); thus, the RHS of the vibration Eq. (4.37) equals zero.

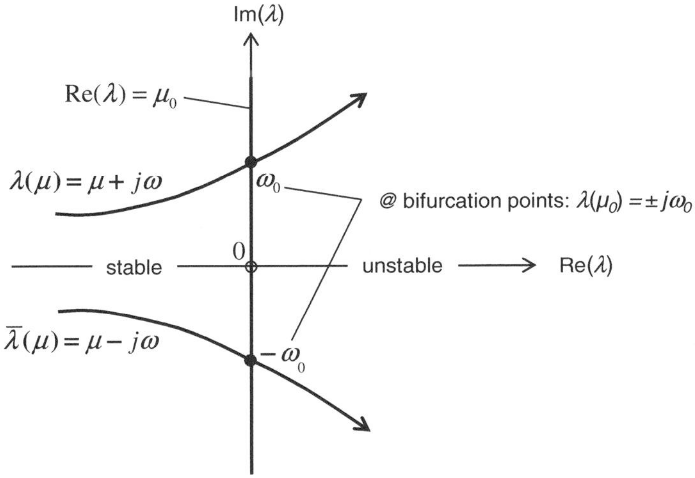

Fig. 4.7 Hopf bifurcation in the λ plane

$$\mathbf{A}^*\mathbf{u}^* + \mathbf{b}^*(\mathbf{u}^*, \Omega, \theta) = \mathbf{0} \tag{4.38}$$

The bifurcation parameter μ is a function of the rotor speed Ω and other characteristics of the vibration system, e.g., stiffness and damping coefficients K and C, oil temperature T, and oil viscosity v, etc.

The critical rotor speed Ω_0 at the bifurcation point results as

$$\mu = \mu(\Omega, K, C, T, v, \ldots) = \mu_0$$
$$\Rightarrow \Omega_0 = f(K, C, T, v, \ldots)$$

At the bifurcation point $\mu_0 = 0$, the system state changes from the equilibrium into periodic vibrations with the initial frequency ω_0. The equilibrium points are also called fixed or singular points. Obviously, the critical rotor speed (threshold of instability) depends on the system characteristics nonlinearly [5, 6].

(b) Bifurcation Point

Bifurcation point Ω_0 is the intersection of the trivial branch and nontrivial half-branches; their solutions correspond to the bifurcation point. At this point, the loss of stability happens and the vibration response of the system changes from the equilibrium state (trivial branch) to periodic vibrations (nontrivial half-branches), as shown in Fig. 4.8.

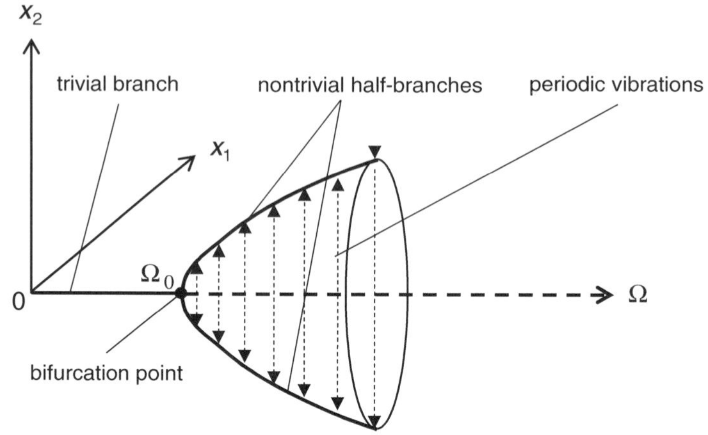

Fig. 4.8 Vibration behaviors near the bifurcation point

Fig. 4.9 Node and focus in the phase plane

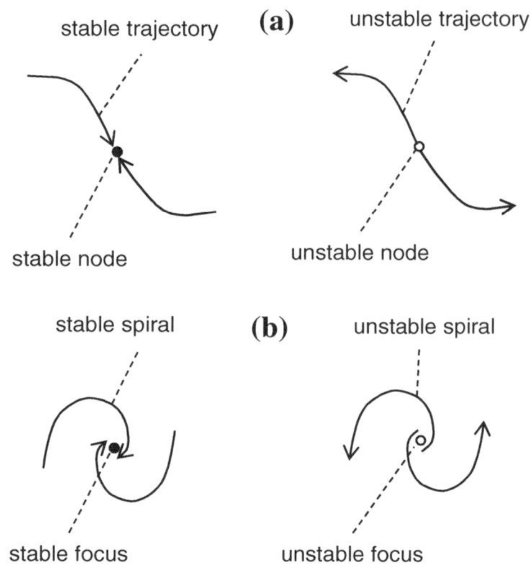

(c) Node and Focus

- Node has no oscillation ($\omega = 0$) where the stable trajectories approach the stable node and the unstable trajectories move away from the unstable node (see Fig. 4.9a).
- Focus contains a vibration with a nonzero frequency ω where the spirals approach the stable focus with time. They are called stable spirals whose speeds are negative. At an unstable focus, the spirals move away from it (see Fig. 4.9b). Hence, the spirals are unstable; their speeds are positive. As a result, the spiral amplitudes increase with time. Note that the focus is also called spiral point.

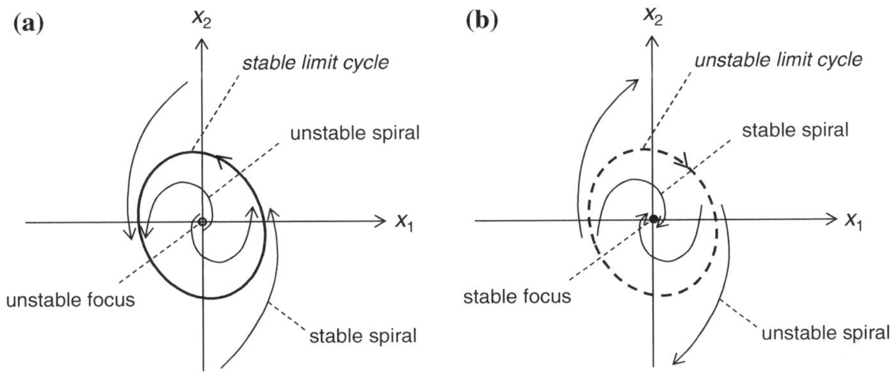

Fig. 4.10 Stable and unstable limit cycles on the phase plane

(d) Limit Cycles

Limit cycle is a closed orbit, which the trajectories asymptotically move to (stable limit cycle), away from (unstable limit cycle), or both cases (half-stable limit cycle). In general, the limit cycle has an elliptic shape on the phase plane.

There are three kinds of the limit cycle in nonlinear rotordynamics:

- Stable limit cycle encircles an unstable focus from that the inner unstable spiral moves away. The inner and outer spirals are limited on the limit cycle (see Fig. 4.10a). Thus, the limit cycle is defined as *stable*.
- Unstable limit cycle encircles a stable focus to that the inner stable spiral moves. The inner and outer spirals move away from the limit cycle (see Fig. 4.10b). As a result, the limit cycle is defined as *unstable*.
- Half-stable limit cycle encircles a stable focus to that the inner stable spiral moves. On the one hand, the inner stable spiral moves away from the limit cycle. On the other hand, the outer stable spiral coming from outside is limited on the limit cycle (see Fig. 4.11a). Hence, the limit cycle is defined as *half-stable*.
- Half-stable limit cycle encircles an unstable focus from that the inner unstable spiral moves away. On the one hand, the inner unstable spiral is limited on the limit cycle. On the other hand, the outer unstable spiral moves away from the limit cycle to outside (see Fig. 4.11b). Therefore, the limit cycle is defined as *half-stable*.

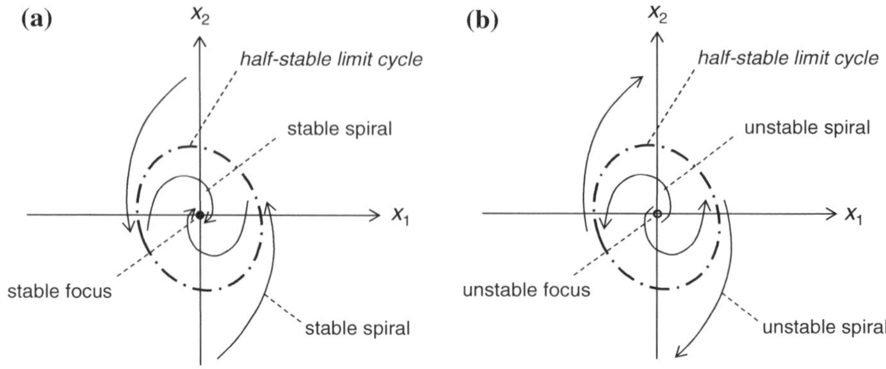

Fig. 4.11 Half-stable limit cycles on the phase plane

(e) Definitions of Lyapunov Stability
The dynamic behavior of **u** is *Lyapunov stable* at the equilibrium $\mathbf{u}^*$ when the perturbation ε remains smaller than the limit value $\varepsilon_0 > 0$ as time t goes to infinity.

$$\lim_{t\to\infty} \mathbf{u}(t) = \lim_{t\to\infty} (\mathbf{u}^* + \varepsilon) = \mathbf{u}^* + \varepsilon;\ |\varepsilon| < \varepsilon_0$$

The dynamic behavior of **u** is *asymptotically stable* to the equilibrium $\mathbf{u}^*$ when the perturbation ε approaches zero as time t goes to infinity.

$$\lim_{t\to\infty} \mathbf{u}(t) = \lim_{t\to\infty} (\mathbf{u}^* + \varepsilon) = \mathbf{u}^*;\ |\varepsilon| = 0$$

The dynamic behavior of **u** is *unstable* when the perturbation ε increases to an arbitrary $\delta > 0$ as time t goes to infinity where δ is larger than the limit value $\varepsilon_0 > 0$.

$$\lim_{t\to\infty} \mathbf{u}(t) = \lim_{t\to\infty} (\mathbf{u}^* + \varepsilon) = \mathbf{u}^* + \delta;\ |\delta| > \varepsilon_0$$

4.3.4 Classification of Hopf Bifurcation

Hopf bifurcation is classified into transcritical, supercritical, and subcritical bifurcations, as shown in Fig. 4.12. The bifurcations are classified into the transcritical bifurcations when the quantitative characteristics change along the trivial branch; into the supercritical bifurcations when two stable nontrivial branches encircle the unstable trivial branch; and into the subcritical bifurcations when two unstable nontrivial branches encircle the stable trivial branch.

(a) Transcritical Bifurcation
The loss of stability on the trivial branch occurs when the rotor speed Ω arrives at the critical value Ω_0 at the bifurcation point.

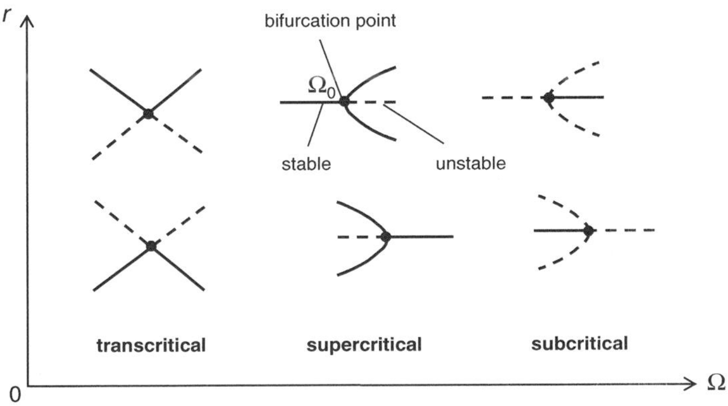

Fig. 4.12 Classification of Hopf bifurcations

(b) Supercritical Pitchfork Bifurcation
Two stable nontrivial half-branches encircle the unstable trivial branch starting from the bifurcation point. The supercritical bifurcations are called subtle bifurcations because the amplitude of the periodic vibrations is limited on the limit cycle after the soft loss of stability at the bifurcation point Ω_0 (see Figs. 4.8 and 4.10a). As long as the limit cycle remains within the bearing clearance, they are harmless and not dangerous for the bearings.

(c) Subcritical Pitchfork Bifurcation
Two unstable nontrivial half-branches encircle the stable trivial branch starting from the bifurcation point. The subcritical bifurcations are much more dangerous and harmful than the supercritical bifurcations because the amplitude of the periodic vibrations moves away from the unstable limit cycle outwards (see Fig. 4.10b). They are called the hard loss of stability at the bifurcation point Ω_0. Such jumps to higher amplitudes could often happen near the bifurcation point where the rotor behavior changes from the equilibrium state to unstable periodic vibrations. In this case, bearing wears and damage could occur when the journal amplitude exceeds the bearing clearance.

4.3.5 Coordinate Transformation in the Bifurcation

The vibration equations of the nonlinear rotor is written as

$$\dot{\mathbf{u}}(t) = \mathbf{A}^*\mathbf{u} + \mathbf{b}^*(\mathbf{u}, \Omega, \theta) \equiv \mathbf{f}(\mathbf{u}, \Omega, \theta) \tag{4.39}$$

For simplicity, the vector $\mathbf{u}(t)$ is considered in the two dimensions of $x_1(t)$ and $x_2(t)$. The coordinates of x_1 and x_2 are transformed into the phase plane r-ϕ as

$$\begin{aligned} r(t) &= \sqrt{x_1^2(t) + x_2^2(t)}; \\ \phi(t) &= \arctan\left(\frac{x_2(t)}{x_1(t)}\right) \end{aligned} \tag{4.40}$$

The rotor response is written at a given time parameter θ as

$$\begin{aligned} \dot{r}(t) &= f(r, \Omega, \theta); \\ \dot{\theta}(t) &= 1 \Rightarrow \theta = t_0 + t \end{aligned} \tag{4.41}$$

where t_0 is an integrating constant.

The bifurcation behavior of Eq. (4.41) is displayed in the bifurcation plane r-Ω (see Fig. 4.13).

The rotor locus of the supercritical pitchfork bifurcation at the rotor speed $\Omega_k > \Omega_0$ is studied in the phase plane r-ϕ (see Fig. 4.14).

Varying the bifurcation parameter of the rotor speed Ω, Hopf bifurcation takes place when the rotor speed reaches the critical value Ω_0 that is dependent on the rotor characteristics. After the bifurcation point, the stable rotor vibration changes from the steady-state equilibrium into the unstable periodic or quasiperiodic vibrations. At the rotor speed of Ω_k, the unstable periodic vibration moves away from the unstable branch and is finally limited on the limit cycle. The response amplitude $r(t)$ and the rotor locus at the rotor speed $\Omega_k > \Omega_0$ are displayed on the

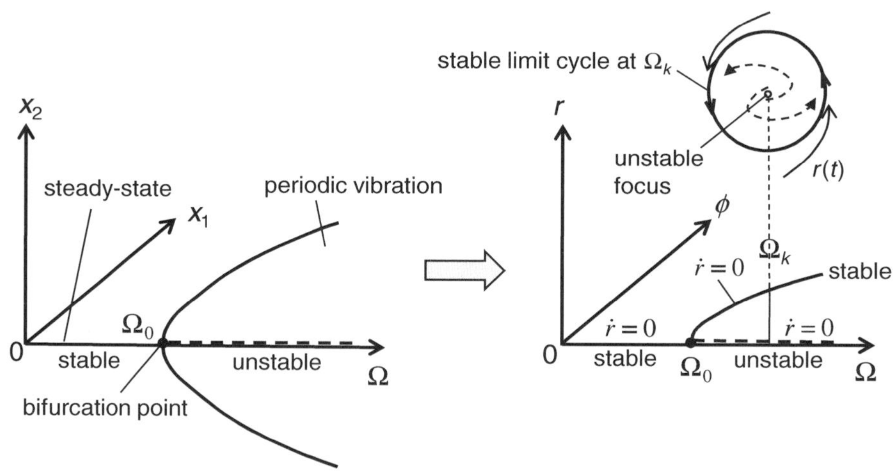

Fig. 4.13 Coordinate transformation in the bifurcation plane

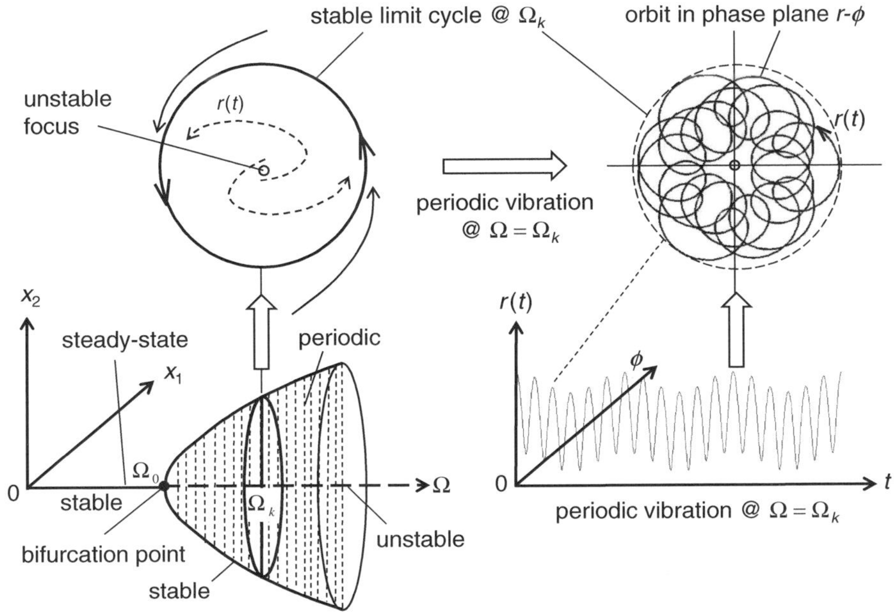

Fig. 4.14 Supercritical pitchfork Hopf bifurcation

phase plane r-ϕ. The rotor orbit stabilizes on the limit cycle that corresponds to the nontrivial stable half-branches. Obviously, the radius of the limit cycle varies with the rotor speed Ω.

4.3.6 Jacobian Matrix of the Vibration Equations

In order to find the eigenvalues of the equilibrium points (singular points), the Jacobian matrix is necessary. The Jacobian matrix $\mathbf{J}$ results from the first partial derivatives of the vibration Eq. (4.39) as

$$\mathbf{J} \equiv \begin{bmatrix} \frac{\partial f_1}{\partial x_1} & \frac{\partial f_1}{\partial x_2} & \cdot & \cdot & \frac{\partial f_1}{\partial x_n} \\ \frac{\partial f_2}{\partial x_1} & \frac{\partial f_2}{\partial x_2} & \cdot & \cdot & \frac{\partial f_2}{\partial x_n} \\ \cdot & \cdot & \cdot & \cdot & \cdot \\ \cdot & \cdot & \cdot & \cdot & \cdot \\ \frac{\partial f_n}{\partial x_1} & \frac{\partial f_n}{\partial x_2} & \cdot & \cdot & \frac{\partial f_n}{\partial x_n} \end{bmatrix} \tag{4.42}$$

Setting the characteristic determinant D of the matrix $(\mathbf{J} - \lambda\mathbf{I})$ at the bifurcation point x^* to zero, the corresponding eigenvalues at the bifurcation point result as

$$D = \begin{vmatrix} \left(\frac{\partial f_1}{\partial x_1} - \lambda\right) & \frac{\partial f_1}{\partial x_2} & . & . & \frac{\partial f_1}{\partial x_n} \\ \frac{\partial f_2}{\partial x_1} & \left(\frac{\partial f_2}{\partial x_2} - \lambda\right) & . & . & \frac{\partial f_2}{\partial x_n} \\ . & . & . & . & . \\ . & . & . & . & . \\ \frac{\partial f_n}{\partial x_1} & \frac{\partial f_n}{\partial x_2} & . & . & \left(\frac{\partial f_n}{\partial x_n} - \lambda\right) \end{vmatrix}_{x=x^*} = 0 \tag{4.43}$$

In fact, Eq. (4.43) is the characteristic equation of the vibration system at the bifurcation point x^*. If the real part of the complex eigenvalue λ is negative, the solutions of the rotor response are stable. Hence, the trivial branch and nontrivial half-branches are stable in the supercritical pitchfork bifurcations. Otherwise, the rotor responses are unstable with positive real parts of λ at the trivial branch and nontrivial half-branches in the subcritical pitchfork bifurcations (see Fig. 4.12).

4.3.7 A Study Case of the Subcritical Hopf Bifurcation

Having considered a rotor with two DOF of x_1 and x_2, the vibration equations are written as

$$\begin{aligned} \dot{x}_1 &= -x_2 + x_1 \cdot \left(\mu + \left(x_1^2 + x_2^2\right)\right) \\ \dot{x}_2 &= +x_1 + x_2 \cdot \left(\mu + \left(x_1^2 + x_2^2\right)\right) \end{aligned} \tag{4.44}$$

Using the coordinate transformation in Eq. (4.40), one obtains the vibration equations of the rotor response.

$$\dot{r}(t) = r(t) \cdot \left(\mu + r(t)^2\right) \tag{4.45}$$

where μ is the bifurcation parameter; $r(t)$ the radius.

(a) Finding the Equilibrium Points

Three equilibrium points result from Eq. (4.45) at the equilibrium as

$$\begin{aligned} r_1^* &= 0 \ for \ \forall \mu \\ r_{2,3}^* &= \pm\sqrt{-\mu} \ for \ \forall \mu < 0 \end{aligned} \tag{4.46}$$

(b) Finding the Corresponding Eigenvalues

The eigenvalues result from setting determinant D at the equilibrium points as

$$\begin{aligned} D &= \det(\mathbf{J} - \lambda \mathbf{I}) \\ &= \left(3r^2 + \mu - \lambda\right)_{r_i^*} = 0 \end{aligned} \tag{4.47}$$

Solving Eq. (4.47), one obtains three eigenvalues at the bifurcation points.

$$\begin{aligned} \lambda_1 &= \mu \leq 0 \text{ or } > 0 \; at \; r_1^* = 0, \; \forall \mu \\ \lambda_{2,3} &= -2\mu > 0 \; at \; r_{2,3}^* = \pm\sqrt{-\mu}, \; \forall \mu < 0 \end{aligned} \tag{4.48}$$

(c) Bifurcation Plane

The first eigenvalue has only a real part μ that can be negative or positive depending on the bifurcation parameter μ. In case of $\mu < 0$, one has the stable trivial branch; otherwise, the unstable trivial branch. The second and third eigenvalues have also only a real part of -2μ that is always positive because the bifurcation parameter μ must be negative at the singular solutions in Eq. (4.46). Hence, the two nontrivial half-branches are unstable in the bifurcation plane, as displayed in Fig. 4.15.

(d) Studying Behavior of Stability

The speeds of the inner and outer spirals are calculated from Eq. (4.45) for various radii r that depend on the bifurcation parameter μ.

$$\begin{aligned} \dot{r}(t) &\leq 0 \; for \; 0 < r \leq \sqrt{-\mu}, \; \forall \mu < 0 \\ \dot{r}(t) &> 0 \; for \; r > \sqrt{-\mu}, \; \forall \mu < 0 \end{aligned} \tag{4.49}$$

The behavior of rotor stability in the subcritical Hopf bifurcation is shown in Fig. 4.16. As long as the rotor locus is inside the limit cycle, the rotor behavior is always stable and tends to move to the stable focus due to the negative speed. The stable focus corresponds to the stable trivial branch at $\mu < 0$.

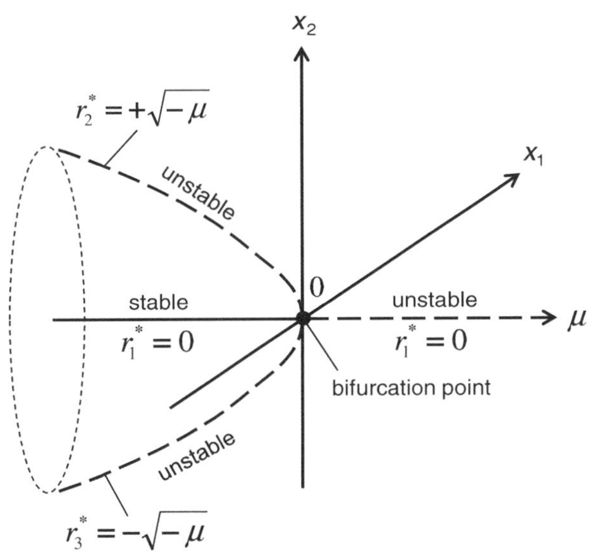

Fig. 4.15 Subcritical pitchfork Hopf bifurcation

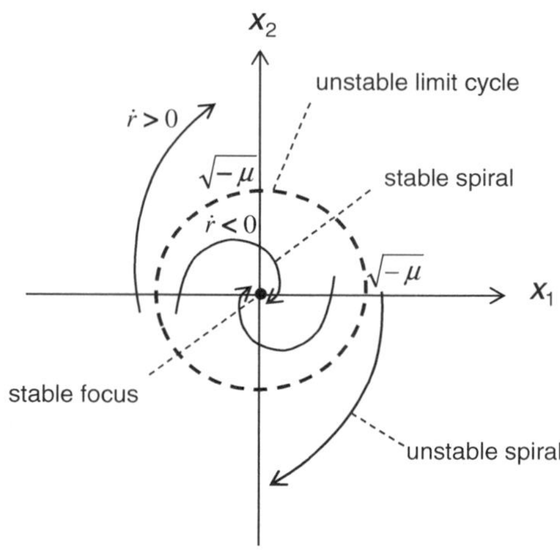

Fig. 4.16 Behavior of the rotor locus at $\mu < 0$

When the rotor locus exceeds the limit cycle, the response behavior becomes unstable. The rotor moves away from the limit cycle outwards while the response amplitude at the positive speed increases with time. In this case, the limit cycle corresponds to the unstable nontrivial half-branches at $\mu < 0$ (see Fig. 4.15). Such unstable response causing the jump from the unstable limit cycle in which the amplitude exceeds the bearing clearance leading to damage of the radial bearings.

4.3.8 Stability with Neimark–Sacker Torus Bifurcations

(a) Classification of Neimark–Sacker Bifurcations

Hopf bifurcation occurs when the dynamic behavior changes from the equilibrium state to periodic vibrations and loses its stability at the first bifurcation point $\Omega_{0,1}$. On the contrary, to the Hopf bifurcation, the periodic vibrations at the Neimark–Sacker bifurcation lose their stabilities to the torus at the second bifurcation point $\Omega_{0,2}$. Therefore, it is a kind of the secondary Hopf bifurcation or generalized Hopf bifurcation [7, 9, 10].

The torus can be attractive or repelling with the periodic vibrations. In case of attractive torus (stable torus), the trajectories move toward the torus surface. In case of repelling torus (unstable torus), the trajectories move away from it. The first one is called supercritical torus bifurcation; the latter is the subcritical torus bifurcation. The periodic or quasiperiodic vibration amplitudes are limited on the surface of the limit torus in the Neimark–Sacker bifurcation that is called torus bifurcations. Their quasiperiodic vibrations often occur at the beginning of the chaotic vibrations.

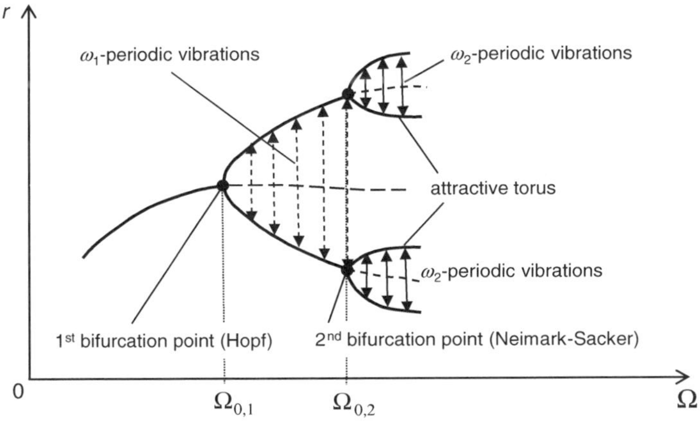

Fig. 4.17 A supercritical Neimark–Sacker bifurcation

Figures 4.17 and 4.19 display the super- and subcritical Neimark–Sacker bifurcations. In the supercritical torus bifurcation, the trajectories of periodic vibrations with frequency ω_2 or quasiperiodic vibrations move on the limit surface of the attractive torus. The resulting vibrations of the Neimark–Sacker bifurcation with two incommensurate frequencies of ω_1 and ω_2 are quasiperiodic and on the verge of becoming chaos, as shown in Fig. 4.17.

At the supercritical torus bifurcation in the Poincaré map, the vibration response consists of two frequencies ω_1 and ω_2 in the longitudinal and latitudinal directions, respectively [7]. Depending on the frequency ratio, the vibration is quasiperiodic if the ratio is irrational (incommensurate frequencies ω_1 and ω_2). The orbit of the quasiperiodic vibration response that wraps on the torus surface intersects the Poincaré map at the point P on the invariant curve C. Due to quasiperiodicity, the orbit never cuts the Poincaré map at the same point P after repeating the period T at time $t = T, 2T, \ldots, nT$. Instead, the orbit cuts the Poincaré map at different points $P', P'', \ldots, P^{(n)}$ on the invariant curve C. Therefore, the Poincaré map is usually applied to study the periodic vibrations, especially aperiodic vibrations of nonautonomous systems. The quantitative change of stability in the Neimark–Sacker bifurcation can be investigated in the Poincaré map [7, 10], see Fig. 4.18.

Contrary to the supercritical torus bifurcation, the trajectories of the subcritical torus bifurcation are repelled from the repelling inner torus (see Fig. 4.19). Just before the second bifurcation point $\Omega_{0,2}$, the periodic vibrations with the frequency ω_2 jump from the inner (repelling) torus into the outer (attractive) torus directly. Shortly after the turning point Ω_T between the inner and outer tori, the response

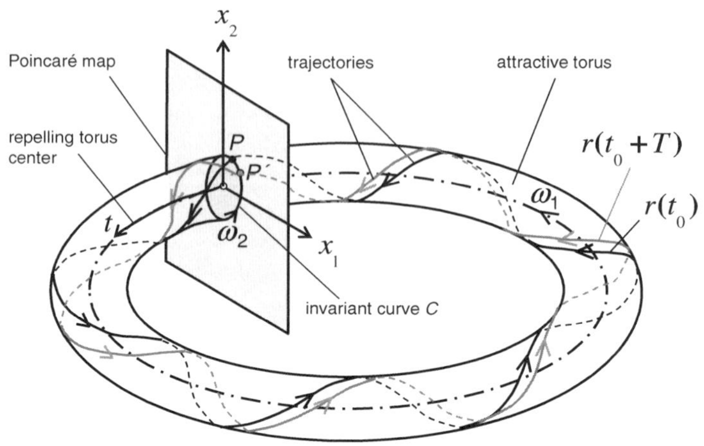

Fig. 4.18 A supercritical Neimark–Sacker bifurcation on an attractive torus

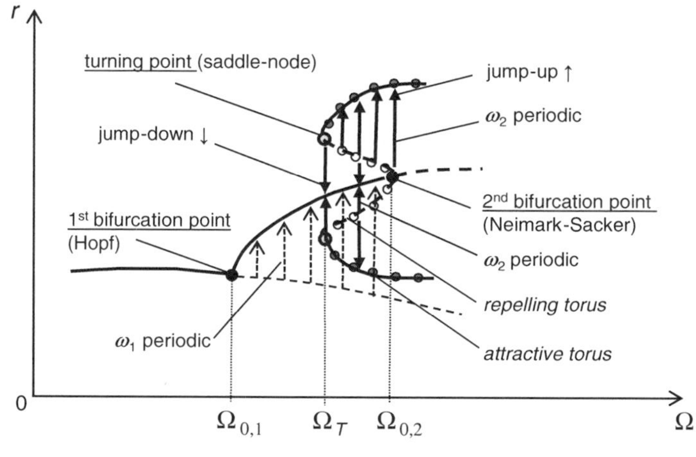

Fig. 4.19 Behavior of the subcritical Neimark–Sacker bifurcation

amplitude jumps from the turning point into the inner torus center. Between the second bifurcation and turning point, the outer trajectories are repelled from the inner torus surface to the outer torus surface. The inner trajectories are repelled from the inner torus surface into the nontrivial half-branches of the Hopf bifurcation.

The vibrations that are superimposed by the periodic vibrations with the frequency ω_1 (Hopf bifurcation) and the periodic vibrations with the frequency ω_2 (Neimark–Sacker bifurcation) could contain other periodic or quasiperiodic vibrations with rational and irrational frequencies. In case of the irrational frequency ratio, ω_1 and ω_2 are incommensurate. The vibrations could change from the periodic vibration of the Hopf bifurcation into the quasiperiodic, and finally into the chaotic vibration in the Neimark–Sacker bifurcation. Such sudden jumps of the journal in the radial bearings with large amplitudes are very dangerous and harmful for the radial bearings and destructively damage them.

(b) Stability in Neimark–Sacker Bifurcations

In the following section, the stability of a two-DOF problem in the Neimark–Sacker bifurcation is studied in the λ plane [10].

Having developed the Taylor series at the vicinity of the equilibrium point x^*, one obtains the vibration solutions in the first order at the time iteration step $i + 1$.

$$\begin{aligned} x_{1,i+1} &= x_1^* + \varepsilon_{1,i+1} \\ x_{2,i+1} &= x_2^* + \varepsilon_{2,i+1} \end{aligned} \tag{4.50}$$

where ε_1 and ε_2 are the disturbances of x_1 and x_2, respectively.

At small disturbances, they can be written in terms ε of the time step i.

$$\begin{aligned} \varepsilon_{1,i+1} &\approx a\varepsilon_{1,i} + b\varepsilon_{2,i} \\ \varepsilon_{2,i+1} &\approx c\varepsilon_{1,i} + d\varepsilon_{2,i} \end{aligned} \tag{4.51}$$

where a, b, c, and d are the coefficients.

According to [10], the disturbances in Eq. (4.51) can be transformed into the new coordinates of u_1 and u_2.

$$\begin{aligned} u_{1,i+1} &= \alpha u_{1,i} - \beta u_{2,i} \\ u_{2,i+1} &\approx \alpha u_{1,i} + \beta u_{2,i} \end{aligned} \tag{4.52}$$

where α and β are the real and imaginary parts of the eigenvalues λ, respectively.

Equation (4.52) is further transformed into the polar coordinate system (r,θ) with using the relation of α and β with the complex eigenvalues of λ_1 and λ_2.

$$\begin{aligned} u_{1,i} &= r_i \cos\theta_i \\ u_{2,i} &= r_i \sin\theta_i \end{aligned} \tag{4.53}$$

and

$$\begin{aligned} \lambda_1 &= \alpha + j\beta = \rho e^{j\phi} = \rho\cos\phi + j\rho\sin\phi \\ \lambda_2 &= \alpha - j\beta = \rho e^{-j\phi} = \rho\cos\phi - j\rho\sin\phi \end{aligned} \tag{4.54}$$

Thus,

$$\alpha = \rho\cos\phi;\ \beta = \rho\sin\phi \tag{4.55}$$

Substituting of Eqs. (4.53, 4.52, and 4.55), the vibration solutions at time step $i + 1$ are written as

$$\begin{aligned} r_{i+1} \cos \theta_{i+1} &= \rho r_i \cos(\theta_i + \phi) \\ r_{i+1} \sin \theta_{i+1} &= \rho r_i \sin(\theta_i + \phi) \end{aligned} \tag{4.56}$$

Therefore,

$$\begin{aligned} r_{i+1} &= \rho r_i \\ \theta_{i+1} &= \theta_i + \phi \; for \; \forall i = 0, 1, 2, \ldots, n. \end{aligned} \tag{4.57}$$

- Substituting $i = 0$ into Eq. (4.57), one obtains

$$\begin{aligned} r_1 &= \rho r_0 \\ \theta_1 &= \theta_0 + \phi \end{aligned}$$

- Next, for $i = 1$:

$$\begin{aligned} r_2 &= \rho r_1 = \rho^2 r_0 \\ \theta_2 &= \theta_0 + 2\phi \end{aligned}$$

- Further, for $i = n > 0$:

$$\begin{aligned} r_n &= \rho^n r_o \\ \theta_n &= \theta_0 + n\phi \end{aligned} \tag{4.58}$$

The solution $r(t)$ in Eq. (4.58) is stable if the radius $\rho < 1$ for increasing n. Hence, the stability condition is written as

$$\lim_{\substack{n \to \infty; \\ \rho < 1}} r_n = \rho^n r_0 = 0 \tag{4.59}$$

in which the eigenvalues are

$$\lambda_{1,2} = \alpha \pm j\beta = \rho e^{\pm j\phi}.$$

The dynamic behavior is stable inside the unit circle ($\rho = 1$) and unstable outside the unit circle, according to Eq. (4.59), as shown in Fig. 4.20.

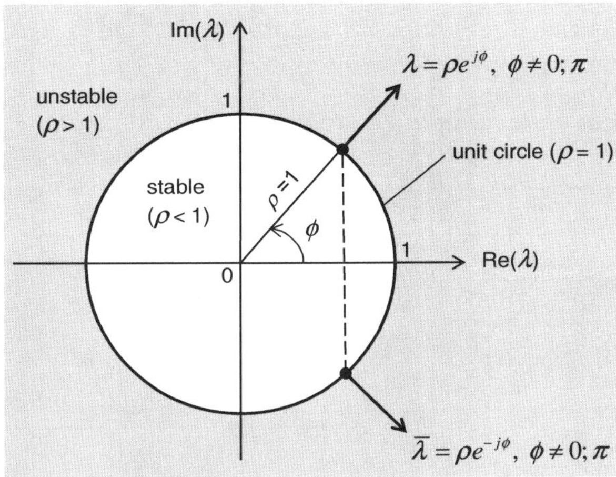

Fig. 4.20 Stability regions of the Neimark–Sacker bifurcation

4.3.9 Vibration Equations of the Nonautonomous Systems

The vibration system is called nonautonomous when its solutions depend on time t explicitly. Hence, the vibration Eqs. (4.36) with the rotor speed Ω is written as

$$\dot{\mathbf{z}} = \mathbf{A}\mathbf{z} + \mathbf{b}(\mathbf{z}, \Omega, t) \tag{4.60}$$

The nonautonomous first-order ODE with two DOF of x_1 and x_2 in Eq. (4.60) can be converted into the autonomous first-order ODE of x_1 and x_2 and one ODE of time t that is considered as a parameter during iteration.

$$\begin{aligned} \dot{x}_1 &= f_1(x_1, x_2, \Omega) \\ \dot{x}_2 &= f_2(x_1, x_2, \Omega) \\ \dot{t} &= 1 \end{aligned} \tag{4.61}$$

Suppose that $x_1(t)$ and $x_2(t)$ are converged, their initial solutions are $x_{1,0}$ and $x_{2,0}$ at $t_0 = 0$ of the starting point P_0 in the phase plane x_1x_2.

The Poincaré mapping M is used in the transformation from P_0 to P_1, P_1 to P_2, and successively into P_k. Finally, one can write the relation of P_k at time t_k in P_0 successively iterating the mapping function M in k times [10].

$$\begin{aligned} P_k &= M^k P_0 \\ at\, t_k &= kT;\ k = 0, 1, 2, \ldots, m \end{aligned} \tag{4.62}$$

where T is the time period of the vibration.

The vibration responses $x_1(t)$ and $x_2(t)$ consist of many periodic frequency components. They result from superimposing sub- and supersynchronous vibrations on the excitation harmonics. Therefore, the vibration responses are periodic or

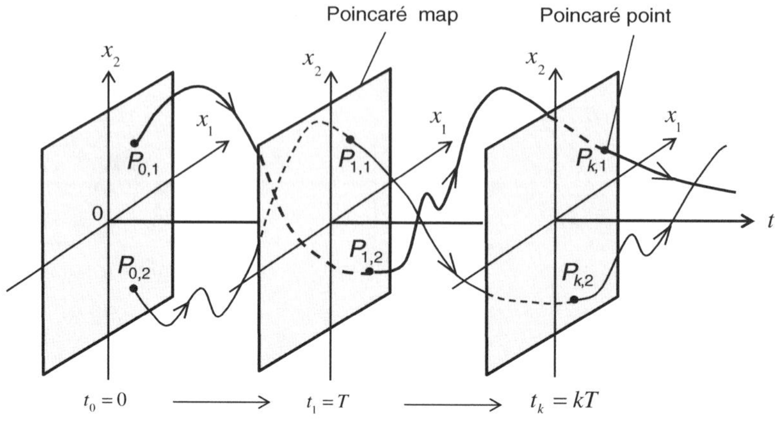

Fig. 4.21 Poincaré points in the Poincaré map at various times

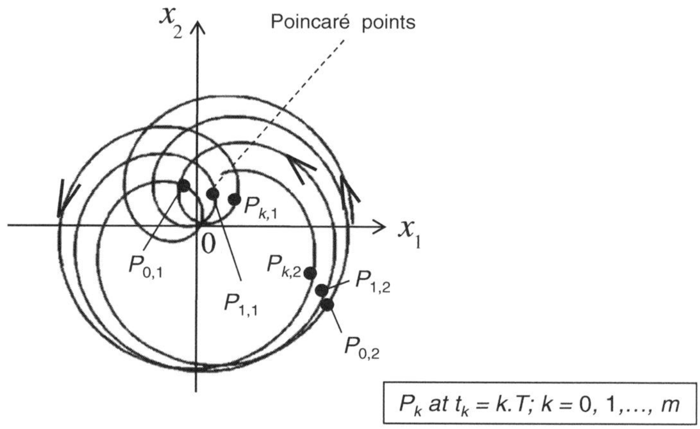

Fig. 4.22 Response orbits at various times in the Poincaré map

quasiperiodic vibrations. In case of subsynchronous frequency components with a frequency order (1/*N*)X, one obtains *N* points in the Poincaré map in every convolution where the whirl orbit completes a cycle. Due to quasiperiodicity, the Poincaré points are never the same points after repeating the time period *T* at time $t = 0, T, 2T, \ldots, mT$ (see Fig. 4.21). Therefore, the transient vibrations will appear rather in scattered dots in the Poincaré map (see Fig. 4.22).

For a vibration equation system of two DOF (x_1 and x_2), the transient solutions are quasiperiodic vibrations containing the unbalance harmonic frequency of 1X and subsynchronous frequency component with the frequency order of nearly (1/2) X (oil whirl with 0.47X). They intersect the Poincaré map at two points $P_{k,1}$ and $P_{k,2}$ at time $t_k = kT$ with $k = 0,1,\ldots,m$ after repeat of *k* times period *T*. The transient response orbits at the various times t_k are plotted in Fig. 4.22.

References

1. Bently, D.E., Hatch, C.: Fundamentals of Rotating Machinery Diagnostics. Bently Pressurized Bearing Press, Minden (2002)
2. Ehrich, F.: Handbook of Rotordynamics. Krieger Publishing Company, Melbourne (2004)
3. Kraemer, E.: Rotordynamics of Rotors and Foundations. Springer, New York (1993)
4. MATCONT: Govaerts, W. et al.: Program of Bifurcation Analysis of Dynamical Systems. Dover Publications Inc., New York (2006)
5. Newland, D.E.: Mechanical Vibration Analysis and Computation. Dover Publications Inc., New York (2006)
6. Randall, R.B.: Vibration-based Condition Monitoring: Industrial, Aerospace and Automotive Applications. Wiley, New York (2011)
7. Seydel, R.: Practical Bifurcation and Stability Analysis, 3rd edn. Springer, New York (2010)
8. Schmidt, G.,Tondl, A.: Nonlinear Vibrations. Cambridge University Press, Cambridge (2009)
9. Strogatz, S.H.: Nonlinear Dynamics and Chaos. Westview, New York (1994)
10. Thompson, J.M.T., Stewart, H.B.: Nonlinear Dynamics and Chaos. Wiley, New york (1986)

Chapter 5
Linear Rotordynamics of Turbochargers

5.1 Introduction

We have thoroughly discussed the rotordynamic stability in Chap. 4. In the following section, it deals with the resonance that is also a harmful effect causing damage of the turbochargers. When the rotor frequency equals its eigenfrequency (undamped natural frequency), the resonance occurs at the critical speed at which the rotor reaches a maximum deflection amplitude [1, 2, 4].

The unbalance force F_{Un} that is proportional to squared rotor speed induces the lateral deflection r of the rotor. That means that the higher the rotor speed is, the larger the unbalance force acts upon the rotor. During operating turbochargers, the unbalance force deflects the rotor outwards in lateral direction against the stiffness force of the rotor shaft and the damping force of the bearing. The larger the rotor amplitude is, the higher the stiffness force acts upon the rotor, and the larger the restoring potential energy induces in the rotor. Additionally, the damping force is proportional to the deflection velocity of the shaft. In other words, the rotor is in balance from the inertia, stiffness, damping, and unbalance forces of the rotor at any rotor speed according to Newton's second law.

The vibration behaviors of the rotor versus rotor speed are analyzed using the balance between the acting forces on the rotor (see Fig. 5.1).

- At subcritical rotor speeds ($\Omega \ll \Omega_{cr}$), both the inertia and damping forces are quite small. As a result, the rotor keeps its balance quasi between the rotor stiffness and unbalance forces at increasing the rotor speeds. The deflection direction is nearly the same direction of the unbalance force. The rotor deflection further increases during speed-up leading to raising the phase angle ϕ of the vibration.
- At the critical speed ($\Omega = \Omega_{cr}$), the direction of the unbalance force is perpendicular to the deflection direction leading to a phase angle $\phi = -\pi/2$. At the resonance, the rotor stiffness force keeps the balance with the inertia force; the damping force keeps the unbalance force in balance. Therefore, the resonance

H. Nguyen-Schäfer, *Rotordynamics of Automotive Turbochargers*,
Springer Tracts in Mechanical Engineering, DOI 10.1007/978-3-319-17644-4_5

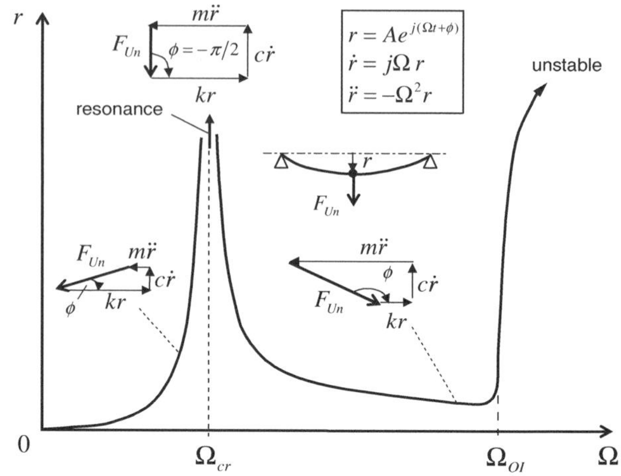

Fig. 5.1 Rotor response behavior versus rotor speed

amplitude strongly depends on the damping force. Equation (5.17) indicates that the rotor amplitude is proportional to the inversion of the damping ratio ζ at the resonance ($\eta = 1$). Thus, if no or very small damping force exists, the deflection rotor amplitude becomes very large. Note that the more the rotor is damped, the smaller its resonance amplitude becomes.

- At supercritical speeds ($\Omega > \Omega_{cr}$), the phase angle ϕ increases due to the self-centering of the rotor in which the deflection direction is nearly opposite to the unbalance force. As a result, the rotor amplitude continuously reduces to the unbalance radius as the rotor speed passed the critical speed.
- At hypercritical speeds ($\Omega \gg \Omega_{cr}$), the stiffness and damping forces become smaller. Therefore, the inertia force keeps the balance with the unbalance force. The rotor mass center tends to approach the bearing center at the phase angle of $\phi \approx -\pi$, as shown in Figs. 5.5 and 5.6.

In the point of view of energy, the resonance amplitude of the rotor is determined by the energy balance between the total kinetic energy of the rotor (translational and rotational), total potential energy of the rotor stiffness (translation and rotation), and dissipated damping energy of the bearings acting upon the rotor. The rotor damping energy dissipates the kinetic energy of the rotor during deflection. The higher the rotor speed is, the larger the dissipation energy induces in the rotor. At increasing rotor speeds, the potential and dissipation energies have been increased in order to balance the kinetic energy of the rotor. When the rotor reaches the equilibrium of the energy balance, the resonance occurs at the maximum rotor deflection where its deflection velocity equals zero. In general, the larger the damping force is, the smaller the resonance amplitude becomes since the translational kinetic energy is partly dissipated by the damping energy during the lateral deflection of the rotor.

After traversing the resonance at the critical speed Ω_{cr}, the rotor speed increases at speed-up of the turbocharger and reaches the onset of instability Ω_{OI} at which the rotor behavior becomes unstable, as discussed in Chap. 4. Therefore, the amplitude sharply increases with time and finally becomes very large in case of oil whip, a subsynchronous self-excited vibration in nonlinear rotordynamics.

At designing the rotordynamic system of automotive turbochargers, the first critical speed should be as high as possible so that the rotor speed at the onset of instability (about twice Ω_{cr}) occurs outside the operating speed range of the turbocharger. It would be very suitable if the operating speed range lay between the resonance and onset of instability.

In fact, automotive turbochargers operate in a wide speed range in the resonance region that varies from small to very large rotor speeds. In order to avoid damage of the turbochargers, a large damping ratio in the bearings is required to reduce the rotor deflection at the resonance in case of linear rotordynamics; and to keep the limit cycle as small as possible in case of nonlinear rotordynamics. However, the large bearing damping forces could induce more friction in the bearings leading to a low efficiency of the turbochargers, especially in low-end torque. Hence, the bearing damping force should be as large as required but not larger than necessary in order to compromise between the resonance amplitude, bearing friction, and reducing noises.

5.2 Vibration Response of the Linear Rotordynamic System

Vibration response of the linear rotordynamic system results from the excitation force of the rotor unbalance. The rotor response results in the product of the system transfer impedance and unbalance excitation force (see Fig. 5.2). The transfer impedance Z of the linear system is defined as the inversion of the complex dynamic stiffness coefficient K_S that contains the stiffness and damping coefficients of the rotor, rotor mass, and rotor speed [2, 5, 7, 8].

In the linear rotordynamic system, the unbalance rotor response is the sum of the rotor responses of N degrees of freedom (DOF).

$$\mathbf{r}(t) = \sum_{i=1}^{N} \mathbf{Z}_i(\Omega) \cdot F_{U,i}(t) \equiv \sum_{i=1}^{N} \mathbf{K}_{S,i}^{-1}(\Omega) \cdot F_{U,i}(t) \tag{5.1}$$

The linear rotordynamic system is only valid in small response amplitudes at low rotor unbalances. Otherwise, at large amplitudes in case of high unbalances, the

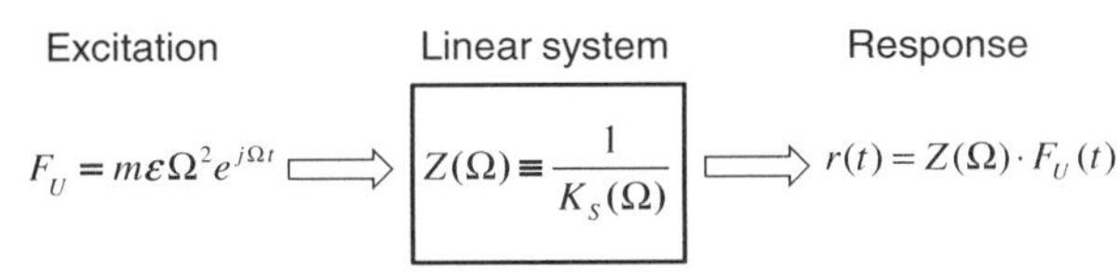

Fig. 5.2 Vibration response of a linear rotordynamic system

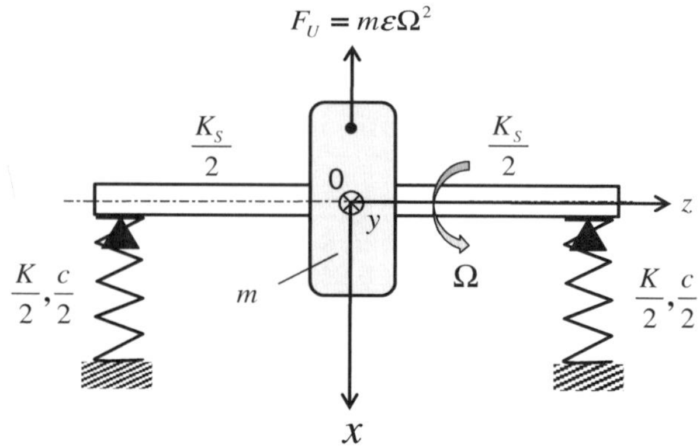

Fig. 5.3 Jeffcott rotor with isotropic bearings

rotor response becomes nonlinear because the stiffness and damping forces of the radial bearings have not been linear any longer.

According to Eq. (5.1), the vibration response results from the complex dynamic stiffness coefficients and excitation unbalance force of the rotor. For the sake of simplicity, the Jeffcott rotor with the isotropic bearings are used to calculate the rotor response in linear rotordynamics, as displayed in Fig. 5.3.

In case of isotropic bearings ($k_1 = k_2 \equiv k$; $c_1 = c_2 \equiv c$), the vibration equations of the rotor are written as

$$\begin{cases} m\ddot{x} + c\dot{x} + kx = m\varepsilon\Omega^2 \cos \Omega t \\ m\ddot{y} + c\dot{y} + ky = m\varepsilon\Omega^2 \sin \Omega t \end{cases} \tag{5.2}$$

where

k is the equivalent stiffness coefficient resulting from K and K_S given in Eq. (4.5);
c is the damping coefficient of the rotor.

Note that both degrees of freedom x and y (DOF = 2) could be combined in one degree of freedom r (DOF = 1) only in case of the isotropic bearings that have the same stiffness and damping coefficients in both directions x and y.

The complex coordinates r and ϕ result from the rotor coordinates as

$$\mathbf{r} = x + jy; \quad \phi = \arctan\left(\frac{y}{x}\right) \tag{5.3}$$

Using Eq. (5.3), the vibration equation of the rotor is written in the complex plane as

$$m\ddot{\mathbf{r}} + c\dot{\mathbf{r}} + k\mathbf{r} = m\varepsilon\Omega^2 e^{j\Omega t} \tag{5.4}$$

The rotor response $\mathbf{r}(t)$ of the linear second-order differential equation consists of the free and forced vibration responses that are the homogeneous and particular solutions of Eq. (5.4), respectively.

$$\mathbf{r}(t) = \mathbf{r}_h(t) + \mathbf{r}_u(t) \tag{5.5}$$

where

$\mathbf{r}_h(t)$ is the free vibration response (homogeneous solutions);

$\mathbf{r}_u(t)$ is the forced vibration response (unbalance response or particular solution).

The generalized free vibration response of the homogeneous Eq. (5.4) can be expressed as

$$\begin{aligned} \mathbf{r}_h(t) &= \sum_{i=1}^{N} \phi_i \left(r_i e^{\lambda_i t} \right) \\ &= \sum_{i=1}^{N} \phi_i \left(r_{i,fw} e^{(\alpha_i + j\omega_i)t} + r_{i,bw} e^{(\alpha_i - j\omega_i)t} \right) \end{aligned} \tag{5.6}$$

where

N is the degrees of freedom (DOF) of the rotor;

λ_i is the eigenvalue corresponding to the eigenfrequency ω_i;

$\boldsymbol{\Phi}_i$ is the eigenvector corresponding to the eigenvalue λ_i;

r_i is the vibration amplitude of the DOF i, and forward whirl (*fw*); backward whirl (*bw*).

The $(N \times N)$ modal matrix $\mathbf{\Phi}$ results from the $(N \times 1)$ matrix of the eigenvectors $\boldsymbol{\Phi}_i$.

$$\mathbf{\Phi} = (\phi_1 \, \phi_2 \ldots \phi_N)$$

The generalized unbalance vibration response of Eq. (5.4) is written as

$$\mathbf{r}_u(t) = \sum_{i=1}^{N} \phi_i p_i e^{j\Omega t} \tag{5.7}$$

where

N is the degrees of freedom (DOF) of the rotor;

$\boldsymbol{\Phi}_i$ is the eigenvector corresponding to the eigenvalue λ_i;

p_i is the response vibration amplitude of the DOF i;

Ω is the rotor speed.

For a stable rotor response, the free vibration response $\mathbf{r}_h(t)$ diminishes in a short time (short-term response), as discussed in Sect. 4.2. Note that the free vibration response (homogeneous solution) is responsible for the rotor stability; the forced vibration response (unbalance response or particular solution) is for the rotor resonance.

Therefore, the rotor response $\mathbf{r}(t)$ approaches the unbalance response $\mathbf{r}_u(t)$ of the $(N \times 1)$ vector in the long-term response.

$$\begin{aligned} \mathbf{r}(t) &= \mathbf{r}_h(t) + \mathbf{r}_u(t) \\ &\approx \mathbf{r}_u(t) = \mathbf{\Phi}\mathbf{p}_u(t) \end{aligned} \tag{5.8}$$

where $(N \times N)$ modal matrix $\mathbf{\Phi}$ of the eigenvectors $\boldsymbol{\Phi}_i$

$$\mathbf{\Phi} = (\ \phi_1\ \phi_2\ \ \ldots\ \ \phi_N\ ) \tag{5.9}$$

and the $(N \times 1)$ modal response vector

$$\mathbf{p}_u(t) = \begin{pmatrix} p_1 \\ p_2 \\ \ldots \\ p_N \end{pmatrix} e^{j\Omega t} \tag{5.10}$$

Now, we turn back to the simple Jeffcott rotor (DOF = 1) with the vibration Eq. (5.4), the unbalance response becomes after a short time

$$r_u(t) = A(\Omega)e^{j\Omega t} \tag{5.11}$$

Inserting r_u, its first, and second derivatives into Eq. (5.4), one obtains the complex response amplitude

$$\begin{aligned} A(\Omega) &= \frac{m\varepsilon\Omega^2}{(k - m\Omega^2) + jc\Omega} \\ &= \left(\frac{m\varepsilon\Omega^2}{(k - m\Omega^2)^2 + (c\Omega)^2}\right) \cdot \left[(k - m\Omega^2) - jc\Omega\right] \end{aligned} \tag{5.12}$$

within the complex dynamic stiffness coefficient K_S of the rotor is given in

$$K_S(\Omega) = (k - m\Omega^2) + jc\Omega \tag{5.13}$$

The amplitude of the rotor response A results from Eq. (5.12)

$$|A| = \frac{m\varepsilon\Omega^2}{\sqrt{(k - m\Omega^2)^2 + (c\Omega)^2}} \tag{5.14}$$

and the phase angle ϕ of the rotor response results in

$$\phi = \arctan\left(\frac{-c\Omega}{k - m\Omega^2}\right) \tag{5.15}$$

Thus,

$$\begin{aligned} r_u(t) &= A(\Omega)e^{j\Omega t} = |A|e^{j\phi}e^{j\Omega t} \\ &= |A|e^{j(\Omega t+\phi)} \end{aligned} \tag{5.16}$$

Using Eqs. (5.14 and 5.15), the amplitude and phase angle of the rotor response are written in the unbalance radius, speed and damping ratio.

$$\begin{aligned} |A| &= \frac{\varepsilon\eta^2}{\sqrt{(1-\eta^2)^2 + 4\zeta^2\eta^2}} \\ \phi &= \arctan\left(\frac{-2\zeta\eta}{1-\eta^2}\right) \end{aligned} \tag{5.17}$$

where

ε is the unbalance radius,

ζ is the damping ratio defined in Eq. (4.16),

η is the speed ratio of the rotor to critical speed ($\eta \equiv \Omega/\Omega_{cr}$).

The amplitude of the vibration response relative to the unbalance radius versus speed ratio η is plotted in Fig. 5.4. At a given damping ratio ζ, the rotor amplitude reaches the resonance amplitude of $\varepsilon/(2\zeta)$ according to Eq. (5.17) since the denominator reaches the minimum of 2ζ at the speed ratio $\eta = 1$. In case of the critically damped vibration at $\zeta = 1$, the rotor critical amplitude equals $\varepsilon/2$, even

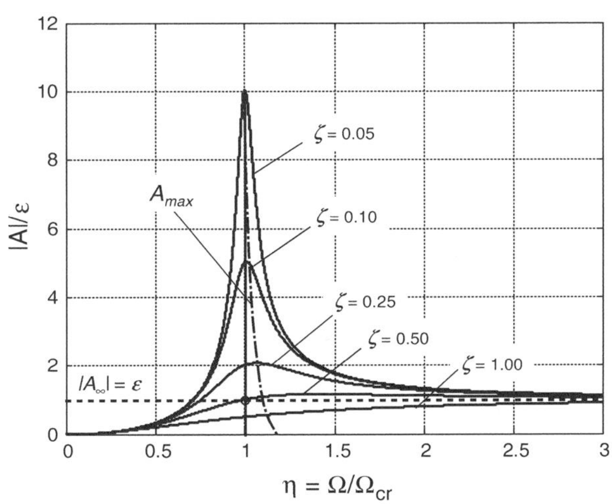

Fig. 5.4 Relative rotor amplitude $|A|/\varepsilon$ of the vibration response

lower than the unbalance radius ε. At $\zeta = 0.5$, the rotor amplitude equals ε at the resonance. The pikes of the resonance amplitudes at various damping ratios are located in the curve A_{max} on the RHS of the critical speed ratio $\eta = 1$. Obviously, the resonance amplitudes change inversely related to ζ, as shown in Fig. 5.4.

Applying the L'Hôpital's rule to Eq. (5.17), the rotor amplitude asymptotically approaches the unbalance radius ε at hypercritical speeds ($\eta >> 1$).

$$\begin{aligned} |A_\infty| &= \lim_{\eta\to\infty} \frac{\varepsilon\eta^2}{\sqrt{(1-\eta^2)^2 + 4\zeta^2\eta^2}} \\ &= \lim_{\eta\to\infty} \frac{2\varepsilon\eta}{2\eta} \\ &= \lim_{\eta\to\infty} \frac{2\varepsilon}{2} = \varepsilon \end{aligned} \tag{5.18}$$

Figure 5.5 shows the self-centering effect of the rotor where the rotor mass center G moves toward the bearing center O at supercritical rotor speeds ($\eta > 1$); at hypercritical rotor speeds of about $3\Omega_{cr}$ ($\eta \approx 3$), the rotor mass center nearly approaches the bearing center ($G \to O$) according to Eq. (5.18), as shown in Fig. 5.4.

The phase angle of the vibration response versus speed ratio is shown in Fig. 5.6a. According to Eq. (5.17), the phase angle begins from zero at subcritical

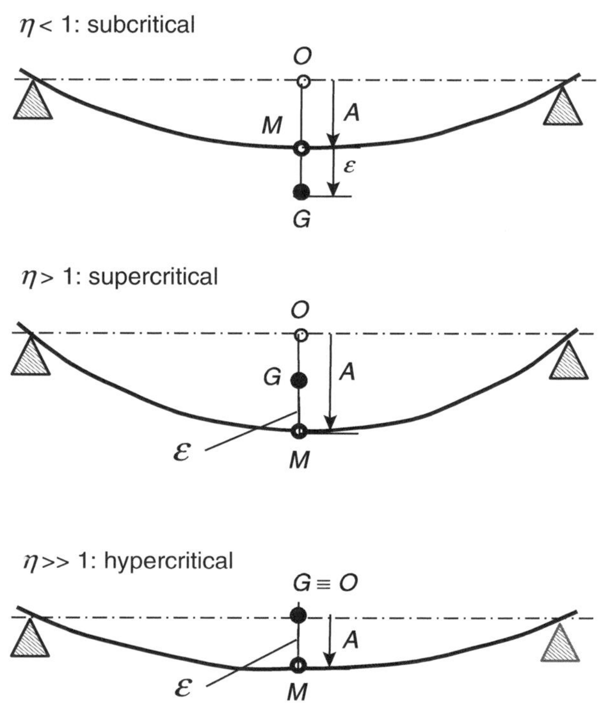

Fig. 5.5 Self-centering of the rotor at super- and hypercritical speeds (O bearing center; M rotor geometric center; G rotor mass center; ε unbalance radius)

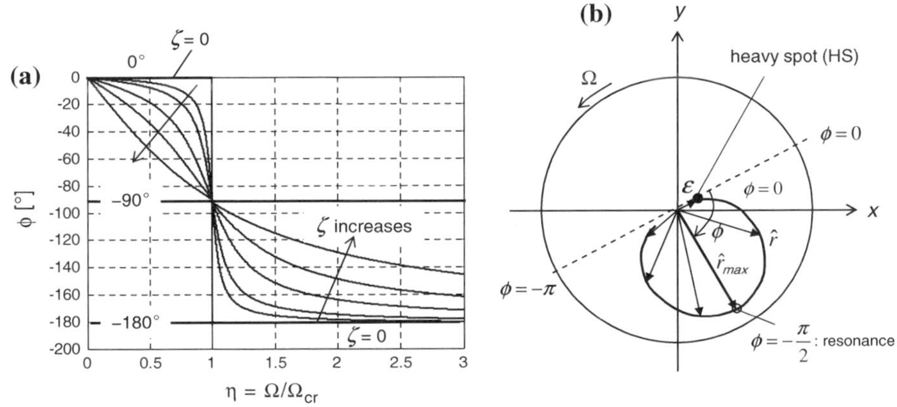

Fig. 5.6 **a** Phase angle of the vibration response; **b** Response vector

rotor speeds ($\eta \ll 1$), decreases to $-90°$ at the critical speed ($\eta = 1$) and changes from -90 to $-180°$ at supercritical speeds ($\eta \gg 1$) [10, 12, 16, 17].

Similarly, the deflection amplitude of the rotor is displayed in the polar coordinate system (r,ϕ) in Fig. 5.6b. At the small rotor speeds, the deflection direction is nearly the same direction of the unbalance force locating at the heavy spot (HS) which is identical to the point G in Fig. 5.5. This behavior is useful at the balancing of the rotor (see Chap. 8). The deflection direction lags behind the unbalance force direction with a phase lag ϕ that grows with the rotor speed $\mathbf{\Omega}$. The amplitude of the response vector reaches the maximum at the resonance with a phase lag of $-90°$, where the rotor deflects perpendicularly to the unbalance force at the critical speed Ω_{cr}. For hypercritical speeds, the rotor amplitude decreases and the rotor deflection changes its direction opposite to the unbalance force with a phase lag larger than $-90°$. At very large hypercritical speeds ($\mathbf{\Omega} \rightarrow \infty$), the phase lag approaches $-180°$ opposite to the unbalance force, and the rotor amplitude converges to the unbalance radius ε, as shown in Fig. 5.4.

5.3 Bearing Force Acting on the Flexible Rotor

The bearing force F_B acting upon the rotor at the steady-state condition is shown in Fig. 5.7. Let k be the effective stiffness coefficient of the rotor shaft. The bearing force that is proportional to the rotor deflection can be written as

$$F_B = k r_M \tag{5.19}$$

Using the relation between k, m, and the critical speed Ω_{cr}, one obtains

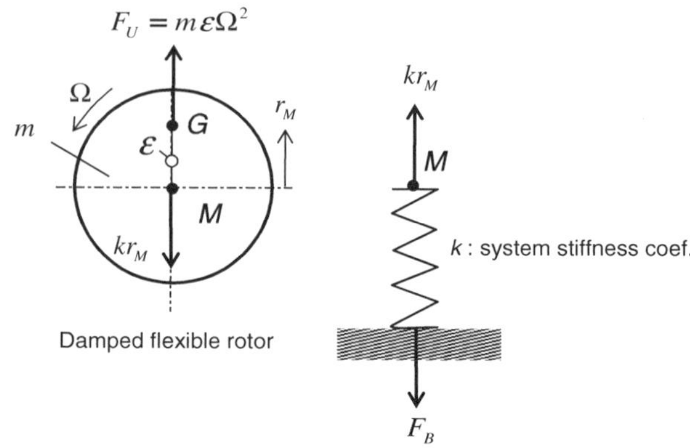

Fig. 5.7 Forces acting upon a damped flexible rotor

$$F_B = m\Omega_{cr}^2 r_M \tag{5.20}$$

Further relations of η and r_M result from Eq. (5.17) as

$$\begin{aligned} \eta &= \frac{\Omega}{\Omega_{cr}} \Rightarrow \Omega_{cr} = \frac{\Omega}{\eta} \equiv \sqrt{\frac{k}{m}} \\ r_M &= \frac{\varepsilon\eta^2}{\sqrt{(1-\eta^2)^2 + 4\zeta^2\eta^2}} \end{aligned} \tag{5.21}$$

Substituting Eq. (5.21) into Eq. (5.20), one obtains the bearing force acting upon the damped flexible rotor.

$$F_{B,flex} = \frac{m\varepsilon\Omega^2}{\sqrt{(1-\eta^2)^2 + 4\zeta^2\eta^2}} = \frac{m\varepsilon\Omega_{cr}^2\eta^2}{\sqrt{(1-\eta^2)^2 + 4\zeta^2\eta^2}} \tag{5.22}$$

In a rigid rotor, the bearing force equals the unbalance force since the rotor stiffness coefficient k is very large ($k \to \infty$).

$$F_{B,rigid} = m\varepsilon\Omega^2 = m\varepsilon\Omega_{cr}^2\eta^2 \tag{5.23}$$

The relative bearing force in Eq. (5.22) versus speed ratio with various damping ratios is plotted in Fig. 5.8. Contrary to the rigid rotor, the bearing force of the flexible rotor reduces after the resonance in supercritical speeds.

Using L'Hôpital's rule, one obtains the bearing force of the damped flexible rotor at a very high rotor speed.

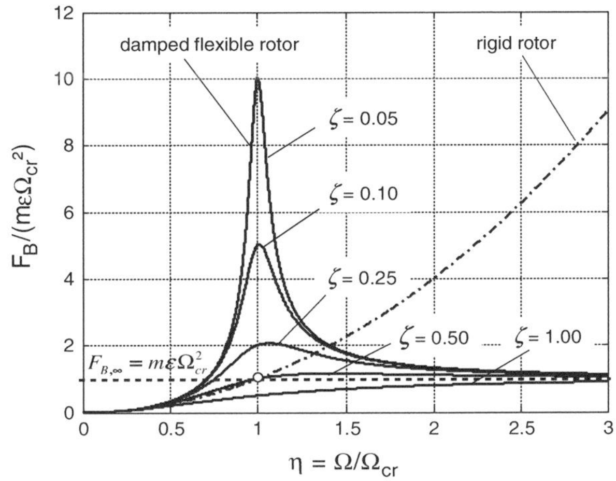

Fig. 5.8 Bearing forces acting on a rigid and damped flexible rotor

$$F_{B,flex,\infty} = \lim_{\eta \to \infty} m\varepsilon\Omega_{cr}^2 \left(\frac{\eta^2}{\sqrt{(1-\eta^2)^2 + 4\zeta^2\eta^2}} \right) = m\varepsilon\Omega_{cr}^2 \quad (5.24)$$

According to Eq. (5.23), the bearing force of the rigid rotor increases without limit at large hypercritical speeds ($\eta \gg 1$), as shown in Fig. 5.8. The discrepancy between both bearing forces is relatively small up to the rotor speed ratio of 0.5 since the rotor deflection is quite small in this speed range. After that, the bearing force of the damped flexible rotor increases very fast to the resonance and reaches the critical force depending on the damping ratio ζ.

$$F_{B,flex,cr} = \frac{m\varepsilon\Omega_{cr}^2}{2\zeta}. \quad (5.25)$$

With a very low damping ratio ζ, the bearing force becomes too large enough to damage the rotor and bearings. At a damping ratio of 5 %, the bearing force of the flexible rotor at the critical speed is 10 times higher than the asymptote unbalance force at $\eta \gg 1$ (see Fig. 5.8).

According to Eq. (5.22), the bearing force depends on two key parameters: the unbalance radius ε and damping ratio ζ. In fact, the high-speed rotor balancing (trim balancing) could be carried out at a high balancing speed to about maximum 1.5 times of the first critical speed and not higher because the turbine wheel is generally driven using the cold air at the balancing, instead of the exhaust gas at high temperatures (large enthalpy). Therefore, the residual unbalance remains high at the operating condition in the supercritical speeds. Normally, the maximal speed of automotive turbochargers is about 2.5 times of the critical speed. Hence, the bearing force in the operating condition of the speed ratios η between 1.5 and 2.5 is relatively large. In this speed range, more damping in the bearings is required to reduce the bearing force. Generally, the damping ratio should be large enough to prevent the bearing from damages and to reduce airborne noises. Fortunately, the engine noise is loud enough in the speed ratios between 1.5 and 2.5 to suppress the unbalance whistle.

5.4 Gyroscopic Effect of the Rotor System

Gyroscopic effect of the large compressor and turbine wheels becomes more important at high rotor speeds in rotordynamics. The gyroscopic effect is proportional to the rotor speed and the wheel polar mass inertia moment that is in turn proportional to mass and squared radius. In the turbochargers, the compressor and turbine wheel diameters are much larger than the rotor shaft diameter (see Fig. 5.9). Therefore, their gyroscopic effects must be taken into account in the rotordynamic computation of the rotor, especially at high rotor speeds.

The gyroscopic effect of the rotor is based on the angular momentum theorem. It indicates that the time change rate of the angular momentum equals all resulting moments acting upon the rotor [15, 17].

$$\mathbf{M}_P = \left(\frac{d\mathbf{L}}{dt}\right)_P \tag{5.26}$$

where

$\mathbf{M}_P$ is the resulting moment vector at a reference point P on the rotor;

$\mathbf{L}_P$ is the angular momentum vector at the point P.

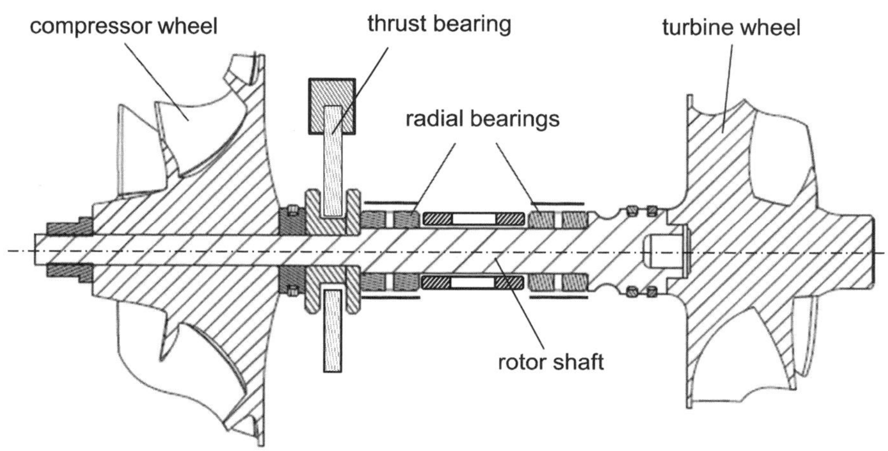

Fig. 5.9 Rotor of a small automotive turbocharger

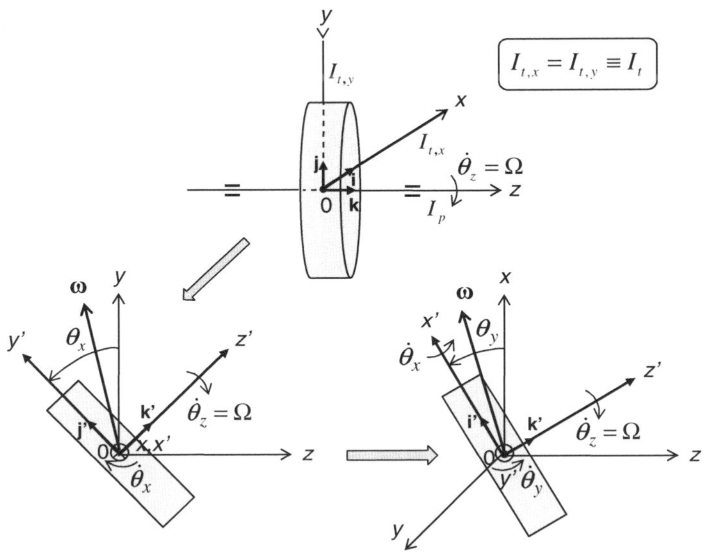

Fig. 5.10 Gyroscopic effect of the spinning wheel

Using the chain rule of differentiation between the inertial and precessing coordinate systems (x, y, z) and (x', y', z'), the time derivative of the angular momentum is written as

$$\mathbf{M} = \frac{d\mathbf{L}}{dt} = \left(\frac{\partial \mathbf{L}}{\partial t}\right)_{(x',y',z')} + (\boldsymbol{\omega} \times \mathbf{L}) \tag{5.27}$$

where $\boldsymbol{\omega}$ is the precessing angular velocity of the spinning wheel at the rotor speed Ω around its axis.

Due to the gyroscopic effect, the precessing angular velocity $\boldsymbol{\omega}$ causes a precession of the spinning wheel around its axis. It can be expressed in the precessing coordinates (x', y', z') with the basis vectors $\mathbf{i}'$ and $\mathbf{j}'$ as (see Fig. 5.10)

$$\boldsymbol{\omega} = \dot{\theta}_x \mathbf{i}' + \dot{\theta}_y \mathbf{j}' \tag{5.28}$$

The angular momentum of the spinning wheel is calculated in the precessing coordinate system (x', y', z') with the basic vectors ($\mathbf{i}'$, $\mathbf{j}'$, $\mathbf{k}'$).

$$\begin{aligned} \mathbf{L} &= I_{t,x}\dot{\theta}_x \mathbf{i}' + I_{t,y}\dot{\theta}_y \mathbf{j}' + I_p \dot{\theta}_y \mathbf{k}' \\ &= I_{t,x}\dot{\theta}_x \mathbf{i}' + I_{t,y}\dot{\theta}_y \mathbf{j}' + I_p \Omega \mathbf{k}' \end{aligned} \tag{5.29}$$

within

$I_{t,x}$, $I_{t,y}$ are the transverse mass moments of inertia in the directions of x' and y';

I_p is the polar mass inertia moment in the direction z';

$\dot{\theta}_x, \dot{\theta}_y$ are the precessing angular velocities of the wheel in the directions x' and y';

Ω is the spinning rotor speed (Ω = const).

Partially differentiating Eq. (5.29) with time, the derivative of the angular momentum in the precessing coordinate system results as

$$\left(\frac{\partial \mathbf{L}}{\partial t}\right)_{(x',y',z')} = I_{t,x}\ddot{\theta}_x\mathbf{i}' + I_{t,y}\ddot{\theta}_y\mathbf{j}' \tag{5.30}$$

Using orthonormality of the basis vectors, the vector cross product of ($\boldsymbol{\omega} \times \mathbf{L}$) in Eq. (5.27) is calculated as

$$\begin{aligned}(\boldsymbol{\omega} \times \mathbf{L}) &= \left(\dot{\theta}_x\mathbf{i}' + \dot{\theta}_y\mathbf{j}'\right) \times \left(I_{t,x}\dot{\theta}_x\mathbf{i}' + I_{t,y}\dot{\theta}_y\mathbf{j}' + I_p\Omega\mathbf{k}'\right)\\ &= I_p\dot{\theta}_y\Omega\mathbf{i}' - I_p\dot{\theta}_x\Omega\mathbf{j}'\end{aligned} \tag{5.31}$$

Substituting Eqs. (5.30 and 5.31) into Eq. (5.27), one obtains the resulting moment acting upon the spinning wheel.

$$\begin{aligned}\mathbf{M} &= \left(I_t\ddot{\theta}_x + I_p\Omega\dot{\theta}_y\right)\mathbf{i}' + \left(I_t\ddot{\theta}_y - I_p\Omega\dot{\theta}_x\right)\mathbf{j}'\\ &\equiv M_x\mathbf{i}' + M_y\mathbf{j}'\end{aligned} \tag{5.32}$$

The basis vectors $\mathbf{i}'$ and $\mathbf{j}'$ of the coordinate system (x', y', z') with the precessing angular velocity $\boldsymbol{\omega}$ become at very small angles θ_x, $\theta_y << 1$ (cf. Appendix A).

$$\begin{aligned}\mathbf{i}' &\approx \mathbf{i} - \theta_y\mathbf{k} \approx \mathbf{i};\\ \mathbf{j}' &\approx \mathbf{j} + \theta_x\mathbf{k} \approx \mathbf{j}\end{aligned}$$

Using the approximation of the basis vectors, the resulting moment $\mathbf{M}$ on the wheel is written in the inertial coordinate system (x, y, z) as

$$\begin{aligned}\mathbf{M} &\approx \left(I_t\ddot{\theta}_x + I_p\Omega\dot{\theta}_y\right)\mathbf{i} + \left(I_t\ddot{\theta}_y - I_p\Omega\dot{\theta}_x\right)\mathbf{j}\\ &\equiv M_x\mathbf{i} + M_y\mathbf{j}\end{aligned} \tag{5.33a}$$

Thus,

$$\begin{aligned}M_x &= I_t\ddot{\theta}_x + I_p\Omega\dot{\theta}_y \Rightarrow \sum M_x \equiv (M_x - I_p\Omega\dot{\theta}_y) = I_t\ddot{\theta}_x\\ M_y &= I_t\ddot{\theta}_y - I_p\Omega\dot{\theta}_x \Rightarrow \sum M_y \equiv (M_y + I_p\Omega\dot{\theta}_x) = I_t\ddot{\theta}_y\end{aligned} \tag{5.33b}$$

Equation (5.33b) shows that the fictitious gyroscopic moments additionally act upon the wheel besides the external moments of M_x and M_y.

As a result, the gyroscopic moment vector $\mathbf{M}_{Gyro}$ is written in the inertial coordinate system (*x*, *y*, *z*) as

$$\mathbf{M}_{Gyro} = (I_p \Omega \dot{\theta}_y)\mathbf{i} - (I_p \Omega \dot{\theta}_x)\mathbf{j} \tag{5.34}$$

The gyroscopic moment stiffens the rotor stiffness of the forward whirls and des-tiffens the rotor stiffness of the backward whirls at increasing rotor speeds. Therefore, the forward whirl eigenfrequencies increase with the rotor speed; the backward whirl eigenfrequencies, decrease with the rotor speed, as shown in Campbell diagram (see Sect. 5.7). Hence, the critical speeds of the turbochargers are higher than the critical speeds of the rotor without the gyroscopic effect at high rotor speeds.

5.5 Vibration Equations of Turbochargers

The vibration equations of turbochargers are mostly derived by two approaches of the momentum theorem and Lagrange's equations.

(a) The momentum approach

• **Lateral vibrations**

Based on the momentum theorem, the vibration equation at the rotor mass center *G* results as

$$\mathbf{F}_M = m\,\ddot{\mathbf{r}}_G \tag{5.35}$$

where

$\mathbf{F}_M$ is the external forces acting upon the rotor geometric center *M*;

$\mathbf{r}_G$ is the vector of the rotor mass center *G*.

The rotor mass center vector results from the vectors of the rotor geometric center and unbalance radius (see Fig. 5.11).

$$\mathbf{r}_G = \mathbf{r}_M + \mathbf{r}_\varepsilon \tag{5.36}$$

The acceleration of the unbalance radius is calculated as

$$\begin{aligned} \mathbf{r}_\varepsilon &= \varepsilon\, e^{j(\Omega t+\beta)} \\ \Rightarrow \ddot{\mathbf{r}}_\varepsilon &= -\varepsilon\, \Omega^2 e^{j(\Omega t+\beta)} \end{aligned} \tag{5.37}$$

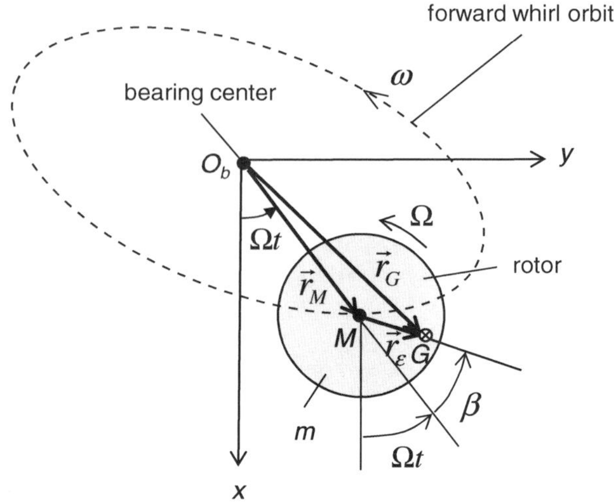

Fig. 5.11 Response vector $\mathbf{r}_M$ of the unbalanced rotor

Inserting the external forces acting at the rotor center M consisting of the stiffness and damping forces into Eq. (5.35), the vibration equation of the rotor is calculated using Eq. (5.37) as

$$\begin{aligned} \mathbf{F}_M &= -c\dot{\mathbf{r}}_\mathbf{M} - k\mathbf{r}_\mathbf{M} = m(\ddot{\mathbf{r}}_\mathbf{M} + \ddot{\mathbf{r}}_\varepsilon) \\ &\Rightarrow m\ddot{\mathbf{r}}_\mathbf{M} + c\dot{\mathbf{r}}_\mathbf{M} + k\mathbf{r}_\mathbf{M} = -m\ddot{\mathbf{r}}_\varepsilon \\ &\Rightarrow m\ddot{\mathbf{r}}_\mathbf{M} + c\dot{\mathbf{r}}_\mathbf{M} + k\mathbf{r}_\mathbf{M} = m\varepsilon\Omega^2 e^{j(\Omega t+\beta)} \end{aligned} \tag{5.38}$$

where $\mathbf{r}_M$ is the vibration response of the rotor; the RHS term is the excitation unbalance force.

- **Rotational vibrations**

The rotational vibration equations are derived from the angular momentum theorem at the reference point P on the rotor.

$$\mathbf{M}_P = \left(\frac{d\mathbf{L}}{dt}\right)_P \tag{5.39}$$

where $\mathbf{M}_P$ is the resulting torques acting upon the rotor; $\mathbf{L}$ is the angular momentum.

Using Eq. (5.33a), the resulting torque vector results as

$$\mathbf{M}_P = (I_t\ddot{\theta}_x + I_p\Omega\dot{\theta}_y)\mathbf{i} + (I_t\ddot{\theta}_y - I_p\Omega\dot{\theta}_x)\mathbf{j}$$

The resulting torque vector $\mathbf{M}_P$ contains the circulant damping, stiffness moments, and as well as the unbalance moment.

- The circulant damping moment is described

$$\mathbf{M}_c = -(c_{r,x}\dot{\theta}_x\mathbf{i} + c_{r,y}\dot{\theta}_y\mathbf{j}) \tag{5.40}$$

- The circulant and interaction stiffness moment results in

$$\mathbf{M}_k = -(k_{r,x}\theta_x + K_{tr}x)\mathbf{i} - (k_{r,y}\theta_y + K_{tr}y)\mathbf{j} \tag{5.41}$$

- The unbalance moment $\mathbf{M}_U$ at the mass center G is caused by the misalignment between the wheel and rotation axis due to the couple unbalance, see Eq. (5.49).

$$\mathbf{M}_U = \begin{bmatrix} (I_t - I_p)\alpha\,\Omega^2 \cos\Omega t \\ (I_t - I_p)\alpha\,\Omega^2 \sin\Omega t \end{bmatrix} \tag{5.42}$$

where

I_t, I_p are the transverse and polar mass moments of inertia at the mass center G;
α is the amplitude of the couple unbalance;
Ω is the rotor speed.

(b) Calculating the couple unbalance moment $\mathbf{M}_\mathrm{U}$

Due to the production process, the geometric axis of the turbine wheel is misaligned with the rotation axis of the shaft at a very small angle α (see Fig. 5.12) leading to the dynamic unbalance that induces the unbalance moment acting on the rotor. The unbalance moment excites the rotor by an excitation moment, similar to the excitation unbalance force. Both excitations cause large rotor response amplitudes in the cylindrical and conical modes.

Using the angular momentum theorem and the chain rule of differentiation, the unbalance moment $\mathbf{M}_U$ induced by the dynamic unbalance like the gyroscopic moment is considered as external moment acting upon the rotor.

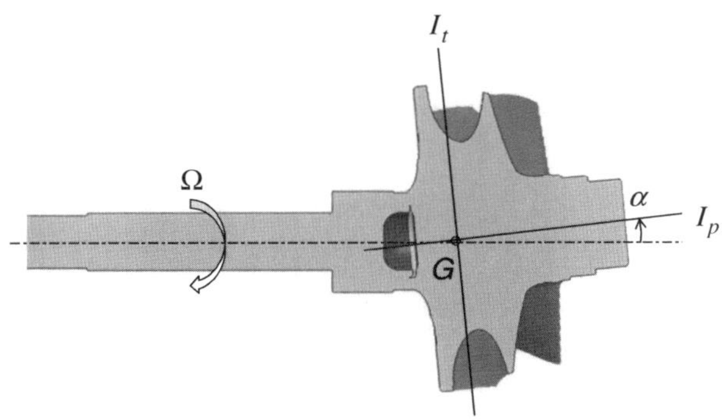

Fig. 5.12 Couple unbalance at a misalignment angle α

The angular momentum balance of the misaligned rotor is written in the inertial coordinate system (x, y, z), cf. Eqs. (5.33a, 5.33b and 5.34).

$$\mathbf{M} = \frac{d\mathbf{L}}{dt} = \left[\left(\frac{\partial \mathbf{L}}{\partial t} \right)_{(x',y',z')} + (\boldsymbol{\omega} \times \mathbf{L}) \right]$$
$$= -\mathbf{M}_U$$

Thus,

$$\mathbf{M}_U = -\left[\left(\frac{\partial \mathbf{L}}{\partial t} \right)_{(x',y',z')} + (\boldsymbol{\omega} \times \mathbf{L}) \right] \quad (5.43)$$

where $\mathbf{L}$ is the angular momentum of the misaligned rotor.

The angular velocity of the rotating coordinate system (x', y', z') is expressed as

$$\boldsymbol{\omega} = \Omega \mathbf{k} \quad (5.44)$$

The angular momentum vector of the turbine wheel is written in the rotating coordinate system (x', y', z'), as shown in Fig. 5.13.

$$\mathbf{L} = -(I_t \Omega \sin \alpha)\mathbf{j}' + (I_p \Omega \cos \alpha)\mathbf{k}' \quad (5.45)$$

The relations between the basis vectors $\mathbf{j}$, $\mathbf{k}$, $\mathbf{j}'$, $\mathbf{k}'$ of two coordinate systems are calculated as

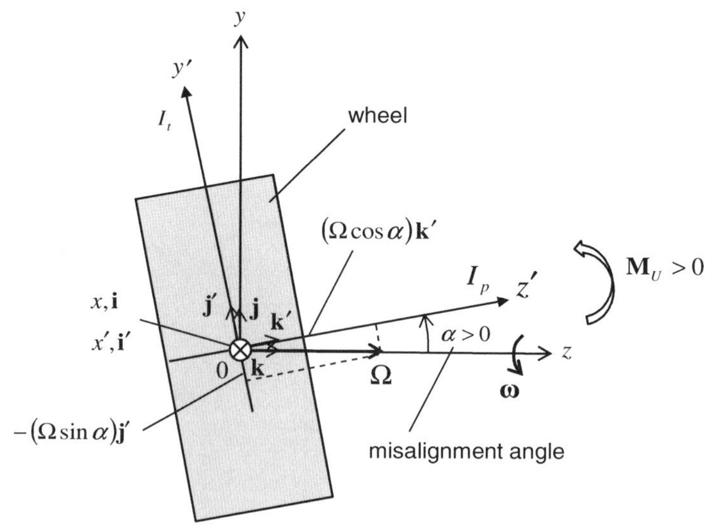

Fig. 5.13 Unbalance moment caused in a misaligned rotor with $I_t > I_p$

$$\begin{aligned} \mathbf{j}' &= (\cos\alpha)\,\mathbf{j} - (\sin\alpha)\,\mathbf{k} \\ \mathbf{k}' &= (\sin\alpha)\,\mathbf{j} + (\cos\alpha)\,\mathbf{k} \\ &\Rightarrow \mathbf{k} = -(\sin\alpha)\mathbf{j}' + (\cos\alpha)\mathbf{k}' \end{aligned} \tag{5.46}$$

Substituting Eq. (5.46) into Eq. (5.45), one obtains the angular momentum vector of the turbine wheel in the inertial coordinate system (x,y,z).

$$\mathbf{L} = (-I_t + I_p)\Omega \sin\alpha \cos\alpha \mathbf{j} + (I_t \sin^2\alpha + I_p \cos^2\alpha)\Omega\mathbf{k}$$

Substituting Eq. (5.46) into Eq. (5.44), the angular velocity $\boldsymbol{\omega}$ can be written in the rotating coordinate system (x', y', z') as

$$\begin{aligned} \boldsymbol{\omega} &= \Omega\,\mathbf{k} \\ &= -(\Omega\sin\alpha)\,\mathbf{j}' + (\Omega\cos\alpha)\,\mathbf{k}' \end{aligned} \tag{5.47}$$

Substituting Eqs. (5.45) and (5.47) into Eq. (5.43), one obtains the unbalance moment vector in the stationary coordinate system (x, y, z). For a very small misalignment angle $\alpha << 1$, it results $\mathbf{i} \approx \mathbf{i}'$ (cf. Appendix A).

$$\begin{aligned} \mathbf{M}_U &= -\left[\left(\frac{\partial \mathbf{L}}{\partial t}\right)_{(x',y',z')} + (\boldsymbol{\omega}\times\mathbf{L})\right] = -(\boldsymbol{\omega}\times\mathbf{L}) \\ &= -(I_t - I_p)\sin\alpha\cos\alpha\,\Omega^2\mathbf{i}' \\ &\approx -(I_t - I_p)\sin\alpha\cos\alpha\,\Omega^2\mathbf{i} \end{aligned} \tag{5.48}$$

where $\left(\frac{\partial \mathbf{L}}{\partial t}\right)_{(x',y',z')} = \mathbf{0}$ because Ω and α in Eq. (5.45) are independent of time.

The misalignment angle α of the couple unbalance is generally very small $\alpha << 1$ leading to $\sin\alpha \approx \alpha$ and $\cos\alpha \approx 1$. Hence, Eq. (5.48) can be written in α (radian).

$$\begin{aligned} \mathbf{M}_U &\equiv \mathbf{U}_\alpha\Omega^2 \approx -(I_t - I_p)\alpha\,\Omega^2\,\mathbf{i} \\ &\Rightarrow M_U = |\mathbf{M}_U| = |I_t - I_p|\alpha\Omega^2 = U_\alpha\Omega^2 \end{aligned} \tag{5.49}$$

Note that the unbalance moment vector $\mathbf{M}_U$ is opposite to the basis vector $\mathbf{i}$ for $I_t > I_p$, as shown in Fig. 5.13.

The moment unbalance U_α caused by the unbalance moment $\mathbf{M}_U$ is defined as

$$\begin{aligned} \mathbf{U}_\alpha &= -(I_t - I_p)\alpha\,\mathbf{i} \\ &\Rightarrow U_\alpha = |\mathbf{U}_\alpha| = |I_t - I_p|\alpha \end{aligned} \tag{5.50}$$

In the automotive turbochargers, the transverse mass moment of inertia I_t of the rotor is mostly larger than the polar inertia moment I_p $(I_t >> I_p)$. Therefore, according to Eq. (5.49), the direction of the unbalance moment $\mathbf{M}_U$ is the same direction of the misalignment angle α that is opposite to the direction **i** (see Fig. 5.14).

During the rotor spins with the rotor speed Ω, the misalignment angle α takes turns changing from the positive to negative direction and vice versa. Thus, the unbalance moment direction is changed with the direction of α. As a result, the unbalance moment $\mathbf{M}_U$ excites the rotor in the third and fourth equations of Eq. (5.52). The rotor amplitude of the conical vibration mode is strongly amplified leading to wear in the bearings or damages of the turbocharger.

Note that the misalignment angle cannot be removed by the low-speed balancing (shop balancing). Moreover, it is very difficult to balance the unbalance moment $\mathbf{M}_U$ of the turbine shaft at the shop balancing in case of $I_t >> I_p$ and a large misalignment angle α. The useful way to reduce this unbalance moment is trying to produce a good turbine shaft by well centering and setting the shaft possibly perpendicular to the turbine wheel at welding and grinding. As a result, it could provide the rotor with a small misalignment angle for a small dynamic unbalance and a low eccentricity (low static unbalance) in order to prevent the rotor from the vibration excitations induced by the unbalance moments and forces, as shown in Eq. (5.52).

The vibration equations of the linear rotordynamic turbocharger with N degrees of freedom (DOF) resulted from the lateral and rotational vibrations are written as

$$\mathbf{M}\ddot{\mathbf{q}} + (\mathbf{C} + \mathbf{G})\dot{\mathbf{q}} + \mathbf{K}\mathbf{q} = \mathbf{Q}(t) \tag{5.51}$$

where

- N degrees of freedom of the turbocharger;
- $\mathbf{M}$ $(N \times N)$ diagonal mass and inertia moment matrix;
- $\mathbf{C}$ $(N \times N)$ lateral and circulant damping matrix;
- $\mathbf{G}$ $(N \times N)$ gyroscopic moment matrix;
- $\mathbf{K}$ $(N \times N)$ lateral and circulant stiffness matrix;
- $\mathbf{q}$ $(N \times 1)$ DOF vector;
- $\mathbf{Q}(t)$ $(N \times 1)$ unbalance force and moment vector.

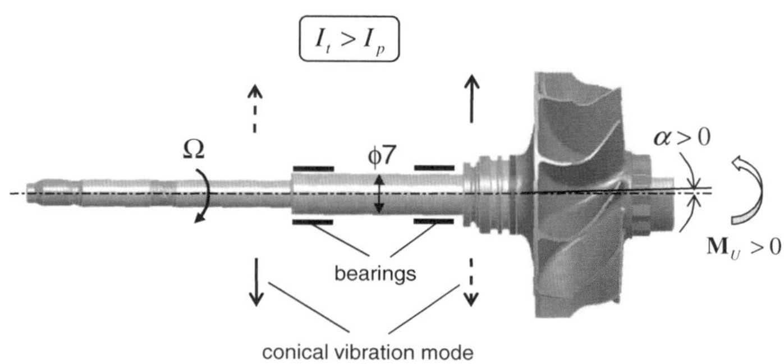

Fig. 5.14 Influence of the unbalance moment on the turbine shaft

Using the momentum approach in Sect. 5.5, the vibration equations for the Jeffcott rotor with four DOF are derived in the inertial coordinate system (x, y, z). In this case, the gyroscopic effect, misalignment angle, unbalance forces, and dynamic unbalance moments of the rotor are taken into account in the vibration equations (see Fig. 5.15).

$$\begin{aligned} m\ddot{x} + c_{l,x}\dot{x} + k_{l,x}x + K_{tr}\theta_x &= m\varepsilon\Omega^2 \cos \Omega t \\ m\ddot{y} + c_{l,y}\dot{y} + k_{l,y}y + K_{tr}\theta_y &= m\varepsilon\Omega^2 \sin \Omega t \\ I_t\ddot{\theta}_x + I_p\Omega\dot{\theta}_y + c_{r,x}\dot{\theta}_x + K_{tr}x + k_{r,x}\theta_x &= (I_t - I_p)\alpha\Omega^2 \cos \Omega t \\ I_t\ddot{\theta}_y - I_p\Omega\dot{\theta}_x + c_{r,y}\dot{\theta}_y + K_{tr}y + k_{r,y}\theta_y &= (I_t - I_p)\alpha\Omega^2 \sin \Omega t \end{aligned} \tag{5.52}$$

where

k_l, k_r are the lateral and circulant system stiffness coefficients;
K_{tr} is the interaction stiffness coefficient of lateral and circulant vibrations;
c_l, c_r are the lateral and circulant system damping coefficients;
ε is the unbalance radius of the rotor;
α is the misalignment angle.

Vibration Eq. (5.52) are written in the matrix formulation.

$$\begin{aligned} &\begin{bmatrix} m & 0 & 0 & 0 \\ 0 & m & 0 & 0 \\ 0 & 0 & I_t & 0 \\ 0 & 0 & 0 & I_t \end{bmatrix} \cdot \begin{bmatrix} \ddot{x} \\ \ddot{y} \\ \ddot{\theta}_x \\ \ddot{\theta}_y \end{bmatrix} + \begin{bmatrix} c_{l,x} & 0 & 0 & 0 \\ 0 & c_{l,x} & 0 & 0 \\ 0 & 0 & c_{r,x} & I_p\Omega \\ 0 & 0 & -I_p\Omega & c_{r,y} \end{bmatrix} \cdot \begin{bmatrix} \dot{x} \\ \dot{y} \\ \dot{\theta}_x \\ \dot{\theta}_y \end{bmatrix} \\ &+ \begin{bmatrix} k_{l,x} & 0 & K_{tr} & 0 \\ 0 & k_{l,y} & 0 & K_{tr} \\ K_{tr} & 0 & k_{r,x} & 0 \\ 0 & K_{tr} & 0 & k_{r,y} \end{bmatrix} \cdot \begin{bmatrix} x \\ y \\ \theta_x \\ \theta_y \end{bmatrix} = \begin{bmatrix} m\varepsilon\Omega^2 \cos \Omega t \\ m\varepsilon\Omega^2 \sin \Omega t \\ (I_t - I_p)\alpha\Omega^2 \cos \Omega t \\ (I_t - I_p)\alpha\Omega^2 \sin \Omega t \end{bmatrix} \end{aligned} \tag{5.53}$$

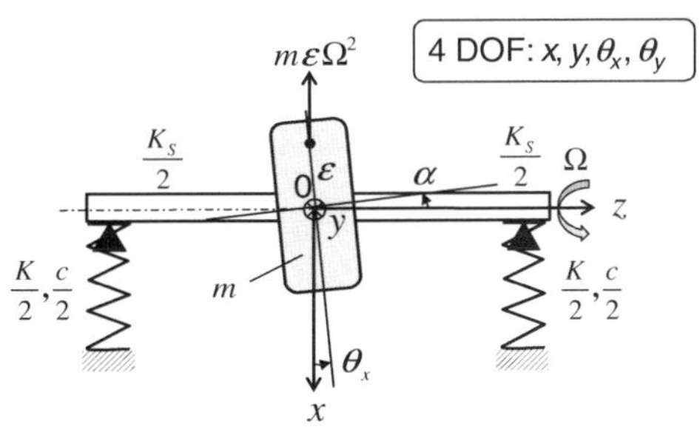

Fig. 5.15 Jeffcott rotor with four DOF

(c) The Lagrange's approach

The vibration equations can be derived from the *Lagrange's equations* based on the kinetic and potential energies of the system with N degrees of freedom (DOF) [15, 18, 19].

Let m particles (mass points) have the coordinates x_i; $i = 1,2,\ldots,N$ in a three-dimensional coordinate system. As a result, the system has N degrees of freedom ($N = 3\ m$).

There is k independent kinematical conditions (k constraints) imposed. Therefore, the mechanical system only depends on the independent parameters q_j; $j = 1,2,\ldots,n$. They are called the generalized coordinates of which the total number results in

$$\begin{aligned} n &= 3m - k \\ &\equiv N - k \end{aligned}$$

The coordinates of the system can be written in functions of the generalized coordinates as

$$\mathbf{r}_i = \begin{pmatrix} x_1 = x_1(q_1, q_2, \ldots, q_n) \\ x_2 = x_2(q_1, q_2, \ldots . q_n) \\ \ldots \\ x_N = x_N(q_1, q_2, \ldots, q_n) \end{pmatrix}$$

The coordinate transformation can be expressed as

$$\mathbf{r}_i = \mathbf{r}_i(q_j), \quad i = 1, 2, \ldots, N; j = 1, 2, \ldots, n \leq N \tag{5.54}$$

The Lagrange's equations for the mechanical system of N degrees of freedom (DOF) are written in the generalized coordinates [18, 19].

$$\frac{d}{dt}\left(\frac{\partial L}{\partial \dot{q}_j}\right) - \frac{\partial L}{\partial q_j} = Q_j; j = 1,\ 2, \ldots, n \leq N \tag{5.55}$$

in which L is the Lagrangian; Q_j are the generalized forces that result from nonconservative external forces acting upon the system. Thus, the generalized forces cannot be derived from a potential function. Generally, they depend not only on the generalized coordinates but also on the generalized velocities.

The Lagrangian L in Eq. (5.55) is defined as

$$L \equiv T - V$$

where T is the kinetic energy and V is the potential energy of the system.

The generalized forces Q_j in the independent generalized coordinates q_j; $j = 1, 2,\ldots,n$ are calculated from the virtual work of the nonconservative (nonpotential) external forces $\mathbf{F}_i$; $i = 1,2,\ldots,N$.

$$\delta W = \sum_{i=1}^{N} \mathbf{F}_i \delta \mathbf{r}_i = \sum_{j=1}^{n \le N} Q_j \delta q_j \tag{5.56a}$$

Using the chain rule of differentiation of Eq. (5.54), the virtual displacement of the coordinates $\mathbf{r}_i$ at a frozen time ($dt = 0$) is expressed as

$$\delta \mathbf{r}_i = \sum_{j=1}^{n \le N} \frac{\partial \mathbf{r}_i}{\partial q_j} \delta q_j \text{ for } i = 1, 2, \ldots, N$$

Substituting it into Eq. (5.56a), one obtains the virtual work of the system.

$$\begin{aligned} \delta W &= \sum_{i=1}^{N} \mathbf{F}_i \left(\sum_{j=1}^{n \le N} \frac{\partial \mathbf{r}_i}{\partial q_j} \delta q_j \right) = \sum_{j=1}^{n \le N} \left(\sum_{i=1}^{N} \mathbf{F}_i \cdot \frac{\partial \mathbf{r}_i}{\partial q_j} \right) \delta q_j \\ &= \sum_{j=1}^{n \le N} Q_j \delta q_j \end{aligned}$$

Thus, the generalized forces result from the nonconservative external forces and the partial derivatives of the coordinates $\mathbf{r}_i$ with respect to q_j.

$$\begin{aligned} Q_j &= \frac{\delta W}{\delta q_j} \\ &= \sum_{i=1}^{N} \mathbf{F}_i \cdot \frac{\partial \mathbf{r}_i}{\partial q_j} \text{ for } j = 1, 2, \ldots, n \le N \end{aligned} \tag{5.56b}$$

in which $\mathbf{F}_i$ are all nonconservative external forces acting upon the system with N DOF, such as friction forces, damping forces, and gyroscopic moments as well.

The kinetic energy T of the rotor includes the translational and rotational kinetic energies.

$$\begin{aligned} T &= T_{trans} + T_{rot} \\ &= \frac{1}{2} \sum_{i=1}^{N} m_i \dot{x}_i^2 + \frac{1}{2} \sum_{i=1}^{N} I_i \dot{\theta}_i^2 \end{aligned} \tag{5.57a}$$

Analogously, the potential energy V of the rotor includes the translational and rotational potential energies.

$$\begin{aligned} V &= V_{trans} + V_{rot} \\ &= \frac{1}{2} \sum_{i=1}^{N} k_i x_i^2 + \frac{1}{2} \sum_{i=1}^{N} c_i \theta_i^2 \end{aligned} \tag{5.57b}$$

The coordinates of the rotor mass center G is written in the angular rotor speed Ω (cf. Fig. 5.11).

$$\begin{aligned} x_G &= x + \varepsilon\cos(\Omega t + \beta) \\ y_G &= y + \varepsilon\sin(\Omega t + \beta) \end{aligned}$$

Differentiating x_G and y_G with respect to time t, one obtains the translational velocity of the mass center G.

$$\begin{aligned} \dot{x}_G &= \dot{x} - \varepsilon\Omega\sin(\Omega t + \beta) \\ \dot{y}_G &= \dot{y} + \varepsilon\Omega\cos(\Omega t + \beta) \end{aligned}$$

Substituting them into Eq. (5.57a), the translational kinetic energy of the rotor is written as

$$\begin{aligned} T_{trans} &= \frac{1}{2}m\left(\dot{x}_G^2 + \dot{y}_G^2\right) \\ &= \frac{1}{2}m\left[\dot{x}^2 + \dot{y}^2 + \varepsilon^2\Omega^2 + 2\varepsilon\Omega(-\dot{x}\sin(\Omega t + \beta) + \dot{y}\cos(\Omega t + \beta))\right] \end{aligned} \tag{5.57c}$$

The rotational kinetic energy of the rotor in the inertial coordinate system (x,y, z) results in

$$T_{rot} = \frac{1}{2}I_t\left(\dot{\theta}_x^2 + \dot{\theta}_y^2\right) + \frac{1}{2}I_p\dot{\theta}_z^2 = \frac{1}{2}I_t\left(\dot{\theta}_x^2 + \dot{\theta}_y^2\right) + \frac{1}{2}I_p\Omega^2 \tag{5.57d}$$

Using the coordinate transformation between the inertial and rotating coordinate systems (cf. Appendix A), one obtains the angular velocity $\boldsymbol{\omega}$ written in the rotating coordinate system (x', y', z') at small translational and rotational displacements.

$$\begin{aligned} \boldsymbol{\omega} &\approx \omega_x\mathbf{i}' + \omega_y\mathbf{j}' + \omega_z\mathbf{k}' \\ &\approx \left(\dot{\theta}_y\sin\theta_z + \dot{\theta}_x\cos\theta_z\right)\mathbf{i}' + \left(\dot{\theta}_y\cos\theta_z - \dot{\theta}_x\sin\theta_z\right)\mathbf{j}' + \left(\Omega - \dot{\theta}_y\theta_x\right)\mathbf{k}' \end{aligned}$$

The rotational kinetic energy in Eq. (5.57d) in the rotating coordinate system (x', y', z') becomes

$$\begin{aligned} T_{rot} &= \frac{1}{2}I_t\left(\omega_x^2 + \omega_y^2\right) + \frac{1}{2}I_p\omega_z^2 \\ &\approx \frac{1}{2}I_t\left(\dot{\theta}_x^2 + \dot{\theta}_y^2\right) + \frac{1}{2}I_p\left(\Omega^2 - 2\Omega\dot{\theta}_y\theta_x\right) \end{aligned}$$

As a result, the Lagrangian L is calculated as

$$
\begin{aligned}
L &= T - V \\
&= \frac{1}{2} m \left[\dot{x}^2 + \dot{y}^2 + \varepsilon^2 \Omega^2 + 2\varepsilon\Omega(-\dot{x}\sin(\Omega t + \beta) + \dot{y}\cos(\Omega t + \beta)) \right] \\
&+ \frac{1}{2} \left[I_t \left(\dot{\theta}_x^2 + \dot{\theta}_y^2 \right) + I_p \left(\Omega^2 - 2\Omega \dot{\theta}_y \theta_x \right) \right] \\
&- \frac{1}{2} \left(\sum_{i=1}^{N} k_i x_i^2 + \sum_{i=1}^{N} c_i \theta_i^2 \right)
\end{aligned}
\tag{5.58}
$$

Using the coordinate transformation in Eq. (5.58) and substituting Eqs. (5.56b and 5.58) into Eq. (5.55), one obtains the Lagrange's equations for N DOF written in the generalized coordinates. The vibration equations of the rotor result from the Lagrange's equations.

5.6 Transient Response at the Run-Up

At run-up, the rotor is accelerated with a constant angular acceleration a_φ. Its transient response will be studied with a Jeffcott rotor with two DOF of x and y (see Fig. 5.16).

The coordinates of the mass center G of the rotor are written in the inertial coordinate system (x, y) as

$$
\begin{aligned}
\mathbf{r}_G &= \mathbf{r}_M + \mathbf{r}_\varepsilon \\
&= [x + \varepsilon \cos(\varphi + \beta)]\,\mathbf{i} + [y + \varepsilon \sin(\varphi + \beta)]\,\mathbf{j}
\end{aligned}
\tag{5.59}
$$

where φ is the rotational angle of the rotor.

The rotor speed results from

$$
\dot{\varphi} = \ddot{\varphi} t \equiv a_\varphi t \tag{5.60}
$$

within a_φ is the constant angular acceleration of the rotor.

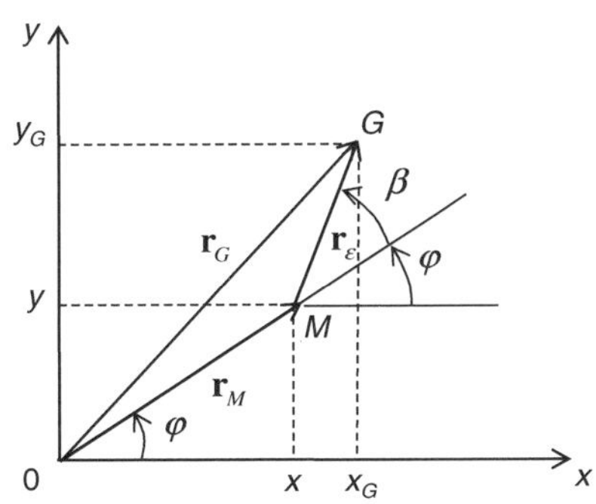

Fig. 5.16 Coordinates of the geometric and mass center of the rotor

The rotation angle of the rotor results from integrating Eq. (5.60).

$$\varphi = \int \dot{\varphi} dt = \int a_\varphi t dt = \frac{1}{2} a_\varphi t^2 \tag{5.61}$$

After twice differentiating Eq. (5.59) with respect to time t, one obtains the acceleration of the mass center G.

$$\begin{aligned} \ddot{x}_G &= \ddot{x} - \varepsilon\left(\dot{\varphi}^2 \cos(\varphi + \beta) + \ddot{\varphi} \sin(\varphi + \beta)\right) \\ \ddot{y}_G &= \ddot{y} - \varepsilon\left(\dot{\varphi}^2 \sin(\varphi + \beta) - \ddot{\varphi} \cos(\varphi + \beta)\right) \end{aligned} \tag{5.62}$$

The first term in the brackets in Eq. (5.62) is much larger than the second term with increasing time t since

$$\frac{\left|\dot{\varphi}^2 \cos(\varphi + \beta)\right|}{\left|\ddot{\varphi} \sin(\varphi + \beta)\right|} = \frac{\left|\dot{\varphi}^2\right|}{\left|\ddot{\varphi}\right|} = \left|a_\varphi\right| t^2 > > 1 \tag{5.63}$$

Thus,

$$\begin{aligned} \ddot{x}_G &= \ddot{x} - \varepsilon\dot{\varphi}^2 \cos(\varphi + \beta) \\ \ddot{y}_G &= \ddot{y} - \varepsilon\dot{\varphi}^2 \sin(\varphi + \beta) \end{aligned} \tag{5.64}$$

The vibration equations of the two DOF Jeffcott rotor with anisotropic bearings, as shown Fig. 5.17, are written at the rotor center M (x,y) at $\beta = 0$.

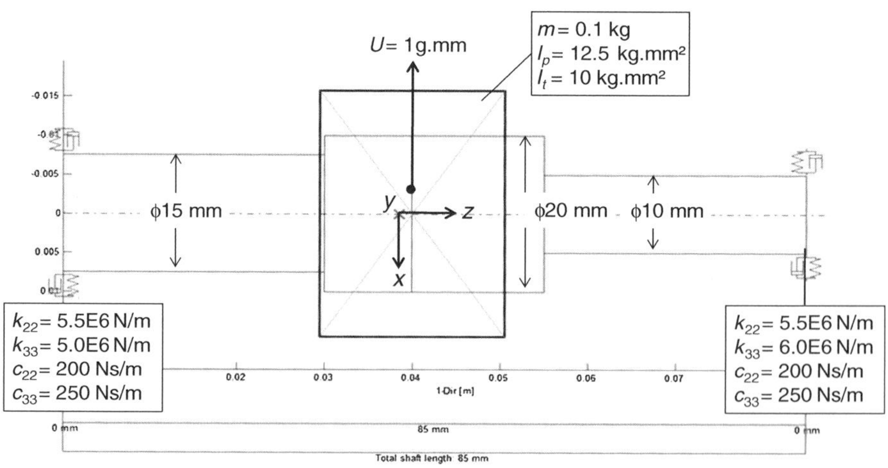

Fig. 5.17 Computational model of the Jeffcott rotor

$$\begin{aligned} m\ddot{x} + c_{22}\dot{x} + k_{22}x &= m\varepsilon\dot{\varphi}^2(t)\cos\varphi \\ m\ddot{y} + c_{33}\dot{y} + k_{33}y &= m\varepsilon\dot{\varphi}^2(t)\sin\varphi \end{aligned} \tag{5.65}$$

where

c_{22}, c_{33} are the diagonal damping coefficients in the directions x and y;
k_{22}, k_{33} are the diagonal stiffness coefficients in the directions x and y.

The Jeffcott rotor has a mass of m = 0.1 kg, the polar and transverse mass moments of inertia I_p = 12.5 kg mm^2 and I_t = 10 kg mm^2, respectively. The total length of the rotor is about l = 85 mm and its various lengths and diameters are given in Fig. 5.17. The residual unbalance U = 1 g mm lies at the mass center of the wheel and induces the excitation unbalance force acting on the rotor. The rotor rotates with accelerated rotational speeds Ω in the axial direction z and deflects in the lateral directions x and y due to the unbalance force. In the computation, it is assumed that the anisotropic bearings are located at each end of the rotor with the constant stiffness and damping coefficients (see Fig. 5.17).

To compute the rotor vibration response of Eq. (5.65), the computational program of rotordynamics [9] based on the finite element method is applied at a steady-state operating condition and various accelerations during speed-up. At steady-state (a_φ = 0), the rotor deflection versus the rotor speed is displayed in Fig. 5.18. Due to the unbalance force, the rotor deflects at increasing rotor speeds and reaches the resonance at the critical speed of nearly 59,100 rpm. The maximum rotor deflection at the resonance is about 27 μm.

After passing the resonance, the rotor amplitude continuously decreases to the unbalance radius ε of approximately 5 μm because of self-centering effect of the rotor at the supercritical speeds, as discussed in Sect. 5.2.

At the run-up, the rotor speed is accelerated with an acceleration a_φ; therefore, the unbalance force is also increased with time according to Eqs. (5.60) and (5.65).

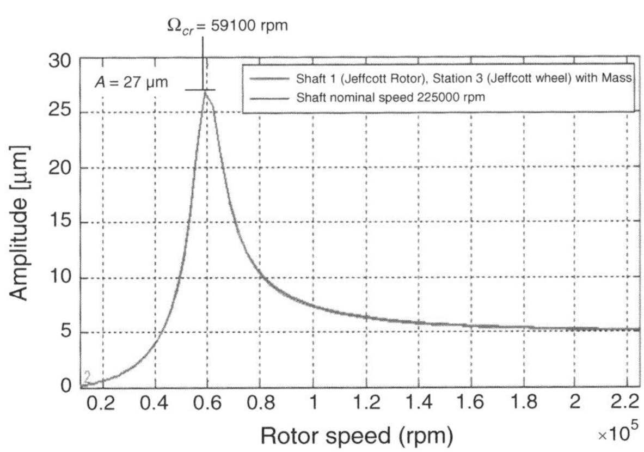

Fig. 5.18 Deflection rotor amplitude versus rotor speed at steady state

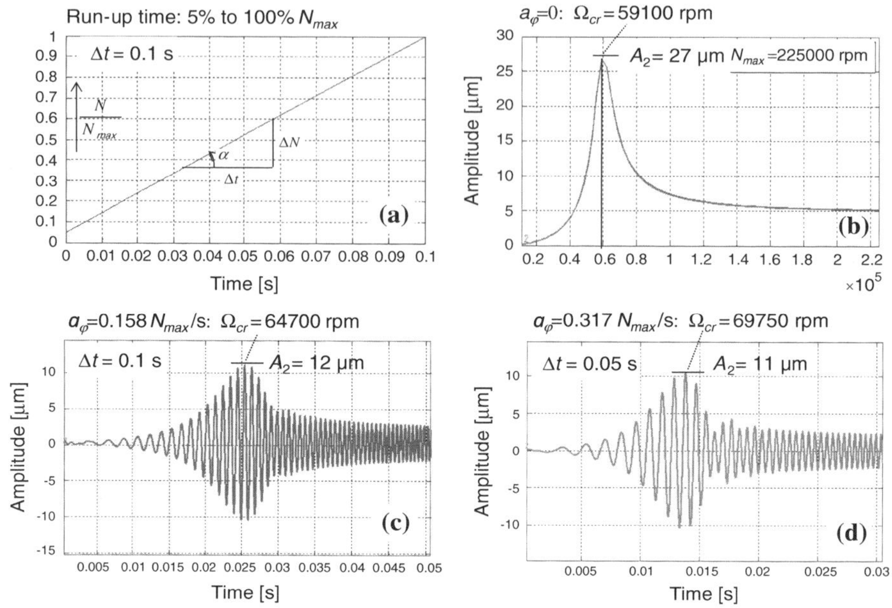

Fig. 5.19 Rotor responses at various run-up accelerations

Figure 5.19a shows a time interval Δt that is needed to run-up the rotor from 5 to 100 % of the maximum rotor speed $N_{\max}$. Note that the shorter the time interval is, the higher the engine acceleration is. At the acceleration with time interval of 0.1 s, the resonance at $a_\varphi = 0.158\ N_{\max}$/s occurs at 64,700 rpm higher than the steady-state critical speed, but the resonance amplitude is reduced from 27 to 12 µm, as shown in Fig. 5.19c. For further increasing accelerations with a time interval of 0.05 s, the resonance at $a_\varphi = 0.317\ N_{\max}$/s takes place much later at the critical speed of about 69,750 rpm where the resonance amplitude of 11 µm is smaller than the earlier case, but the critical speed is higher (see Fig. 5.19d). Contrary to the stationary operating condition, the rotor deflection at the run-up vibrates with reducing amplitudes.

The computing results show that the critical speeds increase with the run-up acceleration compared to the critical speed in case of no acceleration. However, the resonance amplitudes decrease with the acceleration (see Fig. 5.19). Therefore, it is usually to speed-up the rotor over the critical speeds at high accelerations in order to avoid the resonance damage, especially in turbomachines of the power plants.

5.7 Frequency Analysis in Campbell Diagram

In linear rotordynamics, only the response of the rotor has been studied at the resonance. The characteristics of the rotor system are designed in such way that the resonances are possible to occur outside the operating speed range of the rotor.

However, the resonance cannot be always outside the operating speed range in the automotive turbochargers. In this case, more damping is necessary to keep the resonance rotor amplitude as small as possible and therefore to prevent the bearings from fatal damage [11, 13, 14].

According to Eq. (5.17), the resonance amplitude increases drastically when the rotor speeds equal the eigenfrequencies (natural frequencies) of the rotor in case of small damping ratio. In fact, it is not easy to calculate the natural frequencies depending on many rotor characteristics like gyroscopic effects analytically solving the nonlinear equation of the rotor eigenfrequencies and rotor speed.

$$\omega_n(m, I_t, I_p, \Omega, k, c) = \Omega \tag{5.66}$$

However, Eq. (5.66) can be graphically solved in Campbell diagram (whirl speed diagram) by means of the intersections between the eigenfrequencies and the synchronous excitation line 1X where ω_n equals Ω. The critical speeds result from the cutting points in Campbell diagram. The gyroscopic effect stiffens the rotor eigenfrequencies of the forward whirls and destiffens the rotor eigenfrequencies of the backward whirls at increasing rotor speeds, as shown in Fig. 5.20. Therefore, the critical speeds of the forward whirls are higher than the critical speeds without the gyroscopic effect.

The rotational kinetic energy of the rotor excites only the forward whirls because both have the same rotational direction; therefore, the amplitudes are strongly amplified at the resonance. On the contrary, the backward whirl direction is opposite to the rotational direction of the rotor. Hence, the rotational kinetic energy of the rotor is dissipated, instead exciting the backward whirls at the resonance.

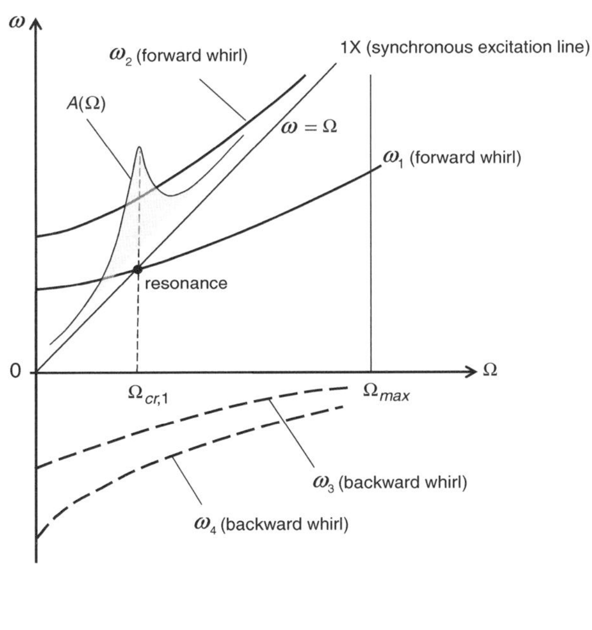

Fig. 5.20 Campbell diagram of the eigenfrequencies versus rotor speed

As a result, the unbalance forces normally excite only the forward whirls at the resonances. However, the unbalance force could excite the backward whirls in case of asymmetric rotors or oscillating rotor speeds. Therefore, the intersections between the eigenfrequencies of the forward whirls and the synchronous excitation line 1X determine the critical speeds of the rotor in Campbell diagram.

The first forward whirl eigenfrequency ω_1 cuts the synchronous excitation line 1X at the first bending resonance; the maximum amplitude A takes place at the critical speed $\Omega_{cr,1}$ ($\omega_1 = \Omega_{cr,1}$). Any point in the excitation line 1X satisfies Eq. (5.66) with $\omega = \Omega$. The eigenfrequency ω_2 may cut the excitation line 1X at a higher rotor speed that is outside the operating speed range. The backward whirls ω_3 and ω_4 have not been normally excited by the unbalance force, and therefore they do not cause resonances at the symmetric rotors.

In the turbochargers, it has to design the rotor having its first bending critical speed as high as possible so that the onset of instability speed at the oil whip is outside the operating speed range of the turbochargers. Moreover, the rotor unbalance induces the unbalance whistle, one of the unwanted airborne noises in the automotive turbochargers. The unbalance whistle has the same frequency of the rotor and disturbs the driver and passengers during driving. However, the unbalance whistle induced at high critical speeds is generally suppressed by the engine noise at high motor speeds. That is also another reason why the first critical frequency should be as high as possible.

For simplicity, the Campbell diagram is sometimes displayed as in Fig. 5.21 where the backward whirls (dashed lines) are in the same side of the forward whirls (solid lines). While the eigenfrequencies of the forward whirls increase with the rotor speed, the backward whirls eigenfrequencies decrease to zero. Even the backward whirls frequencies cut the excitation line 1X; however, the resonances do not occur in the backward whirls.

In the following section, the eigenmodes of the rigid rotor in Fig. 5.22 are computed, as shown in the Campbell diagram. The rotor is supported at each end by two isotropic bearings with an effective stiffness coefficient $k/2$ of the rotor. The

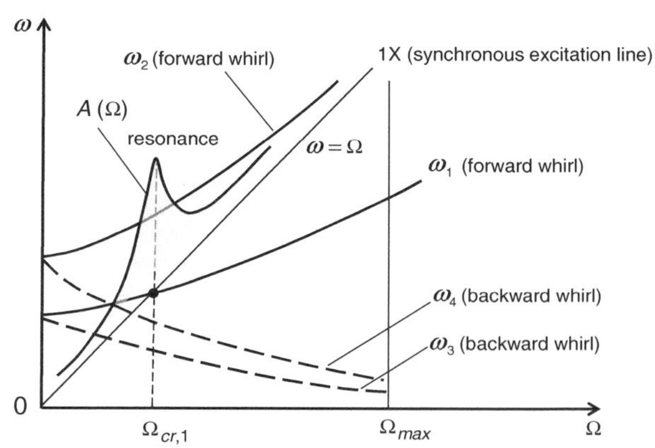

Fig. 5.21 Simplified Campbell diagram

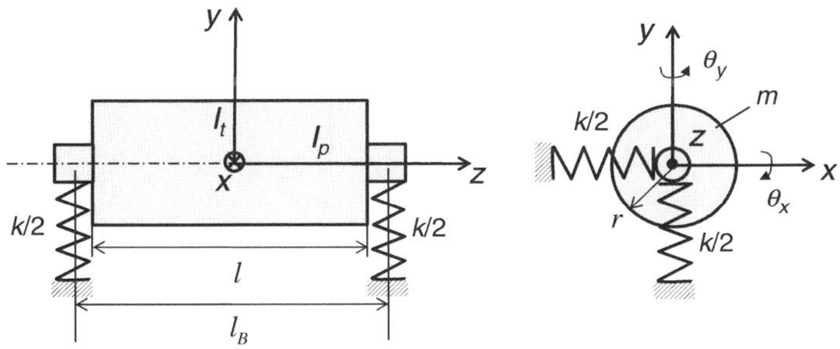

Fig. 5.22 Case study of the rigid rotor

rigid rotor has a translatory vibration in the lateral directions x and y, and a conical vibration in the directions θ_x and θ_y, as shown in Fig. 5.22.

The rotor is assumed as a full cylinder with a radius r and length l. In case of $l^2 < 3r^2$, the polar mass inertia moment is larger than the transverse moment of inertia ($I_p > I_t$) for the short and thick rotor; if $l^2 > 3r^2$, I_p is smaller than I_t ($I_p < I_t$) for the long and slender rotor.

The free vibration equations of the rigid rotor with four DOF of x, y, θ_x, and θ_y are written as

$$\begin{aligned} & m\ddot{x} + kx + k_\delta \theta_x = 0 \\ & m\ddot{y} + ky + k_\delta \theta_y = 0 \\ & I_t\ddot{\theta}_x + I_p\Omega\dot{\theta}_y + k_\theta \theta_x + k_\delta x = 0 \\ & I_t\ddot{\theta}_y - I_p\Omega\dot{\theta}_x + k_\theta \theta_y + k_\delta y = 0 \end{aligned} \tag{5.67}$$

where k and k_θ are the lateral and circulant bearing stiffness coefficients; k_δ is the coupling stiffness coefficient of DOF.

Substituting the free vibration response and its time derivatives

$$\begin{aligned} & r = \hat{r}e^{j\omega t} \\ & \Rightarrow \dot{r} = j\omega r;\ \ddot{r} = -\omega^2 r \end{aligned}$$

into Eq. (5.67), one obtains the characteristic equation

$$D(\omega) = (k - m\omega^2) \cdot (k_\theta + I_p\Omega\omega - I_t\omega^2) - k_\delta^2 = 0 \tag{5.68}$$

The eigenfrequencies result from solving Eq. (5.68). Due to the gyroscopic effect, they are a function of other rotor characteristics and rotor speed Ω. They can be numerically computed at various rotor speeds and are plotted in Figs. 5.23 and 5.24. However, one can analytically calculate the responses depending on the rotor speeds.

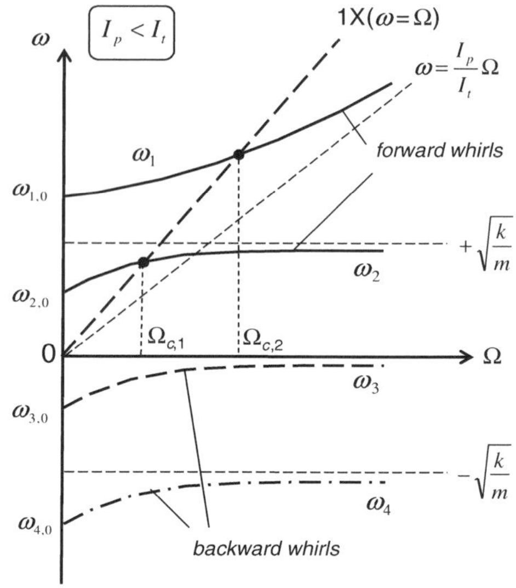

Fig. 5.23 Campbell diagram in case of $I_p < I_t$

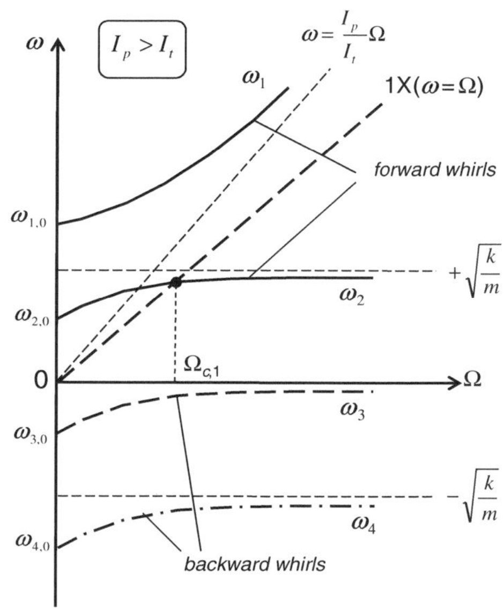

Fig. 5.24 Campbell diagram in case of $I_p > I_t$

The rotor speed is derived from the characteristic Eq. (5.68) and written in the rotor eigenfrequency as [3, 5, 6, 17]

$$\Omega = \frac{mI_t\omega^4 - (kI_t + k_0 m)\omega^2 + \left(kk_0 - k_\delta^2\right)}{mI_p\omega\left(\omega^2 - \frac{k}{m}\right)} \tag{5.69}$$

- At $\Omega = 0$, the numerator of Eq. (5.69) equals zero; hence, four natural frequencies $\omega_{1,0}$, $\omega_{2,0}$, $\omega_{3,0}$, and $\omega_{4,0}$ are determined.
- As Ω goes to ∞, the denominator converges to zero; one obtains

$$\omega_{2;4,\infty} = \pm\sqrt{\frac{k}{m}}; \quad \omega_{3,\infty} \to 0 \tag{5.70}$$

- As ω goes to ∞, the asymptotic relation between Ω and ω according to the L'Hôpital's rule is found as

$$\Omega = \lim_{\omega\to\infty}\left(\frac{mI_t\omega^4 - (kI_t + k_0 m)\omega^2 + \left(kk_0 - k_\delta^2\right)}{mI_p\omega\left(\omega^2 - \frac{k}{m}\right)}\right) = \frac{I_t\omega}{I_p} \tag{5.71}$$

Thus,

$$\omega = \frac{I_p\Omega}{I_t} \tag{5.72}$$

where

- $I_p < I_t$ gives $\omega/\Omega < 1$; its gradient is smaller than 1X ($\omega = \Omega$);
- $I_p > I_t$ gives $\omega/\Omega > 1$; its gradient is larger than 1X ($\omega = \Omega$).

The behaviors of the eigenfrequencies can be plotted by such asymptotic methods in Figs. (5.23 and 5.24). In case of $I_p < I_t$, the gradient of the asymptote ω_1 in Eq. (5.72) is less than 1. There are two critical speeds $\Omega_{c,1}$ and $\Omega_{c,2}$ in the synchronous excitation line 1X (see Fig. 5.23). On the contrary, if $I_p > I_t$, there is only one critical speed $\Omega_{c,2}$ because the gradient of the asymptote ω_1 is larger than 1. Therefore, the eigenfrequency of the forward whirl ω_1 never cuts the excitation line 1X in case of $I_p > I_t$ (see Fig. 5.24).

To investigate the eigenmodes of the eigenfrequencies ω_1 and ω_2, the ratio of critical speeds is used. The critical speed ratio ξ for the symmetric rotor with overhung ends is defined [14] (see Fig. 5.22).

$$\xi \equiv \left(\frac{\Omega_{c,1}}{\Omega_{c,2}}\right)^2 = \frac{4(I_t - I_p)}{ml_B^2} = \left(\frac{l - 3r^2}{3l_B^2}\right) \tag{5.73}$$

where l is the length of the rotor; r is the rotor radius; l_B is the bearing span.

There are two cases of ξ:

- $\xi < 1$ at $(l - 3r^2) < 3l_B^2$, the cylindrical mode occurs at the lower frequency.
- $\xi > 1$ at $(l - 3r^2) > 3l_B^2$, the conical mode occurs at the lower frequency.

5.8 Computations of Linear Rotordynamics

Automotive turbochargers mostly use rotating floating ring bearings (RFRB) with two oil films (inner and outer) which have some nonlinear characteristics of the bearing stiffness and damping coefficients. Therefore, the nonlinear effects of subsynchronous whirls such as oil whirls and supersynchronous whirls like excessively high unbalance forces, rotor misalignment, contact rubs, and frequency modulations must be taken into account in the computations. To compute the rotordynamic behavior of turbochargers, some commercial codes such as Adams, SimPack, DyRoBeS, and MADYN 2000 etc. are used. Unfortunately, the nonlinear characteristics of the oil-film bearings have not been well simulated yet for non-linear rotordynamics because all mutual effects between the fluid-, thermodynamics, and rotordynamics in the oil films are quite difficult to simulate numerically.

However, the modal analysis of turbochargers can be studied using linear rotordynamics. In this case, the journal bearings have constant stiffness and damping coefficients in each direction, instead of the nonlinear bearing forces acting upon the rotor. Nevertheless, the results of the linear rotordynamic computation can provide the critical speeds of the rotor in the Campbell diagram and the rotor eigenmodes of the harmonic vibrations induced by the unbalance excitation forces and moments.

The rotor of the turbocharger is displayed in Fig. 5.25 that consists of the compressor and turbine wheels, and two rotating floating ring bearings. The computational rotor model is transformed into the finite element model where the wheels, shaft, and bearings with all necessary degrees of freedom (DOF) are taken into account in the computation. Each wheel has four DOF of x_2, x_3, θ_2, and θ_3 in the inertial coordinate system (1, 2, 3); the shaft elements and bearings have two DOF of x_2 and x_3. The rotor velocity Ω aligns with the axial direction 1.

The bearings have the stiffness coefficients $k_{22} = 2.0E3$ N/m; $k_{33} = 2.25E3$ N/m and the small damping coefficients $c_{22} = 50$ Ns/m; $c_{33} = 55$ Ns/m, which are used in the computation of the modal analysis. The total mass of the small turbocharger rotor is about 95 g. The polar and transverse mass moments of inertia at the mass center G result in $I_p = 6.4E\text{-}6$ kg m^2 and $I_t = 107E\text{-}6$ kg m^2, respectively.

Using the linear rotordynamics computational program [9], the resonance frequency and vibration eigenmodes over rotor speeds for the turbocharger are computed. Note that linear rotordynamics cannot compute the asynchronous vibrations, such as oil whirls, misalignment. Instead, nonlinear rotordynamics must be used (see Chap. 7).

The first bending resonance is found at the critical speed of nearly 131,900 rpm in Campbell diagram (see Fig. 5.26), in which the forward whirl eigenfrequency ω_4 of the eigenmode #4 cuts the synchronous excitation line 1X. Due to gyroscopic effect of the wheels, the eigenfrequencies of the forward whirls increase with the rotor speed. On the contrary, the eigenfrequencies of the backward whirls decrease with the rotor speed.

Three forward whirls with the eigenmodes #2, #4, and #6 in the frequency range up to 6000 Hz appear in Campbell diagram. However, three backward whirls with

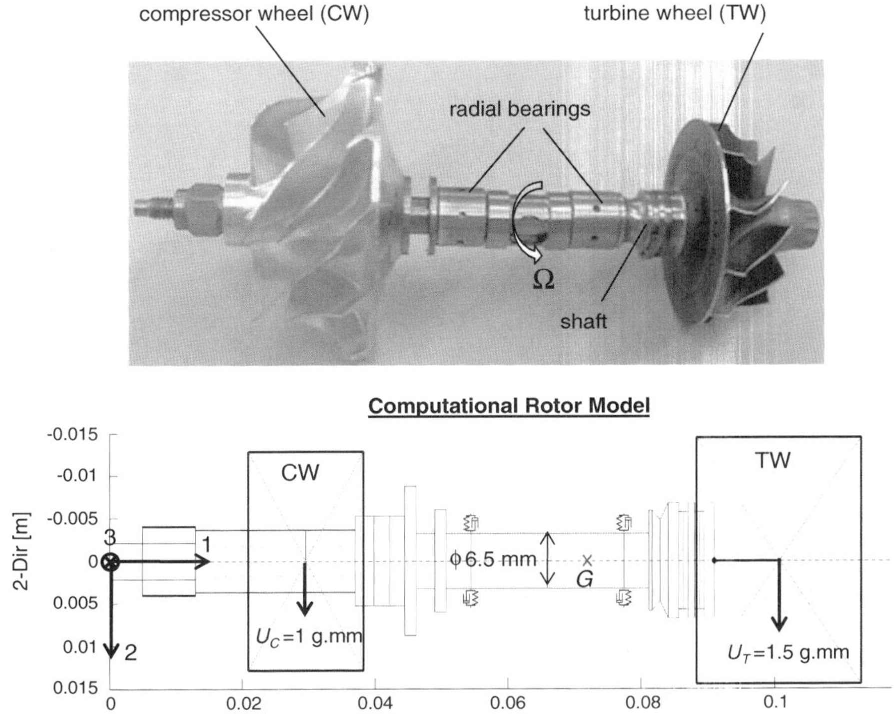

Fig. 5.25 Computational model of a turbocharger rotor

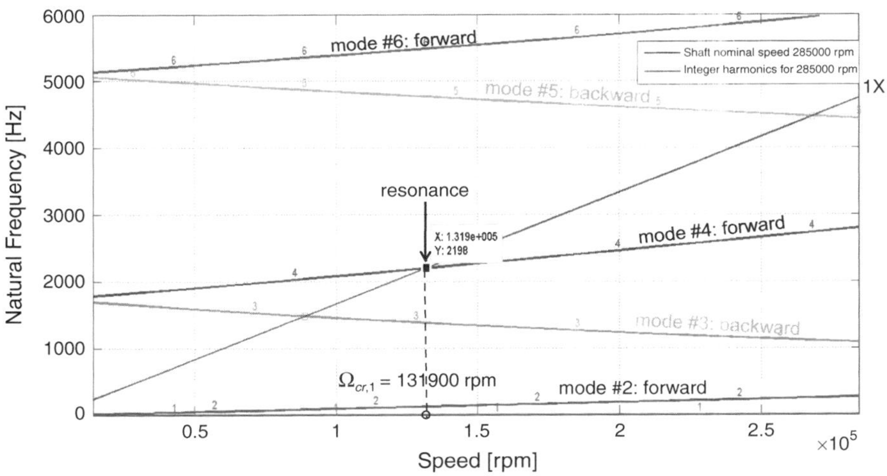

Fig. 5.26 Campbell diagram of the turbocharger rotor

the eigenmodes #1, #3, and #5 do not excite the rotor at the resonances. The second resonance of the eigenmode #6 occurs outside the operating speed range of 285,000 rpm.

The vibration shapes of the eigenmodes of the rotor have been already discussed in Chap. 3. At a low rotor speed of about 7,588 rpm, the rotor is still rigid and its transverse mass moment of inertia $I_t = 107\text{E-}6\ \text{kg m}^2$ is larger than the polar inertia moment $I_p = 6.4\text{E-}6\ \text{kg m}^2$. In this case, the critical speed ratio ξ in Eq. (5.73) is much larger than 1 ($\xi >> 1$); therefore, the conical mode (eigenmode #2) happens at the low frequency, as shown in Fig. 5.27.

Obviously, a long slender rigid rotor ($I_t >> I_p$) tends to tumble with the conical mode shape at low rotor speeds rather than with the cylindrical one. In the conical mode, the displacements of the compressor (CW) and turbine wheels (TW) go out of phase; i.e., their deflection directions are opposite to each other with a phase lag of 180°.

The rotor amplitude of the conical mode is normally larger than the amplitude of the cylindrical because it is amplified by the long arm of the rotor from the rotation center of the conical mode. On the contrary, the rotor amplitude of the cylindrical mode is limited by the small bearing clearance (see Fig. 5.28). At large, the shorter thicker rigid rotor ($I_t << I_p$) is more stable than the long, slender rotor ($I_t >> I_p$) at low rotor speeds since the forward whirl eigenmode is a cylindrical mode. In this case, the displacements of the compressor (CW) and turbine wheels (TW) go in phase and their amplitudes are limited in the bearing clearance.

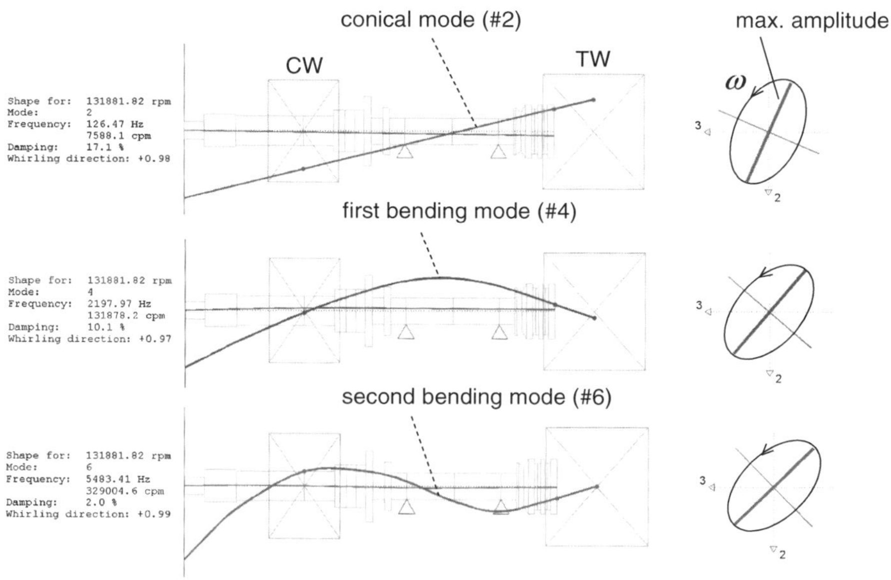

Fig. 5.27 Vibration eigenmodes of the rotor near the resonance

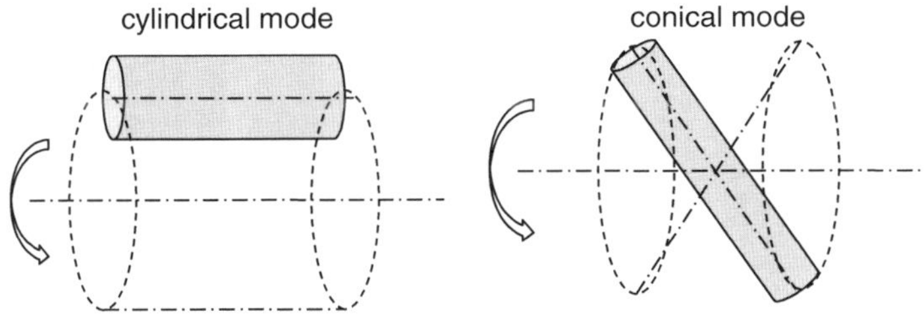

Fig. 5.28 Cylindrical and conical vibration modes of the rotor

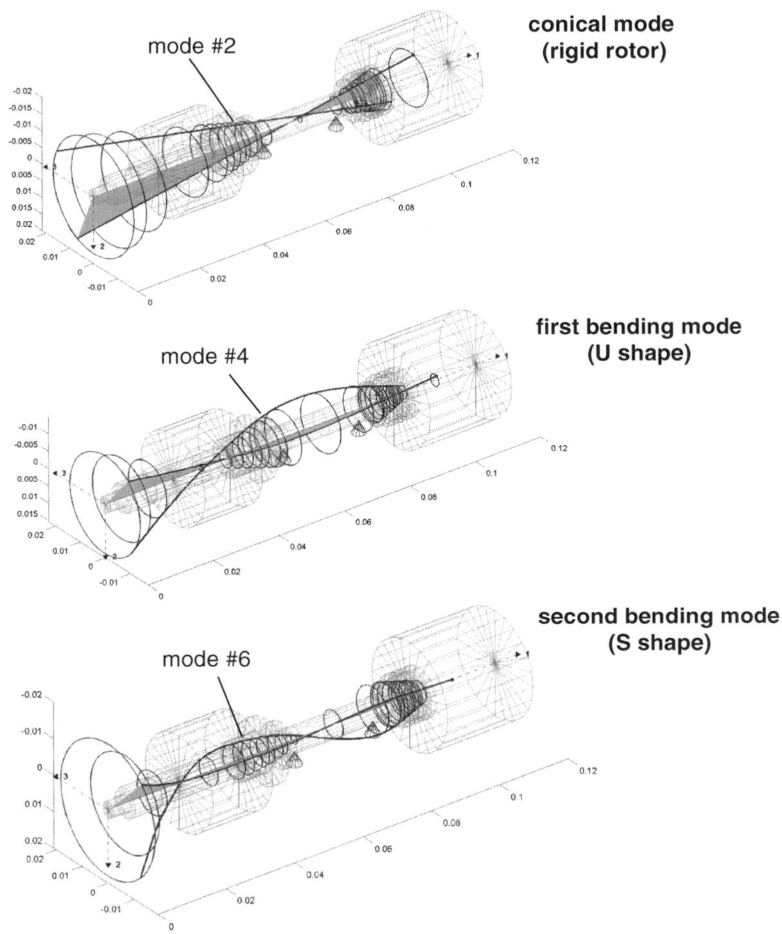

Fig. 5.29 Vibration eigenmodes of the rotor near Ω_{cr} = 132,000 rpm

The whirl orbits of the rotor eigenmodes are illustrated in the phase plane 2–3, in which the amplitude and direction of the minimal and maximal rotor deflections are displayed for each vibration mode (see Fig. 5.27).

The rotor becomes flexible at increasing the rotor speed. It begins to deflect itself due to unbalance and arrives at the maximum deflection at the first resonance with Ω_{cr} = 131,900 rpm. The forward whirl eigenmode #4 of the flexible rotor has a U mode shape where the deflections of the wheels are in phase. At high supercritical speeds, the bending shape of the forward whirl eigenmode #6 changes from the U to S shape and approaches the second resonance at a second critical speed outside the rotor speed range, as shown in Campbell diagram (see Fig. 5.26).

The deflection directions of the wheels are opposite to each other in the S shape of the eigenmode #6. Unless the second resonance is far away from the maximum rotor speed, the rotor amplitude will increase again shortly after the self-centering of the first resonance.

Figure 5.29 displays the shapes of the eigenmodes #2, #4, and #6 of the forward whirls. The elliptical whirl orbit of the rotor has the maximum amplitude whose direction locates in the phase plane 2–3, as shown in Fig. 5.27. In order to measure the rotor eigenmode, four eddy-current sensors should be installed in the ends of the rotor at the compressor and turbine wheels. If the vibration responses at the rotor ends take turns moving in the opposite direction with a phase lag of 180°, the rotor vibration is out of phase, either the conical (rigid rotor) or the second bending mode with an S shape (flexible rotor) depending on the rotor speed is possible.

Otherwise, the rotor vibration is in phase, either the cylindrical (rigid rotor) or the first bending mode with a U shape (flexible rotor) depending on the rotor speed could be the case. However, it is very difficult to measure the rotor deflection at the turbine side due to high temperatures of the exhaust gase at 850 °C in diesel engines and 1050 °C in gasoline engines.

References

1. Adams Jr, M.L.: Rotating Machinery Vibration. Taylor and Francis, CRC (2001)
2. Bently, D.E., Hatch, C.: Fundamentals of Rotating Machinery Diagnostics. Bently Pressurized Bearing Press, Minden (2002)
3. Childs, D.: Turbomachinery Rotordynamics. Wiley, New York (1993)
4. Ehrich, F.: Handbook of Rotordynamics. Krieger Publishing Company, Malabar (2004)
5. Funakawa, M., Tatara, A.: Stability criterion of an elastic rotor in journal bearings. Trans. JSME. **30**(218), 343–352 (1964)
6. Gasch, R., Nordmann, R., Pfützner, H.: Rotordynamik, 2. Springer, Auflage (2006)
7. Genta, G.: Dynamics of Rotating Systems. Springer, New York (2005)
8. Hamill, P.: A Student's Guide to Lagrangians and Hamiltonians. Cambridge University Press, Cambridge (2014)
9. Kraemer, E.: Rotordynamics of Rotors and Foundations. Springer, Berlin (1993)
10. MADYN 2000: Program of Computational Rotordynamics. Delta JS, Zurich (2011)
11. Muszýnska, A.: Rotordynamics. Taylor and Francis, CRC (2005)

12. Newland, D.E.: An Introduction to Random Vibrations, Spectral & Wavelet Analysis, 3rd edn. Dover Publications Inc, New York (2005)
13. Rao, J.S.: Rotordynamics, 3rd edn. Publishers, New Age Intl (2007)
14. Rieger, N.F.: Rotordynamics 2—Problems in Turbomachinery. CISM Courses and Lectures No. 297, New York (1988)
15. Rieger, N.F.: Balancing of Rigid and Flexible Rotors. U.S. DoD, Washington (1986)
16. Taylor, J.R.: Classical Mechanics. University Science Books, Sausalito (2005)
17. Vance, J.: Rotordynamics of Turbomachinery. Wiley, New York (1988)
18. Wells, D.A.: Lagrangian Dynamics. McGraw-Hill, Inc, New York (1967)
19. Yamamoto, T., Ishida, Y.: Linear and Nonlinear Rotordynamics. Wiley, New York (2001)

Chapter 6
Bearing Dynamics of Turbochargers

6.1 Introduction

In the following section, the bearing system is dealt with for the automotive turbochargers operating at very high rotor speeds. To support the rotor during operation, the bearing system including the thrust and radial bearings is necessary for turbochargers. On the one hand, the impulse forces of fluids and pressures acting on the compressor and turbine wheels generate the thrust load on the rotor that depends on various operating conditions. Hence, the thrust load is balanced by the thrust bearing to keep the rotor stable in the axial direction. On the other hand, the radial bearings induce the bearing forces to balance the unbalance forces acting upon the rotor in the radial direction. Additionally, the bearings also deliver the rotor with the bearing stiffness and damping properties that keep the rotor in rotordynamic stability.

Figure 6.1 displays the setup of the bearing system in the core unit CHRA (center housing and rotating assembly) that includes the rotor shaft, turbine and compressor wheels (CW), thrust rings, and radial bearings. The thrust bearing is mounted and fixed in the bearing housing, not rotating with the rotor, and locates in the middle between the thrust rings. The thrust rings fixed on the shaft rotate with the rotor speed to create the thrust acting upon the thrust bearing against the thrust load. The thrust bearing force is induced by squeezing of the oil film between the bearing and thrust rings. Hence, the oil film thickness determines the reacting force of the thrust bearing. Note that the smaller the oil film thickness is, the larger the thrust bearing force becomes, and vice versa.

Oil film bearings rotate and float in the inner and outer oil films; therefore, they are called the rotating floating ring bearings (RFRB). These radial bearings are usually applied to automotive turbochargers in order to reduce the bearing friction compared to the semi-floating ring bearings with two oil films and nonrotating bearing ring, especially in low-end torque (LET). As a result, the response behavior of turbochargers (turbolag) is improved at low rotor speeds.

H. Nguyen-Schäfer, *Rotordynamics of Automotive Turbochargers*,
Springer Tracts in Mechanical Engineering, DOI 10.1007/978-3-319-17644-4_6

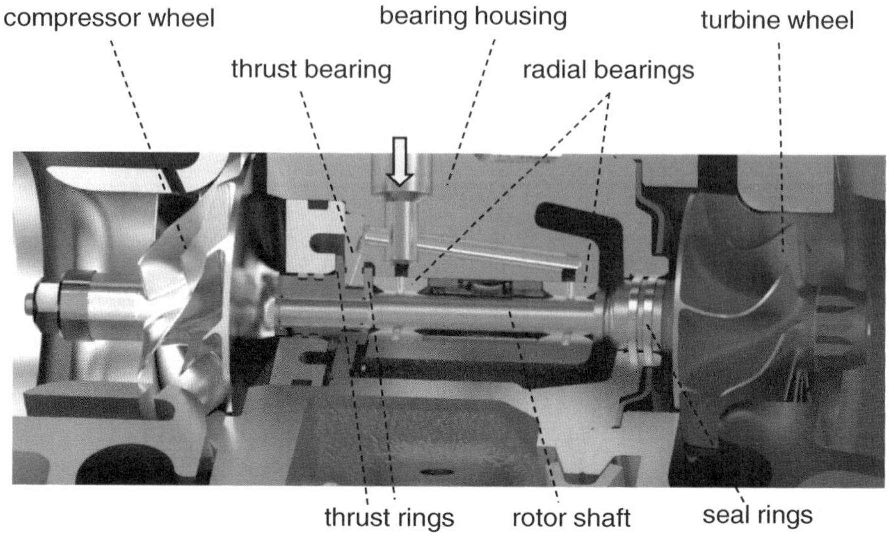

Fig. 6.1 CHRA of a wastegated turbocharger (Courtesy BMTS)

Two kinds of radial bearings are used in automotive turbochargers: hydrodynamic fluid film bearings and rolling element bearings. The latter reduces much more bearing friction at low rotor speeds. Therefore, they are sometimes applied to some automotive turbochargers in order to keep CO_2 emission low according to the current emission law or to improve the transient behavior. At high rotor speeds, the difference of the bearing friction between the fluid film and rolling element bearings is quite small because the oil temperature increases. Hence, the oil viscosity reduces leading to reduction of the bearing friction of the oil film bearings.

The rotor is supported on the bearing system rotating and floating in the bearing housing, as displayed in Fig. 6.2. The bearing system is supplied with pressurized oil at the oil inlet. The oil channels in the bearing housing provide the thrust and radial bearings with the pressurized oil that has two main functions. First, the bearing forces that are generated by the oil hydrodynamic effect keep the rotor stable in the axial and radial directions and damp the rotor vibration to prevent its response from large amplitude at the resonance. Second, fresh supply oil removes the induced friction energy in the bearings by the heat convection in the axial direction, and by the heat conduction between two oil films in the radial direction through the bearing ring. This cooling function keeps the effective oil temperature in the bearings as low as possible and therefore maintains the bearing stiffness and damping coefficients large at the high rotor speeds, in turn large stiffness and damping forces of the bearings [3, 5, 8].

Table 6.1 compares the characteristics between the fluid film and rolling element bearings. Despite many disadvantages, the rolling element bearings have only one main advantage of the small bearing friction at low rotor speeds, especially in LET. As shown in Table 6.1, the friction coefficient of the ball bearings is in the order of

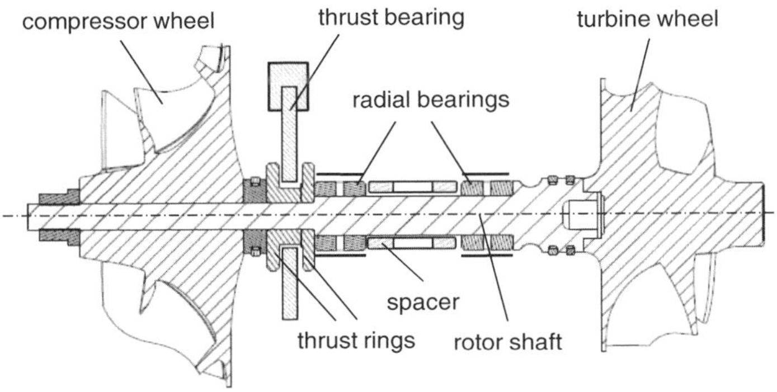

Fig. 6.2 Setup of the bearing system in a turbocharger

Table 6.1 Comparison of bearing characteristics between the oil film and rolling element bearings of turbochargers

Characteristics	Oil film bearings (rotating floating ring bearings)	Rolling element bearings (combined thrust and radial bearings)
Functionality	Separate radial and thrust bearings	Combined radial and thrust bearings
Load direction	Radial and axial	Radial and axial
Volumetric size	Small outer diameter of radial bearings	Outer diameter: 2× larger
Friction coefficient	• in LET: *large* → $\mu = 10^{-2}\ldots10^{-1}$ • at high speeds: small → $\mu = 10^{-3}$	• in LET: *small* → $\mu = 10^{-3}\ldots10^{-2}$ • at high speeds: small → $\mu = 10^{-3}$
Damping coefficient (bearing/rotor)	2× larger because of two oil films: inner bearing clearance (20 μm); outer bearing clearance (70 μm)	Very small, clearance between balls/race (<0.5 μm); 1× oil film damper (50…80 μm)
Noise	Less noisy: unbalance whistle; constant tone (inner oil whirl)	Noisy: unbalance whistle, high harmonics, defected bearing frequencies, sidebands due to frequency modulation
Oil volumetric flow rate	Small	Nearly 2× larger
Lifetime	Unlimited	Limited by high cycle fatigue (HCF)
Damage	Possibly damaged by oil whip (self-excited instability); less or no wear (normal condition)	Ball contacting at high speeds; more wear
Cost	Low	Very high

10 times lower than the oil film bearings at low rotor speeds. However, the friction coefficients of both bearings are nearly equal at high rotor speeds since the effective oil temperature in the bearings has been already high leading to low oil viscosity and low bearing friction at high rotor speeds. However, the bearing efficiency plays a secondary role at high rotor speeds because the turbine power is much larger than the bearing friction power in this case.

The main disadvantage of the rolling element bearings is very loud. Besides the unbalance whistle (1X), the typical induced noises of the ball bearings are high-harmonic noises (2X, 3X, or higher frequency order), defected bearing-related noises, and sideband noises induced by the frequency modulations of the unbalance whistle, high-harmonic noises, and defected bearing-related noises. Compared to the oil film bearings, the so-called constant tone (subsynchronous noise) due to oil whirl does not occur in the ball bearings. However, the ball bearings cost nearly 10 times more than the oil film bearings. When the monetary penalty for exceeding the allowable limit of CO_2 emission is very high, or the transient behavior of the sport vehicles is significantly improved at LET, the ball bearings should be used in the automotive turbochargers.

6.2 Reynolds Lubrication Equation

The Reynolds lubrication equation is used to compute the flow dynamics and induced bearing forces in the oil film bearings. While the Navier–Stokes equations deal with very large convection terms at high Reynolds numbers, the Reynolds equations work with insignificantly small convection terms at low Reynolds numbers. The oil flows in the oil film bearings of automotive turbochargers are normally laminar with low Reynolds numbers between 100 and 200. Therefore, the convection terms are very small corresponding to their low Reynolds numbers. As a result, the Reynolds equations are generally used in the oil film bearings instead of the Navier–Stokes equations [2, 4, 14, 16].

Figure 6.3 shows the hydrodynamic characteristic of the oil film in the oil bearing in which the journal moves with a circumferential velocity U_2; the bearing ring rotates in case of the rotating floating ring bearing with a circumferential velocity U_1.

The oil film can be simplified as a fluid wedge that is surrounded by the bearing and journal moving with the velocities U_1 and U_2, respectively. The ambient pressure of oil at the inlet and outlet is defined as p_0 (see Fig. 6.3). Due to the moving journal and bearing, lubricating oil is drawn into the wedge from the inlet with the inlet pressure p_0 and flows with a velocity $u(x, y)$ through the wedge. The oil film is squeezed in the wedge with a velocity u; hence, the oil film pressure $p(x, z)$ increases in the directions x and z from the inlet to the middle of the fluid wedge. The oil pressure in the wedge remains unchanged in the direction y since the oil film thickness is very small. The Reynolds number of the oil flow in the wedge is very low. Therefore, the flow of the oil film is laminar, and its velocity has a parabolic profile.

The velocity of the journal U_2 contains two components: U_j parallel to the journal surface and V_j perpendicular to the moving direction of U_2. The first velocity is nearly equal to the moving velocity U_2 because the inclination angle α between the journal and bearing is very small ($U_2 \approx U_j$). The second velocity is derived from U_2 and α and written as

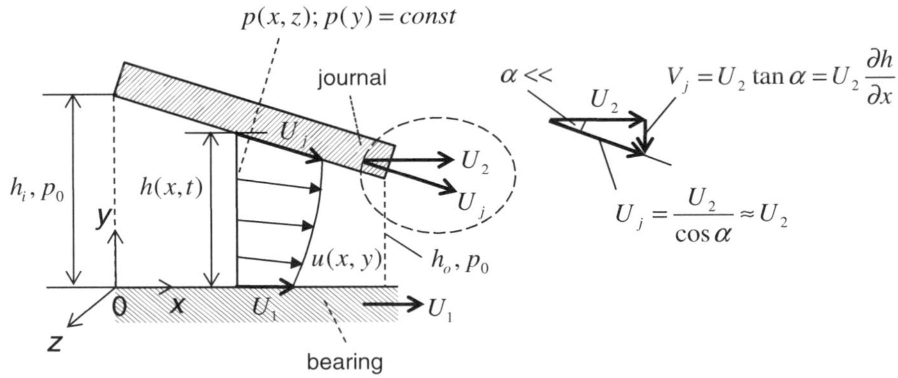

Fig. 6.3 Hydrodynamic oil film in the radial bearing

$$V_j = U_2 \tan \alpha = U_2 \frac{\partial h}{\partial x} < 0 \tag{6.1}$$

The direction of V_j is negative and perpendicular to the oil film. The incompressible oil is squeezed; thus, the oil film pressure increases correspondingly.

Using the Reynolds and continuity equations of the incompressible oil film in the wedge of the bearing, the pressure distribution in the oil film results as [5, 7]

$$\frac{\partial}{\partial x}\left(h^3 \frac{\partial p}{\partial x}\right) + \frac{\partial}{\partial z}\left(h^3 \frac{\partial p}{\partial z}\right) = 6\eta\left[(U_1 - U_2)\frac{\partial h}{\partial x} + 2\left(V_j + \frac{\partial h}{\partial t}\right)\right] \tag{6.2}$$

where

$p(x, z)$ is the oil film pressure;
$h(x, t)$ is the oil film thickness;
t is time;
η is the oil dynamic viscosity.

Substituting Eq. (6.1) into Eq. (6.2), one obtains the Reynolds lubrication equation.

$$\frac{\partial}{\partial x}\left(h^3 \frac{\partial p}{\partial x}\right) + \frac{\partial}{\partial z}\left(h^3 \frac{\partial p}{\partial z}\right) = 6\eta\left[(U_1 + U_2)\frac{\partial h}{\partial x} + 2\frac{\partial h}{\partial t}\right] \tag{6.3a}$$

In a nonrotating radial bearings with $U_1 = 0$, Eq. (6.3a) becomes

$$\frac{\partial}{\partial x}\left(h^3 \frac{\partial p}{\partial x}\right) + \frac{\partial}{\partial z}\left(h^3 \frac{\partial p}{\partial z}\right) = 6\eta U_2 \frac{\partial h}{\partial x} + 12\eta \frac{\partial h}{\partial t} \tag{6.3b}$$

The first terms on the RHS of Eqs. (6.3a and 6.3b) are called the wedge velocity term that is caused by squeezing the oil film in the wedge with the bearing velocities; the last term is called squeeze film term that is generated by the time change rate of the oil film thickness in the wedge. Due to the converged wedge geometry, the change rate of the oil film thickness h in the direction x is always negative. Therefore, the wedge velocity term is negative.

The oil film thickness h in the radial bearing can be calculated as

$$h + e = c_0 \tag{6.4}$$

where

e is the journal displacement inside the bearing clearance;
c_0 the radial bearing clearance (= const).

Thus, the time change rate of the oil film thickness results as

$$\frac{\partial h}{\partial t} = -\frac{\partial e}{\partial t} \tag{6.5}$$

In case of the journal moves from the bearing center outwards due to unbalance; i.e., the time change rate of e is positive. According to Eq. (6.5), the time change rate of the oil film thickness is negative; hence, the squeeze film term is negative.

All terms on the RHS in Eqs. (6.3a and 6.3b) are negative. Therefore, the pressure terms on the LHS (left-hand side) of them must be negative. Note that this term is the second-order derivatives of the oil film pressure. As a result, the oil film pressure $p(x, z)$ has a paraboloid shape with the maximum pressure inside the wedge, as shown in Fig. 6.4.

On the one hand, the wedge velocity terms in Eqs. (6.3a and 6.3b) raise the bearing stiffness coefficients due to the rotation of the bearing with the velocity U_1 and the journal with the velocity U_2 where the oil film is compressed in the wedge by the rotating kinetic energy. On the other hand, the squeeze film term induces the bearing damping coefficients by means of the time change rate of the oil film thickness.

Solving Eqs. (6.3a and 6.3b) with the boundary conditions (BCs), one obtains the pressure distribution of the oil film in the bearing clearance. The total bearing force results from integrating the pressure of the oil film over the bearing surface.

$$F_B = \oint_{A_b} p(x, z)\,\mathrm{d}A = \int_{z_0}^{z_b}\int_{x_0}^{x_b} p(x, z)\,\mathrm{d}x\mathrm{d}z \tag{6.6}$$

where A_b is the pressure surface of the bearing.

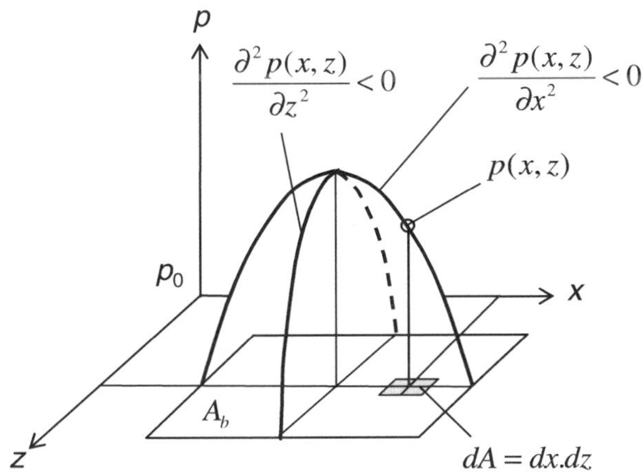

Fig. 6.4 Pressure distribution $p(x, z)$ of the oil film in the wedge

6.3 Lubrication Regimes in the Stribeck Curve

Two rough surfaces 1 and 2 lubricated by the oil film move to each other where the friction force F_t acts on the moving surface 2. Figure 6.5 displays three general lubrication regions of the boundary, mixed, and hydrodynamic lubrications in tribology. We begin to study the lubrication regimes in the Stribeck curve from the RHS to the LHS in Fig. 6.5. The hydrodynamic lubrication occurs when the surfaces are completely separated by the lubricated oil film; i.e., no contact of materials takes place between two surfaces in this regime. In the hydrodynamic lubrication region (full oil film), the friction coefficient μ is proportional to the Hersey dimensionless number according to Eq. (6.11) [6, 7, 14].

The Hersey number is defined as

$$He \equiv \frac{\eta N}{p} \tag{6.7}$$

where

η is the oil dynamic viscosity (Pas);
N is the rotor speed (rps);
p is the oil film pressure between two surfaces (Pa).

The friction force in the hydrodynamic lubrication is calculated

$$F_t = \tau A_t = \eta \left(\frac{\partial u}{\partial h} \right) A_t \approx \eta \left(\frac{U_2}{h} \right) A_t \tag{6.8}$$

The oil velocity U_2 is calculated from the shaft rotational speed and its diameter.

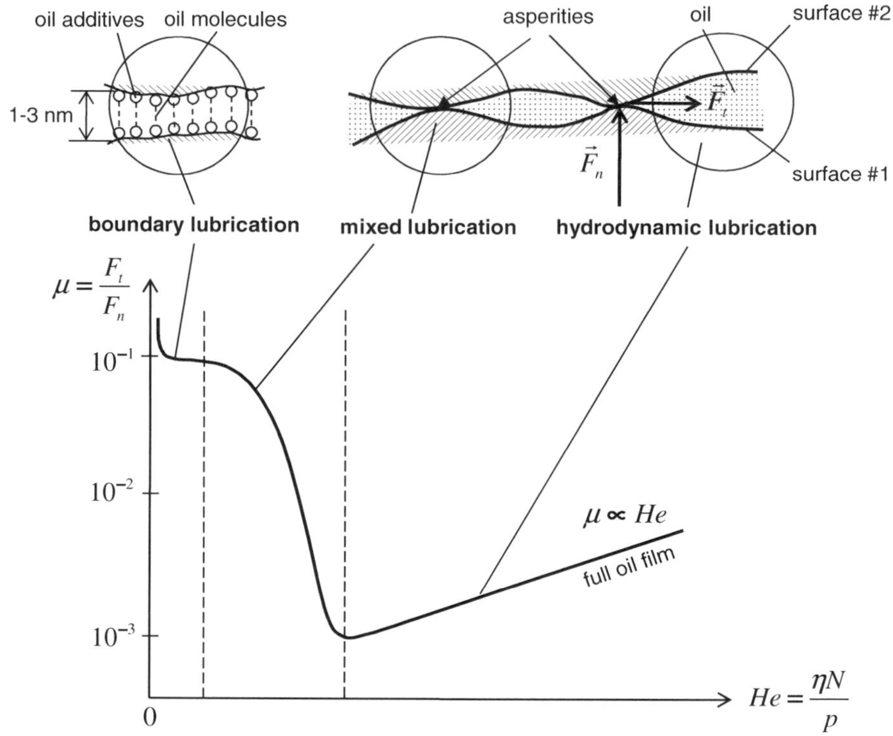

Fig. 6.5 Stribeck curve of the oil lubricated journal bearing

$$U_2 = \Omega R = \pi N D$$

Substituting U_2 into Eq. (6.8), one obtains the friction force

$$F_t = \eta \left(\frac{\pi N D}{h} \right) A_t = \left(\frac{\pi D}{h} \right) A_t \eta N \tag{6.9}$$

where

D is the shaft diameter;
h is the oil film thickness between the surfaces;
A_t is the oil lubricated surface area;
η is the oil viscosity;
N is the rotor speed (rps).

The normal force at the contact surface is calculated as

$$F_n = p A_t \tag{6.10}$$

The friction coefficient is defined as the ratio of the friction to the normal force of Eqs. (6.9 and 6.10).

$$\begin{aligned}\mu \equiv \frac{F_t}{F_n} &= \left(\frac{\pi D}{h}\right)\frac{\eta N}{p} \\ &= k \cdot \left(\frac{\eta N}{p}\right) = k \cdot He\end{aligned} \tag{6.11}$$

Equation (6.11) indicates that the friction coefficient is proportional to the Hersey number in the hydrodynamic lubrication (see Fig. 6.5). It is straightforward that the higher the rotor speed is, the larger the friction force becomes in the oil film.

When the oil film thickness becomes smaller, the micro-contact at the asperities between the surfaces occurs. This lubrication regime is now called mixed lubrication where the adhesion friction mainly causes the friction at the asperities of the surfaces.

The friction force is derived from the shear stress of the softer material at the asperities and contact surface.

$$F_t = \tau A_t \tag{6.12}$$

The normal force is derived from the normal stress at the asperities and contact surface.

$$F_n = \sigma_n A_t \tag{6.13}$$

where σ_n is the plastic normal stress.

Therefore, the friction coefficient in the mixed lubrication results from Eqs. (6.12 and 6.13).

$$\mu \equiv \frac{F_t}{F_n} = \frac{\tau}{\sigma_n} \tag{6.14}$$

The plastic deformation takes place at the vicinity of the contact asperities rather than the elastic deformation that occurs far away from the contact zone. At the plastic deformation, the atoms near the contact zone in the surface slide and move to each other in the form of slip planes and dislocations that correct the errors in the crystal structures of the surfaces. This process is called the work hardening or strain hardening in which the shear stress, normal stress, and hardness of the surface material are strengthened as the plastic deformation continues. Hence, the surface material at the contact zone is hardened by increasing the strength of the plastic shear and normal stress. According to [9], the plastic shear stress increases faster than the plastic normal stress. Therefore, the friction coefficient μ, as given in Eq. (6.14) increases in the mixed lubrication region at reducing the Hersey number. In this region, the plastic deformation occurs at the asperities of the surfaces.

In the boundary lubrication, the surfaces are only lubricated by oil additives connected by the oil molecules in the point of view in nanotribology [9]. In this case, the friction coefficient is nearly unchanged at small Hersey numbers in this regime. At increasing load, the Hersey number reduces since the pressure p in the contact zone becomes larger. At further increasing load or oil temperature, the oil molecule chains break, and therefore the surfaces are now non-lubricating seized. As a result, the friction coefficient jumps up leading to the seizure of the surfaces, as shown in the Stribeck curve at very small Hersey numbers (see Fig. 6.5).

6.4 Thrust Bearings

6.4.1 Working Principle

The working principle of thrust bearings, as shown in Fig. 6.6, is based on the hydrodynamic effect of the lubricating oil described in the Reynolds lubrication equation. The thrust rings fixed to the rotor shaft rotates with the rotor speed Ω and its circumferential velocity U at the mean diameter D_m of the bearing segment (see Fig. 6.7). The bearing segment including the tapered land, flat plateau, and oil groove is mounted and fixed in the bearing housing, and therefore the thrust ring moves relatively to the thrust bearing with the rotor speed Ω. Due to the wedge and squeeze film effects, the bearing force is generated by the oil film pressure acting upon the thrust ring. The bearing force balances the thrust load at a certain oil film thickness to keep the rotor stable in axial direction. At increasing the thrust load on the rotor, the oil film thickness becomes smaller.

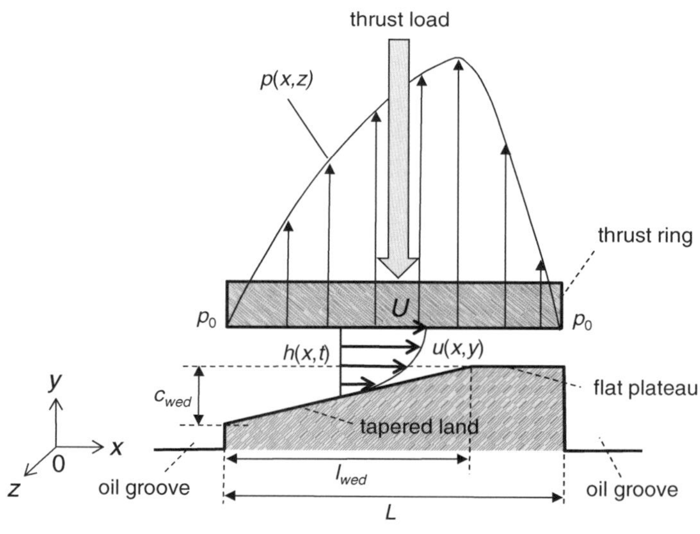

Fig. 6.6 Working principle of a thrust bearing

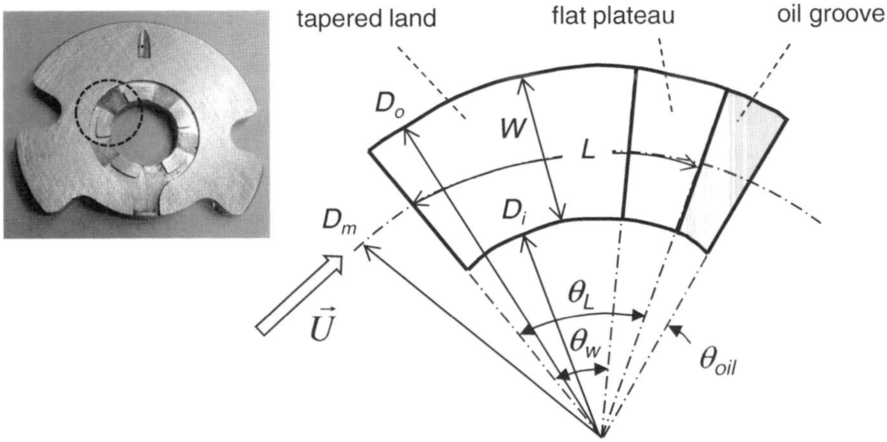

Fig. 6.7 Geometry of the segment of a thrust bearing

In this case, the wedge velocity $U_2 = 0$ and the bearing velocity $U_1 = U$, as indicated in Fig. 6.3 where U is the thrust ring velocity. Therefore, the Reynolds lubrication Eq. (6.3a) for the thrust bearing becomes

$$\frac{\partial}{\partial x}\left(h^3 \frac{\partial p}{\partial x}\right) + \frac{\partial}{\partial z}\left(h^3 \frac{\partial p}{\partial z}\right) = 6\eta U \frac{\partial h}{\partial x} + 12\eta \frac{\partial h}{\partial t} \tag{6.15}$$

where

$p(x, z)$ is the induced pressure in the bearing;
$h(x, t)$ is the oil film thickness in the bearing;
t is time;
U is the circumferential velocity at the mean diameter D_m;
η is the oil dynamic viscosity.

On the one hand, the pressure of incompressible oil in the thrust bearing rises due to the wedge velocity effect described in the first term on the RHS of Eq. (6.15). On the other hand, with increasing the thrust load, the time change rate of the oil film thickness is negative leading to the pressure increase by the squeeze film effect given in the last term on the RHS of Eq. (6.15).

Integrating the pressure $p(x, z)$ over the bearing surface of all bearing segments, one obtains the bearing force that is opposite to the thrust load. At the equilibrium position, the bearing force equals the thrust load at a certain oil film thickness. The oil film thickness must be at least larger than the required minimum oil film thickness, as given in Eq. (6.32) that depends on the surface roughness to prevent the bearing from wear in the mixed lubrication and seizure of the bearing at the thrust rings.

6.4.2 Calculation of the Thrust Load on the Rotor

To design a thrust bearing used in automotive turbochargers, the thrust load must be determined at first. The thrust load is caused by the different pressures acting upon the compressor and turbine wheels and as well as the impulse force induced by the flows in the wheels in the axial direction. Since the turbochargers operate at various speeds at the run-up, slowdown at different traffic situations, the thrust load depends on the rotor speed. Generally, its acting direction could change from one direction to the opposite direction during the operation.

There are two usual ways to compute the thrust load on the rotor: either using CFD (computational fluid dynamics) or applying Newton's second law. The first one provides a precise result but requires a huge computing effort at all operating conditions of turbochargers including the mesh generation of the entire turbocharger, computing time (CPU), and post processing of the computational results. On the contrary, using Newton's second law to compute the axial force is quite simple, but some thermodynamics and turbomachinery backgrounds are required. However, its analytical result is quite good, compared to the CFD numerical results. The discrepancy between them is less than 10 % according to our test computation using both methods at an automotive turbocharger. Moreover, using Newton's second law needs much less CPU to carry out the task for all operating conditions. The result discrepancy between both methods is smaller than the safety tolerance range of the thrust load at the bearing design. As a result, Newton's second law is usually applied to computing the thrust load in automotive turbochargers.

The thrust force $\vec{F}_T$ opposite to the reaction force $\vec{F}_R$ is derived from Newton's second law applied to the control volume (CV). The fluid flows through the CV with a mass flow rate $\dot{m}$, as shown in Fig. 6.8.

The fluid flow is assumed as steady state and the viscous friction at the wall of the CV is negligibly small. According to Newton's second law, the acting forces on the CV are written at the steady state as

$$\sum \vec{F} = \sum_j \left(\vec{F}_{p,j} + \vec{F}_{J,j}\right) + \vec{F}_R = \mathbf{0} \tag{6.16}$$

where $\vec{F}_p$ is the pressure force; $\vec{F}_J$ is the impulse force; $\vec{F}_R$ is the reaction force acting upon the CV.

Integrating pressure p over surface A, the pressure force results as

$$\vec{F}_p = -\oint_A p\vec{n}\mathrm{d}A = -\bar{p}A\vec{n} \tag{6.17}$$

where $\vec{n}$ is the normal vector at the surface; $\bar{p}$ is the average pressure at the surface.

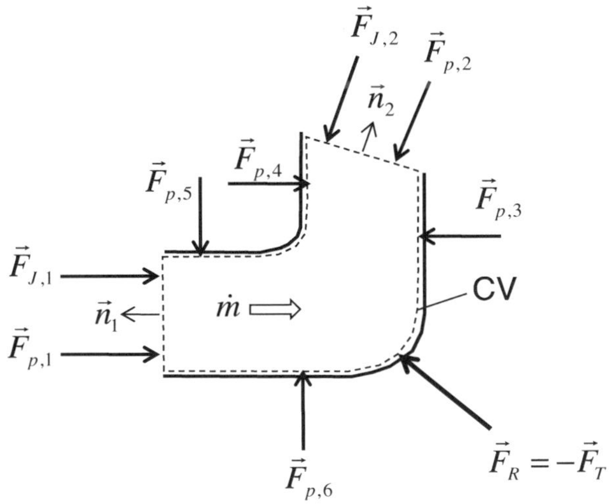

Fig. 6.8 Forces acting upon a contol volume (CV)

The impulse force is calculated by the momentum theorem.

$$\vec{F}_J = -\oint_A \rho\vec{v}\cdot(\vec{v},\,\vec{n})\mathrm{d}A = -\rho\bar{v}^2 A\vec{n} = -\dot{m}\bar{v}\vec{n} \tag{6.18}$$

where

$\dot{m}$ is the mass flow rate through the CV;
$\vec{v}$ is the fluid velocity vector perpendicular to the surface;
ρ is the fluid density at the surface;
$\bar{v}$ is the fluid average velocity at the inlet and outlet flow areas.

Substituting Eqs. (6.16, 6.17, and 6.18), the thrust load acting on the rotor at the steady state results as

$$\begin{aligned}\vec{F}_T = -\vec{F}_R &= \sum_j \left(\vec{F}_{p,j} + \vec{F}_{J,j}\right) \\ &= -\sum_j \left(\bar{p}_j A_j + \dot{m}\bar{v}_j\right)\vec{n}_j\end{aligned} \tag{6.19}$$

Equation (6.19) is the fundamental equation that is used to design the thrust bearing. Due to symmetry of the rotor, the resulting force on the rotor is acting in the axial direction x, which is the thrust load $F_{T,ax}$ (see Fig. 6.9).

The thrust load $F_{T,ax}$ acting upon the rotor results from all forces of the compressor and turbine wheels that are indicated in Fig. 6.9. On the CW (left) $F_{1,C}$ is the

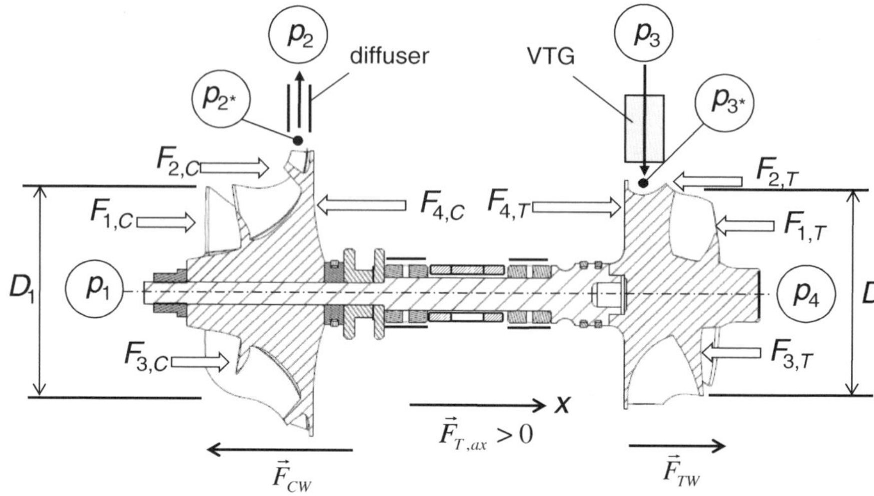

Fig. 6.9 Forces acting on the rotor of an automotive turbocharger

pressure force at the compressor inlet surface; $F_{2,C}$ the pressure force at the shroud surface; $F_{3,C}$ the impulse force in the CW; $F_{4,C}$ the pressure force at the back face. Similarly, the acting forces at the turbine wheel (TW) (right) are $F_{1,T}$, $F_{2,T}$, $F_{3,T}$, and $F_{4,T}$.

The pressure force $F_{1,C}$ is calculated according to Eq. (6.17).

$$F_{1,C} = A_1 p_1 = \frac{\pi D_1^2}{4} p_1 \tag{6.20}$$

where D_1 is the inflow diameter of the compressor; p_1 is the inlet pressure of the ambient air.

Using the mean pressure p_m of the inlet and outlet pressures of the CW, the pressure force $F_{2,C}$ results as

$$F_{2,C} = A_S p_m = A_S \left(\frac{p_1 + p_{2^*}}{2} \right) \tag{6.21}$$

where

A_S is the projected area in the direction x of the shroud surface;
p_1 is the inlet pressure at the CW;
p_{2*} is the outlet pressure at the CW.

The impulse force $F_{3,C}$ results from Eq. (6.18) using the perfect gas equation.

$$F_{3,C} = \dot{m}_C c_{m,1} = \dot{m}_C \left(\frac{\dot{m}_C}{\rho_1 A_{\text{in}}} \right) = \frac{\dot{m}_C^2 R_a T_1}{p_1 A_{\text{in}}} \tag{6.22}$$

where

$\dot{m}_C$ is the air mass flow rate through the CW;
$c_{m,1}$ is the meridional component of air velocity at the compressor inlet;
R_a is the gas constant of air;
T_1 is the inlet temperature of air;
p_1 is the inlet pressure of air;
A_{in} is the cross-sectional area at the inlet of the CW.

The air pressure at the back face of the CW is nearly unchanged when the gap between the bearing housing and the back face of the CW is as large as about 1 mm according to the CFD results. Hence, the pressure force $F_{4,C}$ at the back face is computed with a constant pressure p_{2*} like Eq. (6.17).

$$F_{4,C} = A_{bf,C} p_{2^*} \tag{6.23}$$

where

$A_{bf,C}$ is the acting pressure surface of the back face of the CW;
p_{2*} is the outlet pressure at the CW.

The resulting force on the CW is computed from Eqs. (6.20–6.23).

$$F_{CW} = F_{1,C} + F_{2,C} + F_{3,C} - F_{4,C} \tag{6.24}$$

Analogous to Eq. (6.24) one obtains the resulting on the TW.

$$F_{TW} = -F_{1,T} - F_{2,T} - F_{3,T} + F_{4,T} \tag{6.25}$$

Therefore, the thrust load acting upon the rotor results as

$$F_{T,ax} = F_{CW} + F_{TW} \tag{6.26}$$

Normally, the acting force on the CW is higher than the force on the TW because the CW diameter is generally larger than the TW diameter (see Chap. 10). As a result, the resulting thrust load is normally acting in the direction to the CW. The thrust load is defined as negative when its direction from the turbine to CW ($F_{T,ax} < 0$) and vice versa. At the compressor surge, the thrust load takes turns changing its direction from the turbine to CW ($F_{T,ax} < 0$) and vice versa ($F_{T,ax} > 0$) with a very high frequency (see Fig. 6.9).

The inlet and outlet pressures, temperatures, and mass flow rates of the compressor and turbine are computed using the turbocharged engine processing. However, the necessary pressure p_{2*} between the outlet of the CW and the diffuser, and the pressure p_{3*} between the inlet of the TW and the variable turbine geometry (VTG) are unknown. It is very difficult and time intensive to determine them by means of measurements due to very narrow geometries of the gaps between the wheels and their housings. Therefore, they are estimated using the reaction degrees of the compressor and turbine.

The reaction degree r_C of the compressor is defined as the ratio of the enthalpy increase in the CW to the enthalpy increase of the compressor stage.

$$r_C = \frac{\Delta h_C}{\Delta h_{St}} = \frac{1 - \left(\frac{p_{2^*}}{p_1}\right)^{\frac{\kappa_a - 1}{\kappa_a}}}{1 - \left(\frac{p_2}{p_1}\right)^{\frac{\kappa_a - 1}{\kappa_a}}} \tag{6.27}$$

where κ_a is the isentropic exponent of the charge air.

Having solved Eq. (6.27), the pressure p_{2*} is given as

$$p_{2^*} = p_1 \left[1 + r_C \left(\left(\frac{p_2}{p_1}\right)^{\frac{\kappa_a - 1}{\kappa_a}} - 1\right)\right]^{\frac{\kappa_a}{\kappa_a - 1}} \tag{6.28}$$

where the reaction degree r_C of the compressor is normally between about 55 and 60 % for all operating conditions.

Analogously, the reaction degree r_T of the turbine is defined as the ratio of the enthalpy decrease in the TW to the enthalpy increase of the turbine stage.

$$r_T = \frac{\Delta h_T}{\Delta h_{St}} = \frac{1 - \left(\frac{p_4}{p_{3*}}\right)^{\frac{\kappa_g - 1}{\kappa_g}}}{1 - \left(\frac{p_4}{p_3}\right)^{\frac{\kappa_g - 1}{\kappa_g}}} \tag{6.29}$$

where κ_g is the isentropic exponent of exhaust gas.

The pressure p_{3*} results from Eq. (6.29) as

$$p_{3*} = p_4 \left[1 + r_T \left(\left(\frac{p_3}{p_4}\right)^{-\left(\frac{\kappa_g - 1}{\kappa_g}\right)} - 1\right)\right]^{\frac{-\kappa_g}{\kappa_g - 1}} \tag{6.30}$$

where the reaction degree r_T of the turbine is varied from about 20 % to 90 % depending on the position of the VTG; nearly 50 % at wastegated turbochargers for all positions of the waste gate. The reaction degrees of the compressor and turbine are the key factor to compute the thrust load on the rotor. They should be determined by measurement at the similar type of the designed turbocharger in order to choose an appropriate reaction degree at the first design.

The thrust load acting upon the rotor has been computed using the program CAF (Computing Axial Force) for turbochargers in [10]. The results of the thrust load versus rotor speed are shown in Fig. 6.10.

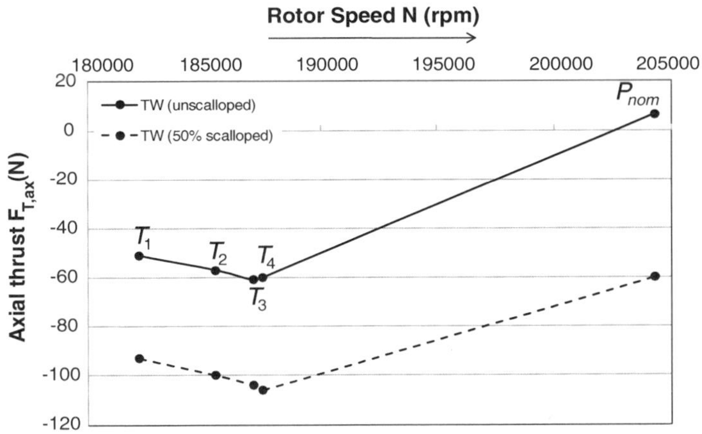

Fig. 6.10 Computed thrust loads versus rotor speed of a small turbocharger

Figure 6.10 shows the characteristic of the thrust load over rotor speeds in a small turbocharger with an unscalloped and a 50 %-scalloped TW. In case of the unscalloped turbine wheel, the thrust load is nearly −50 N in the LET T_1 at a rotor speed of 182,000 rpm; its negative thrust direction is acting to the turbine to CWs, as displayed in Fig. 6.9. At increasing rotor speeds to approximately 188,000 rpm, the turbocharger provides various engine torques T_2, T_3, and T_4 in the full load curve, in which the corresponding thrust loads increase from −50 N to −60 N into the negative thrust direction. After that, the negative thrust load changes at T_4 from −60 N into the positive thrust load of nearly +10 N at the nominal power P_{nom}. Shortly before reaching the nominal power, the negative thrust turns to the positive direction because the acting force on the TW becomes larger than the acting force on the CW. Hence, the negative thrust load changes into the positive direction from the compressor to turbine wheels (see Fig. 6.9).

To improve the response behavior of the turbocharger, the TW is scalloped by 50 %; i.e., the scalloped area of the back face reduces by 50 %. Therefore, the thrust load F_{TW} on the TW is strongly reduced while the thrust load F_{CW} on the CW remains unchanged. As a result, the resulting thrust load $F_{T,ax}$ increases at T_1 from −50 N to −90 N; at T_4 from −60 N to about −105 N; at P_{nom} from +10 N to −60 N into the negative thrust direction. In this case, only negative thrust load ($F_{T,ax} < 0$) acts on the rotor in the negative thrust direction from the turbine to CWs (see Fig. 6.10).

6.4.3 Design of Thrust Bearings

Before designing the geometry of thrust bearings for automotive turbochargers, one has to decide which concept should be applied. Either each thrust bearing for the tailored turbocharger of each application or one thrust bearing for all turbochargers used in various applications. The former offers customers an optimal performance in terms of bearing loads and friction as well because the bearing is tailored for this individual application. However, it is not economical for the turbocharger producers to make each thrust bearing for each individual application. On the contrary, the latter concept is very economical for the turbocharger producers, but the bearing friction is huge at most applications because the overdimensioned bearing has to be designed at the maximum thrust load of the largest turbocharger of the type series in the platform (see Chap. 10).

To make compromise between both extremes, at least two type series of thrust bearings are normally designed for each type series of the turbocharger platform. The type series of the platform is based on the diameter of the rotor shaft. Each type series contains many combinations of various sizes of the compressor and turbine wheels for different nominal powers of the engines.

Usually, it is recommended that at least two type series of thrust bearings should be used for each type series of the turbocharger platform: the small one is applied in the range of the smallest to the middle turbocharger, and the large one for the middle to the largest turbocharger. For an application range in a type series of turbochargers, the maximum thrust load is determined at the combination of the largest CW and the smallest TW at the maximum pressure ratio of the compressor; it has been computed in Sect. 6.4.2 with a safety factor. If the safety factor is chosen too high, the thrust bearing is overdesigned and therefore quite robust against the bearing wear. However, the bearing friction increases; the response behavior of the turbocharger is worse. In case of a small safety factor, the bearing friction reduces, and the response behavior of the turbocharger is improved. However, the bearing could be worn or damaged at some operating conditions in the mixed or boundary lubrication region; in turn, the bearing friction increases significantly. As a result, one should think twice at deciding how large the safety factor should be chosen.

Figure 6.7 shows the typical geometry of a thrust bearing that is derived by experience. At a given thrust load, the minimum oil film thickness of the thrust bearing geometry is computed at the various operating conditions, such as the highest oil inlet temperature, corresponding to rotor speeds using the program TBD [11] based on DIN 31653 [1] so that it satisfies the limit oil film thickness to avoid the mixed lubrication region. Figure 6.11 gives the required minimum oil film thickness that is based on the combined conjunction mean roughness depth of the bearing and thrust ring surfaces surrounded by the oil film (see Table 6.2).

The conjunction mean roughness depth combined the surfaces of the bearing and thrust ring is defined as

$$R_z = \sqrt{R_{z,bearing}^2 + R_{z,disk}^2} \tag{6.31}$$

According to DIN 31653 [1], the limit oil film thickness h_{lim} that depends on the surface roughness R_z and segment mean diameter D_m, is empirically determined as

$$h_{\min} \geq h_{\lim} = 1.25\sqrt{\frac{D_m R_z}{3 \times 10^3}} \tag{6.32}$$

where

D_m is the mean diameter of the bearing segment in meter;
R_z is the combined conjunction mean roughness depth in meter.

To prevent the thrust bearing from operating in the mixed lubrication region, the oil film thickness should be larger than the limit film thickness, as displayed in Fig. 6.11. Otherwise, the thrust bearing operates in the mixed or boundary lubrication regions. In this case, the bearing pressure surface must be enlarged so that the minimum oil film thickness $h_{\min}$ fulfills the requirement given in Eq. (6.32) for all operational conditions.

In the following section, the minimum oil film thickness, oil volumetric flow rate, friction power of the bearing, and effective oil temperature inside the thrust bearing are computed at the various rotor speeds, oil inlet temperatures, and as well as oil types for an application of automotive turbochargers using the program TBD [11].

The dimensionless load number in the hydrodynamic lubrication region is defined as

$$F^* \equiv \frac{F \cdot \left(\frac{h_{\min}}{L}\right)}{F_f} \tag{6.33}$$

where

F is the bearing load;
F_f is the friction force in the bearing;
$h_{\min}$ is the minimum oil film thickness;
L is the bearing segment (pad) length (cf. Fig. 6.7).

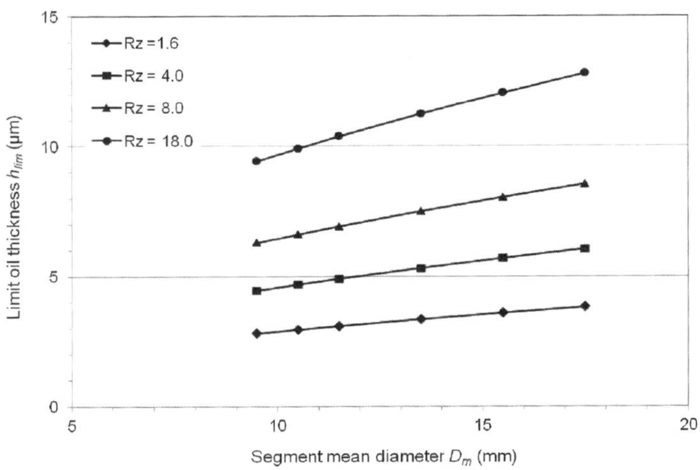

Fig. 6.11 Limit oil film thickness versus segment mean diameter D_m and R_z

Table 6.2 Conjunction mean roughness depths of the various finishing surfaces

Conjunct. roughness R_z (μm)	Finishing methods	Limit oil film thickness $h_{\lim}$ (μm)
0.8–1.6	Grind, lap, honing, polishing	Depends on R_z and the bearing segment mean diameter D_m, as given in Fig. 6.11
1.6–4.0	Grind and lap, honing	
4.0–8.0	Grind and lap	
8.0–18.0	Grind, precision mill, and fine turning	

The friction force results from the bearing theoretical friction power $P_{f,th}$ as

$$F_f \equiv \frac{P_{f,th}}{U} = \eta_{\text{eff}} \frac{U}{h_{\min}} (L \cdot W \cdot Z) \tag{6.34}$$

where

η_{eff} is the effective oil dynamic viscosity;
U is the circumferential velocity at the mean diameter of bearing segment;
W is the width of bearing;
Z is the number of bearing segments.

The dimensionless load number F_B* for a thrust bearing results from F*

$$F_B^* \equiv F^* \left(\frac{c_{\text{wed}}}{h_{\min}}\right)^2 = \frac{F c_{\text{wed}}^2}{\eta_{\text{eff}} U L^2 \cdot W \cdot Z} = F_B^* \left(\frac{h_{\min}}{c_{\text{wed}}}, \frac{W}{L}\right) \tag{6.35}$$

where c_{wed} is the wedge height, as shown in Fig. 6.6.

The dimensionless load number F_B* is a function of $h_{\min}/c_{\text{wed}}$ and W/L given in Eq. (6.35) that is computed solving the Reynolds lubrication equation [1]. At the given bearing geometry and BCs, such as the bearing thrust load F, bearing mean velocity U, the minimum oil film thickness $h_{\min}$ is calculated from F_B* at the initial oil viscosity with the guessed effective oil temperature T_{eff}. The effective oil temperature is computed using the iterative method based on the balance of heat flows in the thrust bearing. The effective oil temperature is converged when the absolute relative change of the oil temperature is smaller than the given convergence value of 10^{-6}.

Similar to Eqs. (6.33 and 6.35), the dimensionless friction number of the thrust bearing is the basics of computing the effective oil temperature at each iteration step.

$$\begin{aligned} f_B^* &\equiv f^* \left(\frac{c_{\text{wed}}}{h_{\min}}\right) = \frac{P_f}{P_{f,th}} \left(\frac{c_{\text{wed}}}{h_{\min}}\right) \\ &= \frac{P_f c_{\text{wed}}}{\eta_{\text{eff}} U^2 L W Z} = f_B^* (c_{\text{wed}}, l_{\text{wed}}, L, W, h_{\min}) \end{aligned} \tag{6.36}$$

where P_f is the real bearing friction power derived from Eq. (6.36).

$$P_f = \frac{f_B^* \eta_{\text{eff}} U^2 LWZ}{c_{\text{wed}}} \tag{6.37}$$

The dimensionless friction number f_B* depends on the bearing geometries and oil film thickness and results from solving the Reynolds lubrication equation [1]. Hence, the real bearing friction power P_f is determined at the effective oil viscosity relating to the iterated oil temperature T_{eff} in the bearing at each iteration step. The oil volumetric flow rates and oil temperature distribution in the bearing are displayed in Fig. 6.12.

Due to the bearing friction, the oil temperature in the bearing increases by a temperature difference

$$\begin{aligned} \Delta T_2 &= T_2 - T_1 \\ &= \frac{P_f}{\rho c_p \left(Q_2 + \frac{Q_3}{2}\right)} = \frac{P_f}{\rho c_p \left(Q_1 - \frac{Q_3}{2}\right)} \end{aligned} \tag{6.38}$$

The volumetric in- and outflow rates of Q_1 and Q_3 are calculated solving the Reynolds equation, which depends on the bearing geometries, rotor velocity, and iterated effective oil viscosity. Therefore, the temperature increase ΔT_2 in the bearing is computed at every iteration step according to Eq. (6.38).

To remove the friction power induced in the bearing, fresh oil with a volumetric flow rate Q at the inlet temperature T_{in} is partly mixed with the oil outflow rate Q_3 from the bearing segment.

$$\rho c_p (MQ + (1 - M)Q_3)\Delta T_1 = \rho c_p Q_2 \Delta T_2 \tag{6.39}$$

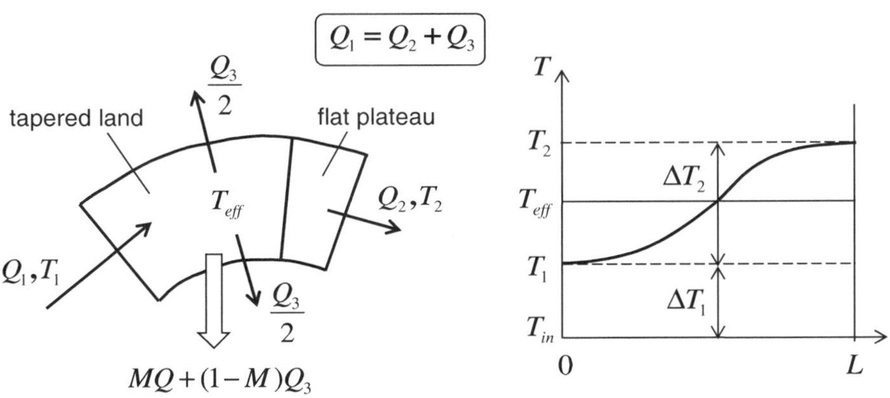

Fig. 6.12 Volumetric flow rates and oil temperatures in a thrust bearing

where

M	is the mixing factor ($0 \le M \le 1$) of the fresh oil (common value 0.5);
$\Delta T_1 = T_1 - T_{in}$,	the temperature increase after the oil mixing;
$\Delta T_2 = T_2 - T_1$,	the temperature increase in the bearing segments;
ρ	is the oil density;
c_p	is the heat capacity at constant pressure.

The temperature increase of the fresh oil due to mixing with the high-temperature outflow oil from the bearing segment results from Eq. (6.39) as

$$\Delta T_1 = T_1 - T_{in} = \frac{Q_2 \Delta T_2}{MQ + (1 - M)Q_3} \tag{6.40}$$

The required fresh oil flow rate Q is calculated from the bearing friction power.

$$Q = \frac{P_f}{\rho c_p (T_{out} - T_{in})} \tag{6.41}$$

The effective oil temperature is defined as the average temperature of T_1 and T_2 at the in- and outlet of the bearing segment.

$$T_{eff} = \frac{T_1 + T_2}{2} = T_{in} + \Delta T_1 + \frac{\Delta T_2}{2} \tag{6.42}$$

The effective bearing temperature in Eq. (6.42) must be iterated as long as the relative solution error reaches the convergence criterion.

$$\frac{\left| T_{eff}^{i} - T_{eff}^{i-1} \right|}{T_{eff}^{i-1}} \le \varepsilon \tag{6.43}$$

where

T_{eff}^{i}	is the effective temperature at iteration step i;
T_{eff}^{i-1}	is the effective temperature at iteration step $(i - 1)$;
ε	is the convergence radius (common value 10^{-6}).

Using the Cameron and Vogel equation, one obtains the effective oil dynamic viscosity η_{eff} at the effective temperature T_{eff}^{i} at the iteration step i.

$$\eta_{eff}^{i} = a \exp\left(\frac{b}{T_{eff}^{i} - c} \right) \tag{6.44a}$$

where a, b, and c are the coefficients determined by three reference points (η_1, T_1), (η_2, T_2), and (η_3, T_3) of the oil type.

Solving the equation system at three reference points, the coefficients a, b, and c result as

$$c = \frac{\beta T_3 - T_2}{\beta - 1}, \; \beta = \frac{\left(\frac{T_1 - T_2}{T_1 - T_3}\right) \cdot \ln\left(\frac{\eta_3}{\eta_1}\right)}{\ln\left(\frac{\eta_2}{\eta_1}\right)}; \tag{6.44b}$$

$$b = \frac{(T_1 - c) \cdot (T_2 - c) \cdot \ln\left(\frac{\eta_2}{\eta_1}\right)}{(T_1 - T_2)}; \tag{6.44c}$$

$$a = \frac{\eta_1}{\exp\left(\frac{b}{T_1 - c}\right)}; \tag{6.44d}$$

where a is in Ns/m^2 (Pas); b, c, and T are in Kelvin (K); β is dimensionless.

The computed dynamic viscosity η of SAE 5W30 oil versus oil temperature T is displayed in Fig. 6.13. The effective oil temperature in the bearing generally lies between 30 and 50 °C above the inlet temperature. The temperature increase caused by the bearing friction depends on many parameters, such as the bearing geometries, oil inlet temperature, rotor speed, minimum oil film thickness. In any case, the effective oil temperature should not exceed the oil coking temperature of about 210 °C for SAE 5W30 oil. When it exceeds the coking temperature, oil will be burnt (coked) in the bearing and remains a hard and black oil layer on the bearing segments leading to a further increase of the effective oil temperature. To avoid the

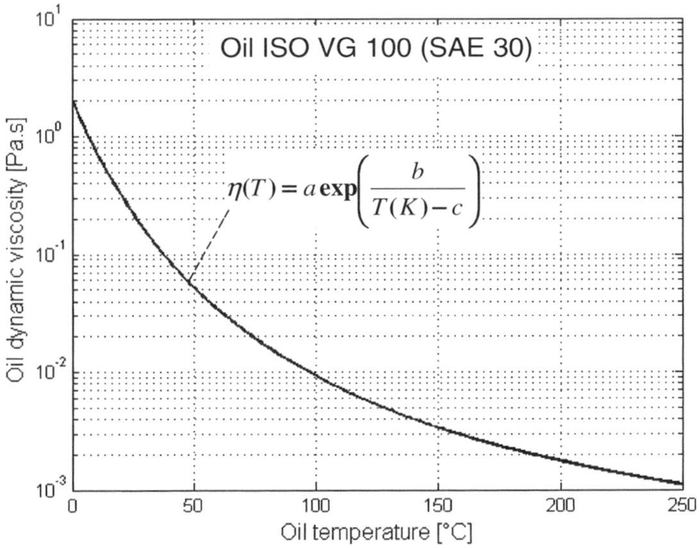

Fig. 6.13 Computed oil dynamic viscosity versus temperature

oil coking in the bearing, the geometry of the bearing segment must be optimized so that the effective oil temperature operates below the oil coking temperature in any operating condition.

The thrust bearing with many segments (pads) is mostly made of brass, a yellowish alloy of copper and zinc including small amounts of manganese and aluminum (CuZnX–MnAl; X is the percentage of zinc). The bearing material of brass has not only excellent tribological characteristics but also very good thermal conductivity that reduces wear and removes the heat flow caused by the bearing friction. Therefore, the effective oil temperature is low and the oil film thickness rises in the thrust bearing (see Fig. 6.14).

Figure 6.15 displays the computed bearing load capacities versus the minimum oil film thickness at various inlet oil temperatures. The minimum oil film thickness in the thrust bearing is found at the generated bearing load capacity of 70 N with the rotor speed of 190,000 rpm and the oil inlet temperatures from 50, 90, 120, and 150 °C. Obviously, the oil film thickness decreases with the thrust load and inlet oil temperature. The bearing geometry must be designed at the worst case with the maximum oil inlet temperature of 150 °C and largest thrust load in the turbocharger. The resulting oil film thickness for this case is about 5 μm (case d) that satisfies the limit oil film thickness with the conjunction surface roughness R_z between 1.6 and 4 μm, as given in Fig. 6.11 according to DIN 31653 [1].

Note that the limit oil thickness is only a recommended value that should be validated by experiments for each application. It is very difficult to predict at which oil film thickness the mixed lubrication occurs in the bearing. In practice, the contaminants in oil cause more wear in form of deep cuts or scratches than wear in the mixed lubrication (see Chap. 9). At reducing the oil inlet temperature from 150 to 120, 90, and 50 °C, the oil film thickness increases from 5 μm to

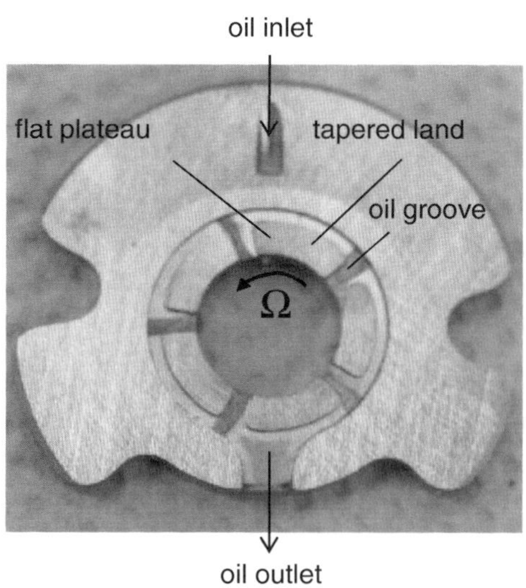

Fig. 6.14 Thrust bearing with five segments (Courtesy BMTS)

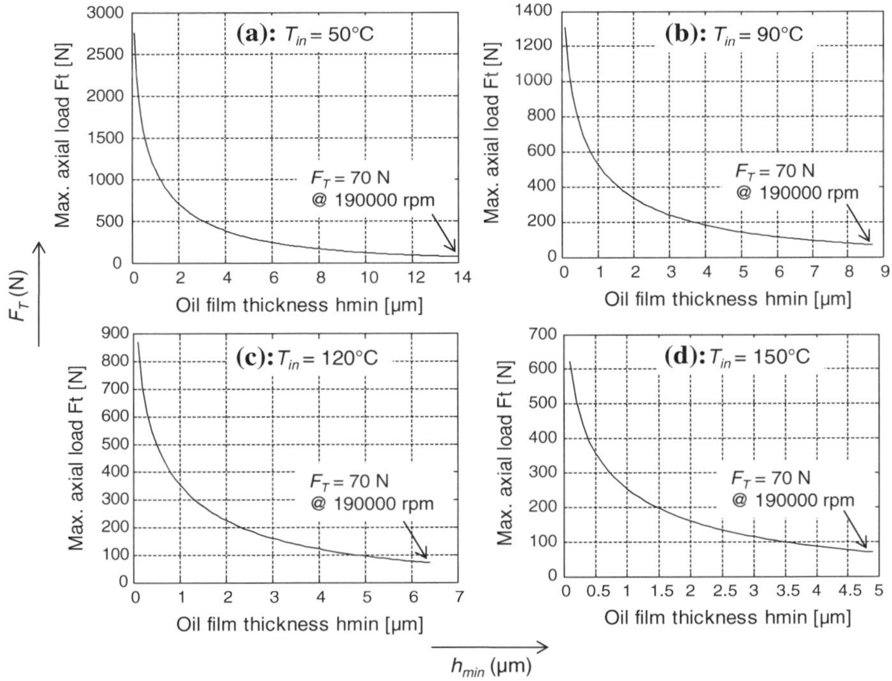

Fig. 6.15 Computed bearing load capacities versus oil film thickness at different T_{in}

approximately 6.5, 8.5, and 14 μm at a thrust load of 70 N and rotor speed of 190,000 rpm for the cases c, b, and a (see Fig. 6.15).

The computed bearing friction power versus the oil inlet temperature at the thrust load of 70 N and rotor speed of 190,000 rpm is shown in Fig. 6.16. At reducing the oil inlet temperature, the minimum oil film thickness increases; hence, the velocity gradient U/h_{min} of the oil film reduces. At the same time, the effective oil viscosity at low temperatures increases faster than the velocity-gradient drops. Therefore, the bearing friction power increases at reducing the oil inlet temperatures. However, the bearing friction power decreases as the oil inlet temperature increases up to about 135 °C, then the friction power increases a little as the oil inlet temperature further increases to 150 °C. The reason for this increase could be that the velocity gradient of the oil film due to strongly reducing the oil film thickness is higher than the decrease of the oil viscosity at the high temperature.

6.4.4 Influential Parameters of Thrust Bearings

To optimize the thrust bearing design in terms of bearing friction, effective temperature in the bearing, and maximum bearing load capacity, some influential

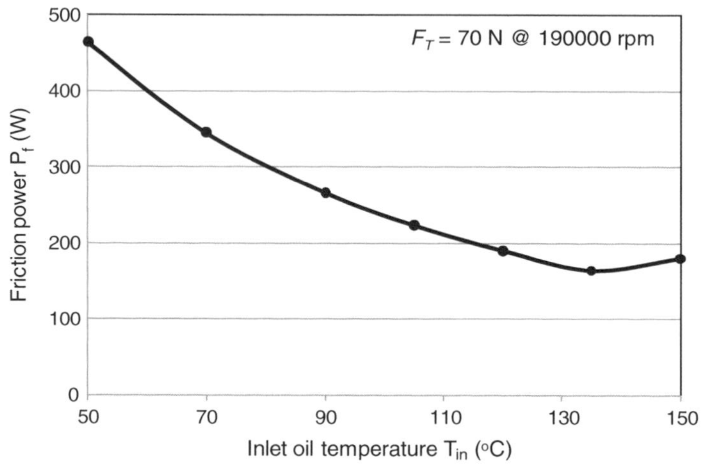

Fig. 6.16 Computed thrust bearing friction power versus oil inlet temperature

parameters should be considered. They are the ratios W/L, l_{wed}/L, c_{wed}/l_{wed}, and D_m, as displayed in Fig. 6.17.

- **Ratio *W/L***

The ratio of the segment width to its length of the bearing should be between 0.5 and 2. On the one hand, at a long bearing segment with the length L much larger

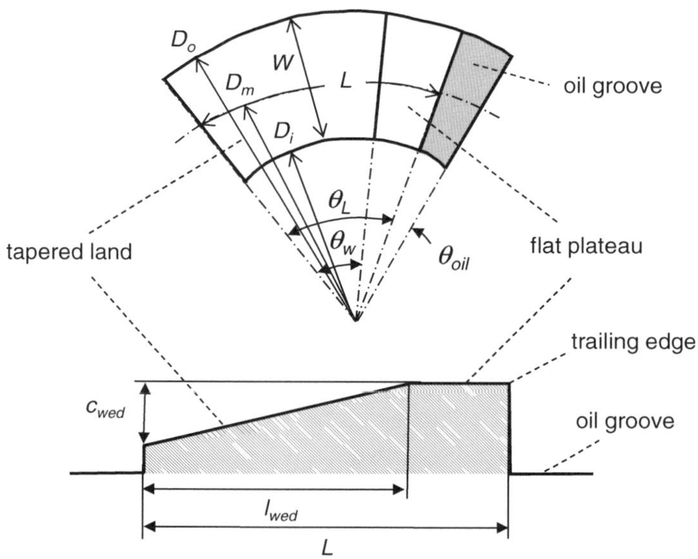

Fig. 6.17 Geometrical parameters of a bearing segment

than its width W ($W/L \ll 1$), the effective oil film is overheated and the effective oil temperature increases due to the long wedge. The oil viscosity reduces at increasing temperature; therefore, the oil film thickness drops so that the operating condition of the bearing occurs in the mixed lubrication.

On the other hand, at the short bearing segment with $W/L \gg 1$, the oil film pressure does not have enough time to build up in the short wedge; hence, the oil film thickness is reduced. Moreover, the side oil outflow Q_3 in the radial direction is smaller due to the large radial flow resistance. It induces an increase of the effective oil temperature in the bearing segment, and in turn the reduction of the oil film thickness. In order to compromise of both extremes of W/L and to optimize the bearing friction, the square bearing segment ($W/L = 1$) would be an optimum geometry for the oil film thickness and maximum bearing load capacity by experience. Therefore, the optimum value for the segment ratio W/L equals 1 at the square bearing segment. However, it is very difficult in practice to design a thrust bearing with square segments that fulfills many predetermined parameters, such as the given bearing outer diameter, given bearing pressure surface, an acceptable number of segments for a small mean diameter, mass production tolerance. In fact, the ratio of W/L between 0.75 and 1 is normally used for a well-designed thrust bearing of the turbochargers.

- **Ratio l_{wed}/L**

The ratio should vary between 0.7 and 0.8 where the optimum value is at 0.75 for a maximum bearing load capacity. If the wedge length l_{wed} is so small that the oil film pressure does not build up enough to balance the thrust load because the hydrodynamic wedge effect is too small. In consequence, the oil film thickness reduces, the bearing friction increases, the effective oil temperature becomes higher leading to the bearing wear and possibly to the seizure of the bearing and thrust rings. In the other case, at $l_{wed} \gg L$, the oil film pressure increases due to large wedge effect; hence, the oil thickness becomes larger and in turn more oil flow rate is required. However, the maximum induced pressure begins dropping in front of the trailing edge sooner than in the normal case with $l_{wed}/L = 0.75$. The early pressure drop is not optimal for the load capacity of the bearing. In order to optimize the bearing load capacity, oil flow rate, oil film thickness, and friction power of the bearing, the ratio l_{wed}/L should be chosen nearly 0.75 for the bearing segments.

- **Ratio c_{wed}/l_{wed}**

This ratio indicates the slope of bearing wedge. At a small slope, the oil film pressure is not increased strongly. Therefore, the oil film thickness decreases, and the effective oil temperature in the bearing segment becomes higher. For the other case, the oil pressure increases faster due to the large wedge effect. As a result, the oil film thickness increases, but much more oil flow rate is necessary.

At small thrust bearings, the bearing slope c_{wed}/l_{wed} varies from 1:75 to 1:150; at large thrust bearings, flatter bearing slopes from 1:200 to 1:400 are usually applied to designing thrust bearings. The reason of using the flatter bearing slopes is to avoid a large pressing force required at manufacturing the thrust bearings. On the

Table 6.3 Influential parameters of thrust bearings

Influential parameters	Normal values	Optimum value
W/L	0.5…2.0	1.0
l_{wed}/L	0.70…0.80	0.75
c_{wed}/l_{wed}	1:75…1:150 (small bearings)	–
	1:200…1:400 (large bearings)	–
D_m	possibly small	–

contrary, steep bearing slopes can be used in the small thrust bearings at which a small pressing force is required for the manufacturing.

- **Mean diameter D_m**

The bearing friction power is proportional to squared mean velocity at the mean diameter of the bearing segment, as shown in Eq. (6.37).

$$P_f \propto \frac{\eta_{\text{eff}} U^2 LWZ}{c_{\text{wed}}} \tag{6.45}$$

with the mean velocity at the mean diameter D_m

$$U = \frac{\Omega D_m}{2} = \frac{\pi N D_m}{60} \tag{6.46}$$

where N is the rotor speed in rpm; U in m/s; D_m in m.

The influential parameters of thrust bearings are summarized in Table 6.3.

In order to reduce the bearing friction, the segment mean diameter should be as small as possible, but it is necessary to check whether

- the minimum oil film thickness is less than the limit oil film thickness;
- the effective oil temperature in the bearing is higher than the oil coking temperature;
- the maximum induced oil pressure in the bearing is higher than the ultimate tensile stress of the bearing material leading to plastic deformation of the bearing.

6.5 Fluid Film Radial Bearings

Some points of view on the design of radial (journal) bearings for turbochargers must be considered, as shown in Fig. 6.18.

First, the main function of the radial bearings is to keep the rotor stable (rotordynamic stability) at all operating conditions of the turbocharger.

Second, the rotor must be well balanced with high-speed balancing so that the residual unbalance is small enough to reduce the unbalance force acting on the

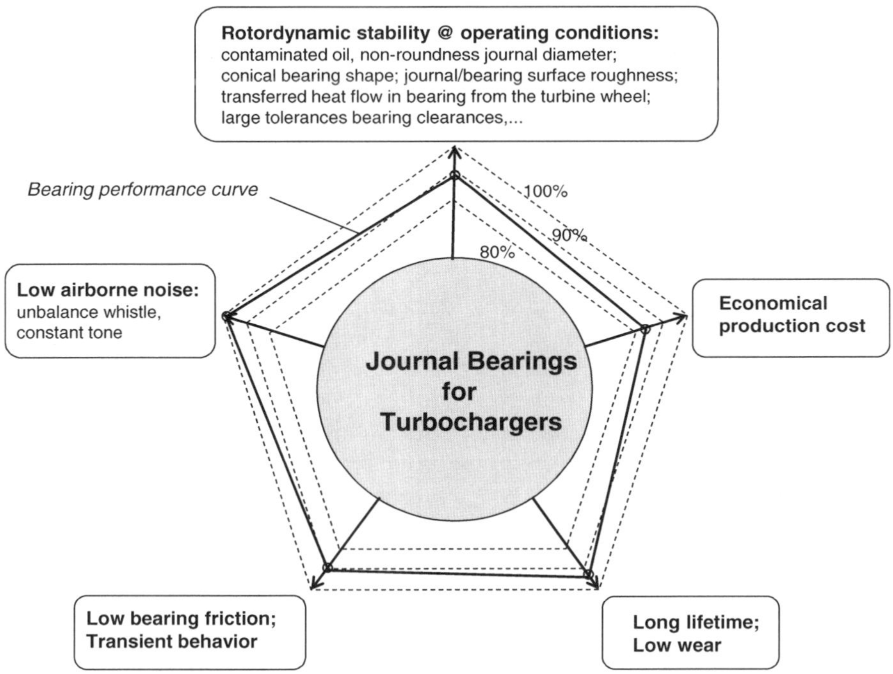

Fig. 6.18 Bearing performance curve of automotive turbochargers

bearing and unbalance whistle. Additionally, radial bearings must have enough damping effect to keep the rotor stable and to suppress noises induced by the unbalance excitation causing the unbalance whistle and the inner oil whirl leading to the constant tone in the automotive turbochargers. The constant tone could be reduced by a suitable geometry of the bearing and the well-chosen inner and outer bearing clearances.

Third, the bearing friction should be reduced as much as possible to improve the transient response of the turbocharger in LETand to reduce CO_2 and NO_x emission.

Fourth, radial bearings should operate in a long lifetime with a minimum or without wear.

Fifth, they should be produced at an economical producing cost with the best quality and functionality for a long lifetime.

However, there is no patent remedy on how to fulfill all five challenging targets perfectly (all 100 %) at the same time. Therefore, the best solution is taking a compromise between them and setting priority of the target with customers. One has to take the compromise between the customer wishes and profit. However, the customers would accept to pay a little more for innovative products that the competitors have not yet brought in the market. Figure 6.18 shows an example of the bearing performance curve for turbochargers. In this case, the customer decides that low airborne noise is the first priority (100 %); long time and low wear are the second priority (95 %); low bearing friction, good transient behavior and rotordynamics stability are the third priority (both 90 %); and in turn, the bearing price could be a little higher (85 %). Note that the bearing performance curve could be changed from one customer to another customer depending on the vehicle segments (regular, sport, premium, and luxury classes).

In the technical point of view, the producing tolerance for the nominal diametral bearing clearance of 20 μm is normally ±6 μm due to saving producing costs. It results from tolerances of the diameters of the bearing and journal. Bearing clearances with large tolerances mostly cause rotor instabilities at high rotor speeds and bearing wear. When the diametral bearing clearance is the minimum of 14 μm, the bearing friction becomes larger; the oil temperature in the bearing increases (coked oil); the oil film thickness reduces (mixed or boundary lubrication); therefore, wear and seizure could occur in the bearing. On the contrary, when the bearing clearance is the maximum of 26 μm, the bearing stiffness and damping coefficients reduce; the rotor instability occurs at some operating conditions (oil whirl or oil whip). Such rotor instabilities cause wear and damage of the rotor and bearings. That is just only one of many technical difficulties at the bearing design, and the goal is always to keep the producing cost as low as possible; hence, the profit as high as possible. In order to save costs of development, testing, and production, only one type of radial bearings is used in each type series for different compressor and turbine wheels. Hence, we have to design one type of radial bearings for different rotors that consist of different diameters of the compressor and turbine wheels operating on various conditions at very high rotor speeds. All of these are the big challenge for the rotordynamic engineers who have to cope with the bearing design in the industry of turbochargers.

6.5.1 Theory of Fluid Film Bearings

The working principle of the radial bearings (journal bearings) shown in Fig. 6.19 is based on the hydrodynamic effect described by the Reynolds lubrication equation. The pressure of the oil film is induced by the wedge contour between the bearing and journal of which the velocities are U_b and U_j, respectively.

Lubricating oil is squeezed in the converged wedge by the rotation of the journal and bearing. Therefore, the oil pressure increases and reaches the maximum pressure in the wedge in front of $\theta = \pi$, as displayed in Fig. 6.20. At the minimum oil film thickness ($\theta = \pi$), the oil film reaches a maximum velocity, and the lowest pressure occurs shortly behind the minimum oil film thickness $h_{\min}$ at the high rotor speeds. Due to the pressure drop, air releasing and cavitation could take place there, and the gas bubbles continue moving in the diverged bearing wedge ($\pi < \theta \leq 2\pi$). The induced pressure of the oil film in the converged wedge generates the bearing forces acting upon the journal to keep the rotor stable in the radial direction.

The Reynolds lubrication equation for radial bearings results from Eq. (6.3a) and it can be written as

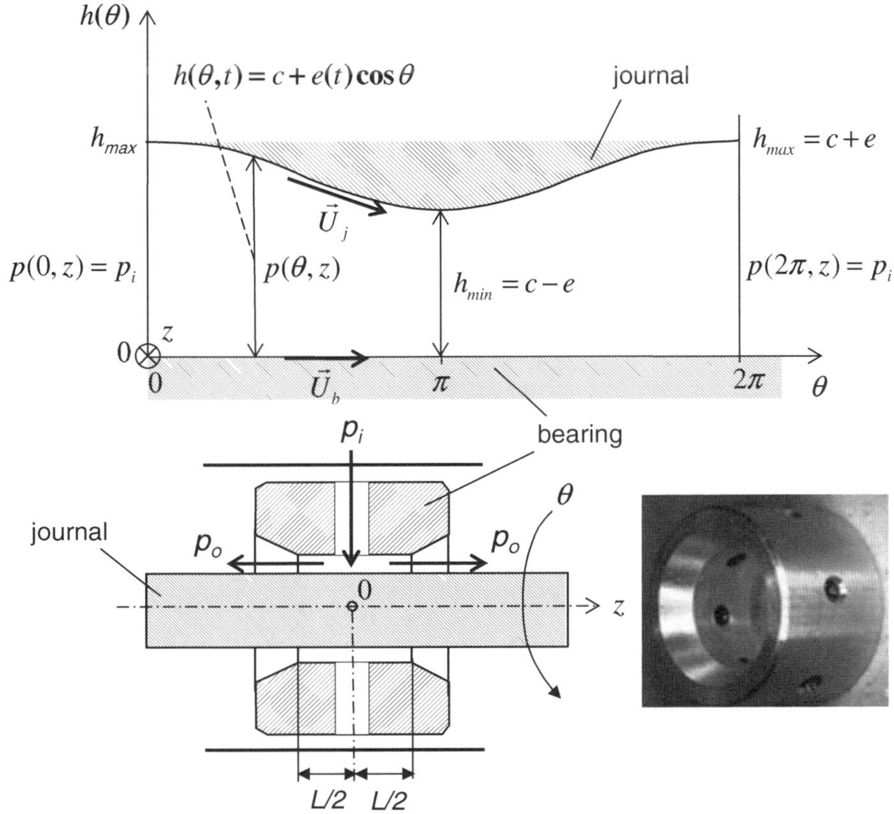

Fig. 6.19 Working principle of a journal bearing

$$\frac{\partial}{\partial x}\left(h^3\frac{\partial p}{\partial x}\right)+\frac{\partial}{\partial z}\left(h^3\frac{\partial p}{\partial z}\right)=6\eta\left[\left(U_j+U_b\right)\frac{\partial h}{\partial x}+2\frac{\partial h}{\partial t}\right] \tag{6.47}$$

where

U_j is the journal velocity;
U_b is the bearing ring velocity;
h is the oil film thickness;
x is the circumferential direction ($x = R\theta$);
z is the axial direction.

At the nonrotating floating ring bearing, the bearing velocity U_b equals zero, and the Reynolds lubrication equation is the same equation for thrust bearings in Eq. (6.15) where U is the journal velocity instead of the velocity of the thrust ring.

Figure 6.20 shows the bearing force induced in the wedge and the pressure distribution in the bearing clearance. To balance the external force $\mathbf{F}_J$ acting upon

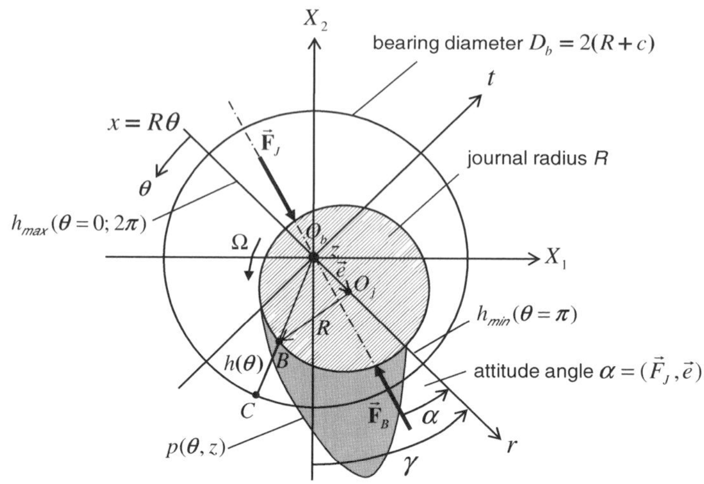

Fig. 6.20 Bearing force vector $\mathbf{F}_B$ acting upon the journal

the rotor, the bearing force $\mathbf{F}_B$ results at the attitude angle α, acting on the journal in the opposite direction of the external force. At the equilibrium position, the minimum oil film thickness at $\theta = \pi$ is determined by the hydrodynamic Reynolds lubrication equation.

The bearing force $\mathbf{F}_B$ results from integrating the pressure distribution over the journal surface in the bearing clearance. The relation between the angular position, journal eccentricity, and bearing clearance allows calculating the oil film thickness in the bearing, as shown in Fig. 6.21.

From the geometries of the journal bearing in Figs. 6.20 and 6.21, one obtains the relation

$$\begin{aligned} O_bC &= (R + c) \approx h + R + e \cdot \cos\varphi \\ &\Rightarrow h(\theta, t) \approx c - e \cdot \cos\varphi = c - e \cdot \cos(\pi - \theta) \\ &= c + e(t) \cdot \cos\theta(t) \end{aligned} \tag{6.48}$$

Fig. 6.21 Geometric relation of the oil film thickness

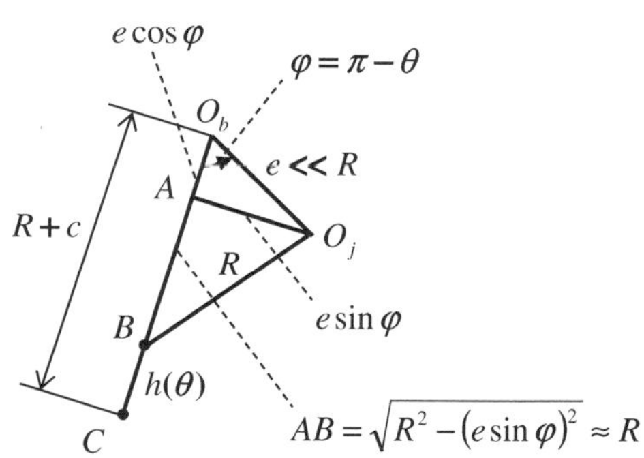

where c is the radial bearing clearance (= const); $e(t)$ is the journal eccentricity depending on time t.

According to Eq. (6.48), the maximum and minimum oil film thicknesses result at $\theta = 0$; 2π and $\theta = \pi$.

$$\begin{cases} h_{\text{max}} = c + e \text{ at } \theta = 0;\ 2\pi \\ h_{\text{min}} = c - e \text{ at } \theta = \pi \end{cases}$$

From Eq. (6.48), one calculates the partial derivatives of the oil film thickness h with respect to t and θ.

$$\begin{aligned} h(\theta, t) &= c + e(t)\cos\theta(t); \\ e(t) &= c\varepsilon(t); \\ \dot{\theta} &= -\dot{\gamma}. \end{aligned} \tag{6.49a}$$

Therefore,

$$\begin{aligned} \frac{\partial h}{\partial t} &= \dot{e}(t)\cos\theta - e(t)\dot{\theta}\sin\theta \\ &= c\dot{\varepsilon}\cos\theta + c\varepsilon\dot{\gamma}\sin\theta; \\ \frac{\partial h}{\partial \theta} &= -e(t)\sin\theta(t) \end{aligned} \tag{6.49b}$$

Moreover, the bearing length x in the circumferential direction θ is written as

$$\begin{aligned} x(\theta, t) &= R\theta(t) \\ \Rightarrow \frac{\partial \theta}{\partial x} &= \frac{1}{R} \end{aligned} \tag{6.49c}$$

Using the chain rule of differentiation (Leibniz rule) and Eqs. (6.49b and 6.49c), one obtains the change rate of the oil film thickness h with respect to x.

$$\begin{aligned} \frac{\partial h}{\partial x} &= \frac{\partial h}{\partial \theta} \cdot \frac{\partial \theta}{\partial x} = -\left(\frac{e}{R}\right)\sin\theta \\ &= -\left(\frac{c}{R}\right)\varepsilon\sin\theta \end{aligned} \tag{6.49d}$$

Substituting Eqs. (6.49a, 6.49b, and 6.49d) into Eq. (6.47), the Reynolds lubrication equation is written in the cylindrical coordinates with a nonrotating bearing ring at $U_b = 0$.

$$\frac{\partial}{\partial\theta}\left((1+\varepsilon\cos\theta)^3\frac{\partial p}{\partial\theta}\right)+R^2\frac{\partial}{\partial z}\left((1+\varepsilon\cos\theta)^3\frac{\partial p}{\partial z}\right)$$
$$=-6\eta\left(\frac{R}{c}\right)^2\cdot[(\Omega-2\dot{\gamma})\varepsilon\sin\theta-2\dot{\varepsilon}\cos\theta] \tag{6.50a}$$

where

$p(\theta, z)$	is the induced pressure in the oil film;
R	is the journal radius (= $D/2$);
Ω	is the angular rotor velocity;
$\dot{\gamma}$	is the whirl velocity of the journal;
ε	is the journal relative eccentricity;
$\dot{\varepsilon}$	is the time change rate of the journal relative eccentricity;
η	is the oil dynamic viscosity.

The BCs for Eq. (6.50a) at the

- inlet: $p(\theta = 0, z) = p(\theta = 2\pi) = p_i$ (6.50b)
- outlet: $p(\theta, z = +\frac{L}{2}) = p(\theta, z = -\frac{L}{2}) = p_o$ (6.50c)

where p_i is the oil inlet pressure; p_o is the oil outlet pressure; L is the bearing width (see Fig. 6.19).

1. Pressure distribution for long bearings (*L/D* > 1)

In the long bearings, the pressure gradient in the circumferential direction θ is much larger than in the axial direction z; thus, the second term on the RHS of Eq. (6.50a) is omitted. The Reynolds lubrication equation is simplified as

$$\frac{\partial}{\partial\theta}\left((1+\varepsilon\cos\theta)^3\frac{\partial p}{\partial\theta}\right)=-6\eta\left(\frac{R}{c}\right)^2[(\Omega-2\dot{\gamma})\varepsilon\sin\theta-2\dot{\varepsilon}\cos\theta] \tag{6.51}$$

Solving Eq. (6.51) with the given BCs, one obtains the pressure distribution $p(\theta)$ in the bearing [16].

$$p(\theta)=p_i+6\eta\left(\frac{R}{c}\right)^2\left(\begin{array}{l}(\Omega-2\dot{\gamma})\frac{\varepsilon(2+\varepsilon\cos\theta)\sin\theta}{(2+\varepsilon^2)(1+\varepsilon\cos\theta)^2}+\\ \frac{\dot{\varepsilon}}{\varepsilon}\left(\frac{1}{(1+\varepsilon\cos\theta)^2}-\frac{1}{(1+\varepsilon)^2}\right)\end{array}\right) \tag{6.52}$$

The bearing forces of the non-cavitating oil film result from integrating the pressure distribution over the journal surface in the radial and tangential directions [5].

$$\begin{aligned} F_r &= -6\eta RL\left(\frac{R}{c}\right)^2\left[\frac{2\varepsilon^2(\Omega-2\dot{\gamma})}{(2+\varepsilon^2)(1-\varepsilon^2)}+\frac{2\dot{\varepsilon}}{(1-\varepsilon^2)^{3/2}}\left(\frac{\pi}{2}-\frac{8}{\pi(2+\varepsilon^2)}\right)\right] \\ F_t &= 6\eta RL\left(\frac{R}{c}\right)^2\left[\frac{\pi\varepsilon(\Omega-2\dot{\gamma})}{(2+\varepsilon^2)(1-\varepsilon^2)^{1/2}}+\frac{4\varepsilon\dot{\varepsilon}}{(2+\varepsilon^2)(1-\varepsilon^2)}\right] \end{aligned} \quad (6.53)$$

2. Pressure distribution for short bearings (*L*/*D* ≤ 1)

In the short bearings, the pressure gradient in the circumferential direction θ is much smaller than in the axial direction z; therefore, the first term on the RHS of Eq. (6.50a) is omitted. The Reynolds lubrication equation is written as

$$\frac{\partial}{\partial z}\left((1+\varepsilon\cos\theta)^3\frac{\partial p}{\partial z}\right)=\frac{-6\eta}{c^2}[(\Omega-2\dot{\gamma})\varepsilon\sin\theta-2\dot{\varepsilon}\cos\theta] \quad (6.54)$$

Solving Eq. (6.54) with the given BCs, one obtains the pressure distribution $p(\theta, z)$ on the journal surface [8].

$$p(\theta,z)=p_0+\frac{3\eta}{c^2(1+\varepsilon\cos\theta)^3}[(\Omega-2\dot{\gamma})\varepsilon\sin\theta-2\dot{\varepsilon}\cos\theta]\left(\frac{L^2}{4}-z^2\right) \quad (6.55)$$

Similarly, the bearing forces of the non-cavitating oil film result from integrating the pressure distribution over the journal surface in the bearing clearance in the radial and tangential directions [5].

$$\begin{aligned} F_r &= -\frac{\eta}{2}\frac{L^3}{R}\left(\frac{R}{c}\right)^2\left[\frac{2\varepsilon^2(\Omega-2\dot{\gamma})}{(1-\varepsilon^2)^2}+\frac{\pi\dot{\varepsilon}(1+2\varepsilon^2)}{(1-\varepsilon^2)^{5/2}}\right] \\ F_t &= \frac{\eta}{2}\frac{L^3}{R}\left(\frac{R}{c}\right)^2\left[\frac{\pi\varepsilon(\Omega-2\dot{\gamma})}{2(1-\varepsilon^2)^{3/2}}+\frac{4\varepsilon\dot{\varepsilon}}{(1-\varepsilon^2)^2}\right] \end{aligned} \quad (6.56)$$

The attitude angle α between the external force and journal eccentricity is calculated from the radial and tangential force components of Eq. (6.56).

$$\alpha=\arctan\left|\frac{-F_t}{F_r}\right|=f(\Omega,\varepsilon,\dot{\varepsilon},\dot{\gamma}) \quad (6.57)$$

where $\dot{\gamma}$ is the whirl velocity of the journal $\dot{\gamma}\equiv\omega$.

6.5.2 Nonlinear Bearing Forces on the Journal

The bearing forces F_r and F_t are based on the rotating coordinate system (r, t) in the journal. In fact, the bearing forces acting upon the journal in the inertial coordinate system (X_1, X_2) are required for the rotordynamic vibration equations. Therefore, the coordinate transformation from the rotating to inertial coordinate system is necessary to compute the bearing forces in the inertial coordinate system (X_1, X_2).

The nonlinear bearing force is transformed to the inertial coordinate system (X_1, X_2).

$$\vec{F}_B = \begin{pmatrix} F_1 \\ F_2 \end{pmatrix} = \begin{pmatrix} \sin\gamma & \cos\gamma \\ -\cos\gamma & \sin\gamma \end{pmatrix} \begin{pmatrix} F_r \\ F_t \end{pmatrix} \tag{6.58}$$

where

F_1 and F_2 are the force components in the inertial coordinates X_1 and X_2;
F_r and F_t are the force components in the rotating coordinates r and t.

Thus,

$$\begin{aligned} F_1 &= +F_r \sin\gamma + F_t \cos\gamma \equiv f_1(\varepsilon, \dot{\varepsilon}, \gamma, \dot{\gamma}, \Omega) \\ F_2 &= -F_r \cos\gamma + F_t \sin\gamma \equiv f_2(\varepsilon, \dot{\varepsilon}, \gamma, \dot{\gamma}, \Omega) \end{aligned} \tag{6.59}$$

where γ is the angular position of the journal (see Fig. 6.22).

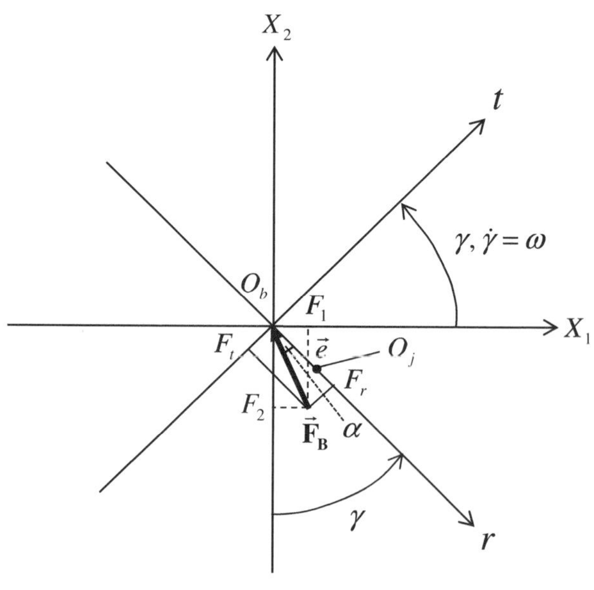

Fig. 6.22 Coordinate transformation between (r, t) and (X_1, X_2)

For a large displacement ε and high whirl velocity $\dot{\gamma}$ of the journal, the bearing force depends on $\varepsilon, \dot{\varepsilon}, \gamma, \dot{\gamma}$, and Ω nonlinearly, as given in Eqs. (6.56 and 6.59). They cannot be separated into the stiffness and damping bearing forces like in linear rotordynamics. As a result, the vibration equation system of the rotor becomes nonlinear due to the nonlinear bearing forces. They change with time, depending on the journal eccentricity ε, its velocity $\dot{\varepsilon}$, the journal angular position γ, the whirl velocity $\dot{\gamma}$, and the rotor speed Ω.

The bearing force given in Eq. (6.56) is valid only for the short bearings of non-cavitating flows. In fact, the radial bearings used in the turbochargers are neither short nor long bearing type, and cavitating flows could occur in the oil film at the high rotor speeds. Hence, two-phase flows must be taken into account in the rotordynamic computation due to cavitation and air releasing [13].

Solving the Reynolds lubrication equation for two-phase cavitating flow of the oil film, one obtains the nonlinear bearing forces for any bearing ratio *L/D*. At a given journal eccentricity ε and angular journal position γ, the bearing forces F_r and F_t are in functions of $\varepsilon, \dot{\varepsilon}, \gamma, \dot{\gamma}$, and the Sommerfeld number *So*.

Using the coordinate transformation in Eq. (6.59), one obtains the bearing forces F_1 and F_2 in the inertial coordinate system (X_1, X_2). To reduce CPU, the transient nonlinear bearing forces are computed by the impedance method, instead of solving the Reynolds lubrication equations at each iteration time step. As a result, the transient bearing forces are linearized at each iteration time step applying the Taylor's series to the computed bearing forces at the journal equilibrium position of ε, γ, and *So*.

Using Taylor's series, the bearing forces in the inertial coordinate system (X_1, X_2) are linearized at the journal equilibrium position.

$$\begin{aligned} F_1 &\approx F_{0,1} + \left(\frac{\partial f_1}{\partial x_1} x_1 + \frac{\partial f_1}{\partial x_2} x_2 + \frac{\partial f_1}{\partial \dot{x}_1} \dot{x}_1 + \frac{\partial f_1}{\partial \dot{x}_2} \dot{x}_2 \right) \equiv F_{0,1} + \Delta F_1 \\ F_2 &\approx F_{0,2} + \left(\frac{\partial f_2}{\partial x_1} x_1 + \frac{\partial f_2}{\partial x_2} x_2 + \frac{\partial f_2}{\partial \dot{x}_1} \dot{x}_1 + \frac{\partial f_2}{\partial \dot{x}_2} \dot{x}_2 \right) \equiv F_{0,2} + \Delta F_2 \end{aligned} \tag{6.60}$$

where

$F_{0,1}, F_{0,2}$ are the static loads acting upon the bearing, cf. Eq. (6.64);
$\frac{\partial f_i}{\partial x_k}$ is the bearing stiffness coefficients; $i, k = 1, 2$;
$\frac{\partial f_i}{\partial \dot{x}_k}$ is the bearing damping coefficients; $i, k = 1, 2$.

The static load F_0 acting on the short bearing results from substituting $\dot{\varepsilon} = \dot{\gamma} = 0$ into Eq. (6.56).

$$\vec{F}_0 = F_{0,r}\vec{e}_r + F_{0,t}\vec{e}_t \tag{6.61a}$$

within

$\vec{e}_r$, $\vec{e}_t$ are the basic vectors in the coordinate system (r, t);
$F_{0,r}, F_{0,t}$ are the force components

$$F_{0,r} = -\frac{\eta}{2}\frac{L^3}{R}\left(\frac{R}{c}\right)^2\left[\frac{2\varepsilon^2\Omega}{(1-\varepsilon^2)^2}\right]; \tag{6.61b}$$

$$F_{0,t} = \frac{\eta}{2}\frac{L^3}{R}\left(\frac{R}{c}\right)^2\left[\frac{\pi\varepsilon\,\Omega}{2(1-\varepsilon^2)^{3/2}}\right] \tag{6.61c}$$

The magnitude of the static load vector F_0 results in

$$\begin{aligned} F_0 &= \sqrt{F_{0,r}^2 + F_{0,t}^2} \\ &= \frac{\eta\Omega}{2}\frac{L^3}{R}\left(\frac{R}{c}\right)^2\left[\frac{\pi\varepsilon}{2(1-\varepsilon^2)^2}\sqrt{(1-\varepsilon^2)+\left(\frac{4\varepsilon}{\pi}\right)^2}\right] \\ &\equiv F_\eta A \end{aligned} \tag{6.62}$$

The reference force F_η in Eq. (6.62) is defined as

$$F_\eta \equiv \frac{\eta\Omega}{2}\frac{L^3}{R}\left(\frac{R}{c}\right)^2 \tag{6.63}$$

Therefore,

$$\begin{aligned} F_{0,1} &= F_{0,r}\sin\gamma + F_{0,t}\cos\gamma = f_{0,1}(\varepsilon,\gamma,\Omega) \\ F_{0,2} &= -F_{0,r}\cos\gamma + F_{0,t}\sin\gamma = f_{0,2}(\varepsilon,\gamma,\Omega) \end{aligned} \tag{6.64}$$

The attitude angle α at the equilibrium condition results from Eq. (6.64).

$$\tan\alpha = -\frac{F_{0,t}}{F_{0,r}} = \frac{\pi\sqrt{1-\varepsilon^2}}{4\varepsilon} \tag{6.65}$$

The static load curve on which the journal center O_j locates at the equilibrium position is displayed in Fig. 6.23. According to Eq. (6.65), the attitude angle α goes to $\pi/2$ ($So \to \infty$) at very small journal eccentricities ($\varepsilon \to 0$); α approaches zero ($So = 0$) at the maximum eccentricity ($\varepsilon \to 1$). Under the static load F_0, the journal center locates at the relative eccentricity ε and attitude angle α. At increasing the bearing load, the journal center moves downwards along the static load curve to the new equilibrium position, locating at the larger eccentricity and smaller attitude angle and vice versa.

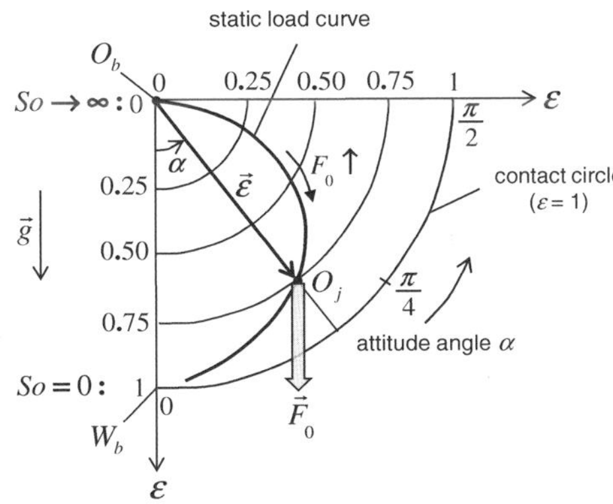

Fig. 6.23 Static load curve of the oil film bearing

At a constant rotor speed Ω, the journal center O_j occurs near the bearing center O_b under a low static load F_0. As the static load increases, the journal center moves downwards with the eccentricity at decreasing the attitude angle along the static load curve to the bearing wall W_b at the contact circle $\varepsilon = 1$ ($So = 0$).

On the contrary, the journal is under the constant load F_0. At first, its center lies near the bearing wall W_b at a low rotor speed Ω. Then, at increasing the rotor speed, the journal center begins moving upwards with the attitude angle at decreasing eccentricity along the load curve to the bearing center O_b at $\varepsilon = 0$ ($So \to \infty$).

The **Sommerfeld number** *So* is defined according to the English convention*. It describes the relation between the bearing geometry, average load pressure, rotor speed, and oil viscosity as

$$So \equiv \frac{\eta N}{\bar{p}} \left(\frac{R}{c}\right)^2 \tag{6.66}$$

where

R is the shaft radius;
c is the radial bearing clearance;
η is the oil dynamic viscosity;
N is the rotor speed (rps) with $N = \Omega/(2\pi)$;
$\bar{p}$ is the average load pressure on the bearing.

(*) In the German convention, the **Sommerfeld number** So^* is defined as

$$So^* \equiv \frac{\bar{p}}{\eta\Omega} \left(\frac{c}{R}\right)^2 = \frac{1}{2\pi\, So}$$

However, the Sommerfeld number *So* according to the English convention in Eq. (6.66) is used in this book.

The average load pressure in the bearing is defined as

$$\bar{p} \equiv \frac{F_0}{DL} \tag{6.67}$$

where

F_0 is the static load on the bearing at equilibrium (cf. Fig. 6.23);
D is the bearing inner diameter;
L is the bearing inner length.

Inserting the average load pressure and rotor speed into Eq. (6.66), the Sommerfeld number becomes

$$So \equiv \frac{\eta \Omega}{2\pi} \frac{DL}{F_0} \left(\frac{R}{c} \right)^2 \tag{6.68}$$

Substituting F_0 of Eq. (6.62) into Eq. (6.68), one obtains the Sommerfeld number for short bearings in the relative journal eccentricity ε.

$$So = \frac{\left(\frac{1}{2\pi\beta^2} \right)}{\frac{\pi\varepsilon}{2(1-\varepsilon^2)^2} \sqrt{(1-\varepsilon^2) + \left(\frac{4\varepsilon}{\pi} \right)^2}} \equiv \frac{\left(\frac{1}{2\pi\beta^2} \right)}{A} \tag{6.69}$$

where $\beta \equiv \frac{L}{D}$ is the length to diameter ratio of the bearing.

Thus,

$$\begin{aligned} \frac{F_0}{F_\eta} \equiv A &= \frac{1}{2\pi\beta^2 So} \equiv \frac{1}{S^*} \\ &= \frac{\pi\varepsilon}{2(1-\varepsilon^2)^2} \sqrt{(1-\varepsilon^2) + \left(\frac{4\varepsilon}{\pi} \right)^2} \end{aligned} \tag{6.70}$$

The **modified Sommerfeld number** S^* results from So and β as

$$\begin{aligned} S^* &\equiv 2\pi\beta^2 So \\ &= \frac{2(1-\varepsilon^2)^2}{\pi\varepsilon\sqrt{(1-\varepsilon^2) + \left(\frac{4\varepsilon}{\pi} \right)^2}} = f(\varepsilon) \end{aligned} \tag{6.71}$$

According to Eq. (6.71), the modified Sommerfeld number S^* is a function of ε (see Fig. 6.24); thus, ε can be expressed in terms of S^* or So.

The impedance method combined with using Eqs. (6.59, 6.60, and 6.64) is used to compute the transient nonlinear bearing forces at each iteration time step that result from the bearing stiffness, damping coefficients, and other parameters.

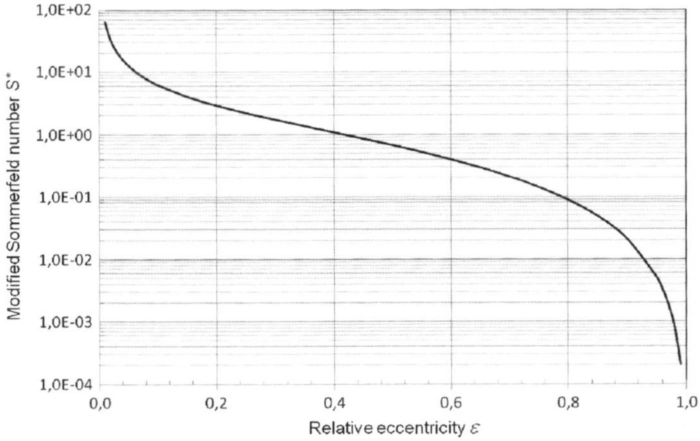

Fig. 6.24 Modified Sommerfeld number S^* versus ε for short bearings

The bearing stiffness coefficients k_{ik} are defined as

$$k_{ik} \equiv -\frac{\partial f_i}{\partial x_k} = \kappa_{ik}\frac{F_0}{c}; \quad i;k = 1,\, 2 \tag{6.72}$$

where

κ_{ik} is the bearing dimensionless stiffness coefficients;
F_0 is the static load at equilibrium;
c is the radial bearing clearance.

Similarly, the bearing damping coefficients d_{ik} are defined as

$$d_{ik} \equiv -\frac{\partial f_i}{\partial \dot{x}_k} = \beta_{ik}\frac{F_0}{c\Omega}; \quad i;k = 1, 2 \tag{6.73}$$

where

β_{ik} is the bearing dimensionless damping coefficient;
F_0 is the static load at equilibrium;
c is the radial bearing clearance;
Ω is the rotor speed.

Since f_1 and f_2 are the functions of $\varepsilon, \dot{\varepsilon}, \gamma, \dot{\gamma}$ in Eq. (6.59), one obtains the partial derivatives using the chain rule of differentiation

$$\begin{aligned} \frac{\partial f_i}{\partial x_k} &= \frac{\partial f_i}{\partial \varepsilon}\frac{\partial \varepsilon}{\partial x_k} + \frac{\partial f_i}{\partial \gamma}\frac{\partial \gamma}{\partial x_k}; \quad i;k = 1,\, 2 \\ \frac{\partial f_i}{\partial \dot{x}_k} &= \frac{\partial f_i}{\partial \dot{\varepsilon}}\frac{\partial \dot{\varepsilon}}{\partial \dot{x}_k} + \frac{\partial f_i}{\partial \dot{\gamma}}\frac{\partial \dot{\gamma}}{\partial \dot{x}_k}; \quad i;k = 1,\, 2 \end{aligned} \tag{6.74}$$

Using the coordinate transformation (see Fig. 6.25), one obtains the relations between the coordinates

$$\begin{pmatrix} dx_1 \\ dx_2 \end{pmatrix} = \begin{bmatrix} \cos\delta & -\sin\delta \\ \sin\delta & \cos\delta \end{bmatrix} \begin{pmatrix} de \\ ed\gamma \end{pmatrix} \tag{6.75}$$

where $\delta = \gamma - \frac{\pi}{2}$.

Thus, Eq. (6.75) becomes

$$\begin{pmatrix} dx_1 \\ dx_2 \end{pmatrix} = \begin{bmatrix} \sin\gamma & \cos\gamma \\ -\cos\gamma & \sin\gamma \end{bmatrix} \begin{pmatrix} de \\ ed\gamma \end{pmatrix} \tag{6.76}$$

Transforming backwards Eq. (6.76), it gives

$$\begin{pmatrix} de \\ ed\gamma \end{pmatrix} = \begin{bmatrix} \sin\gamma & -\cos\gamma \\ \cos\gamma & \sin\gamma \end{bmatrix} \begin{pmatrix} dx_1 \\ dx_2 \end{pmatrix} \tag{6.77}$$

Differentiating with respect to time t both sides of Eq. (6.77) using $e = c\varepsilon$, one obtains

$$\begin{pmatrix} c\dot{\varepsilon} \\ e\dot{\gamma} \end{pmatrix} = \begin{bmatrix} \sin\gamma & -\cos\gamma \\ \cos\gamma & \sin\gamma \end{bmatrix} \begin{pmatrix} \dot{x}_1 \\ \dot{x}_2 \end{pmatrix} \tag{6.78}$$

Partially differentiating Eqs. (6.77 and 6.78), one obtains

$$\frac{\partial\varepsilon}{\partial x_1} = \frac{\partial\dot{\varepsilon}}{\partial\dot{x}_1} = \frac{1}{c}\sin\gamma; \; \frac{\partial\varepsilon}{\partial x_2} = \frac{\partial\dot{\varepsilon}}{\partial\dot{x}_2} = -\frac{1}{c}\cos\gamma; \tag{6.79}$$

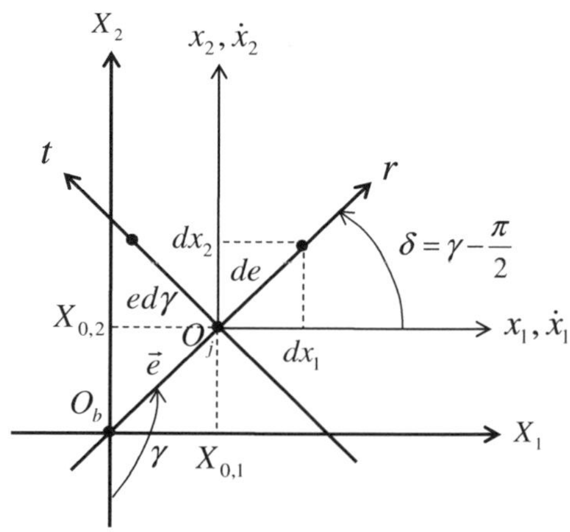

Fig. 6.25 Coordinate transformation between (x_1, x_2) and (X_1, X_2)

$$\frac{\partial\gamma}{\partial x_1} = \frac{\partial\dot{\gamma}}{\partial\dot{x}_1} = \frac{1}{e}\cos\gamma; \ \frac{\partial\gamma}{\partial x_2} = \frac{\partial\dot{\gamma}}{\partial\dot{x}_2} = \frac{1}{e}\sin\gamma \tag{6.80}$$

Substituting Eqs. (6.79 and 6.80) into Eq. (6.74), the bearing dimensionless stiffness and damping coefficients κ_{ik} and β_{ik} result for short bearings [8].

$$\begin{aligned}
\kappa_{11} &= \left[2\pi^2 + (16-\pi^2)\varepsilon^2\right]F(\varepsilon)\\
\kappa_{12} &= \frac{\pi}{4}\left[\frac{-\pi^2 + 2\pi^2\varepsilon^2 + (16-\pi^2)\varepsilon^4}{\varepsilon(1-\varepsilon^2)^{1/2}}\right]F(\varepsilon)\\
\kappa_{21} &= \frac{\pi}{4}\left[\frac{\pi^2 + (32+\pi^2)\varepsilon^2 + 2(16-\pi^2)\varepsilon^4}{\varepsilon(1-\varepsilon^2)^{1/2}}\right]F(\varepsilon)\\
\kappa_{22} &= \left[\frac{\pi^2 + (32+\pi^2)\varepsilon^2 + 2(16-\pi^2)\varepsilon^4}{(1-\varepsilon^2)}\right]F(\varepsilon)
\end{aligned} \tag{6.81}$$

and

$$\begin{aligned}
\beta_{11} &= \frac{\pi(1-\varepsilon^2)^{1/2}}{2\varepsilon}\left[\pi^2 + 2(\pi^2-8)\varepsilon^2\right]F(\varepsilon)\\
\beta_{12} &= \beta_{21} = 2\left[\pi^2 + 2(\pi^2-8)\varepsilon^2\right]F(\varepsilon)\\
\beta_{22} &= \frac{\pi}{2}\left[\frac{\pi^2 + 2(24-\pi^2)\varepsilon^2 + \pi^2\varepsilon^4}{\varepsilon(1-\varepsilon^2)^{1/2}}\right]F(\varepsilon)\\
F(\varepsilon) &\equiv \frac{4}{\left[\pi^2 + (16-\pi^2)\varepsilon^2\right]^{3/2}}
\end{aligned} \tag{6.82}$$

In case of linear or quasilinear rotordynamics, the bearing forces F_1 and F_2 in the coordinate system (X_1, X_2) are linearized from the equilibrium at each iteration time step according to Eq. (6.83) for the small rotor amplitudes.

$$\begin{aligned}
F_1^i &= F_{0,1} + \Delta F_1^i\\
F_2^i &= F_{0,2} + \Delta F_2^i
\end{aligned} \tag{6.83}$$

where ΔF_1^i, ΔF_2^i are the changes of force relating to small displacements and velocities of the rotor.

The changes of force at the equilibrium position result from Eqs. (6.60, 6.72, and 6.73).

$$\begin{bmatrix}\Delta F_1\\ \Delta F_2\end{bmatrix}^i = -\begin{bmatrix}k_{11} & k_{12}\\ k_{21} & k_{22}\end{bmatrix}^i\begin{bmatrix}x_1\\ x_2\end{bmatrix}^i - \begin{bmatrix}d_{11} & d_{12}\\ d_{21} & d_{22}\end{bmatrix}^i\begin{bmatrix}\dot{x}_1\\ \dot{x}_2\end{bmatrix}^i \tag{6.84}$$

where k_{ik} and d_{ik} are the bearing stiffness and damping coefficients, respectively. In case of short bearings, they only depend on the angular position γ and journal relative eccentricity ε, in turn on the Sommerfeld number So and γ.

Solving the Reynolds lubrication equation of the bearings, one obtains the impedance table of the bearing forces that depend on the bearing positions and its characteristics. In case of nonlinear rotordynamics, the nonlinear bearing forces in the entire bearing clearance result from the corresponding bearing stiffness and damping coefficients linearly interpolating the computed values of the bearing characteristics at ε, γ, and So in the impedance table. They contain two force components. First, the bearing rotation force is induced by the rotation with Ω, and it depends on η, ε, and γ. Second, the bearing damping force is caused by the journal velocities $\dot{\varepsilon}, \dot{\gamma}$, and it depends on $\eta, \varepsilon, \gamma, \dot{\varepsilon}, \dot{\gamma}$.

6.5.3 Floating Ring Bearings

Floating ring bearings are usually applied to automotive turbochargers to reduce airborne noises and rotor amplitudes at resonances. There are two common types of the floating ring bearings: semi-floating (SFRB) and RFRB. The common character of both floating ring bearings is having two oil films, the inner and outer oil films. However, they are different because the bearing ring does not rotate in the semi-floating bearing and rotates in the rotating floating ring bearing.

In order to reduce the bearing friction, the RFRB are used in the turbochargers. Due to reduction of the bearing friction at the low rotor speeds, the transient response of turbochargers is improved in LET. Figure 6.26 shows the setup of a rotating floating ring bearing on the rotor shaft. Lubricating oil at the pressure p_i from the inlet flows through the outer bearing clearance in the circumferential and axial directions and leaves the bearing ends at the ambient pressure p_0. By means of the oil holes on the bearing ring, lubricating oil enters the inner bearing clearance. Then, it flows through the inner bearing clearance in the circumferential and axial directions and leaves the bearing ends at the ambient pressure p_0. The oil volumetric flow rate depends on the oil inlet temperature, pressure drop between p_i and p_0, rotor speed, and as well as the rotating speed of the bearing ring.

The inner oil film has a key function to bear the rotor against to the external forces acting upon the rotor. Therefore, the inner radial clearance c_1 of the bearing is relatively small to increase the bearing stiffness. On the contrary, the outer oil film provides the rotor with a large damping coefficient to reduce the rotor deflection at resonances and suppress the airborne noises, such as the unbalance whistle and constant tone caused by rotor unbalance and inner oil whirl in the bearing, respectively. As a result, the bearing outer clearance is larger than the inner clearance.

In the following section, the rotating speed of the bearing ring is calculated using flow characteristics of the oil films and the bearing friction power.

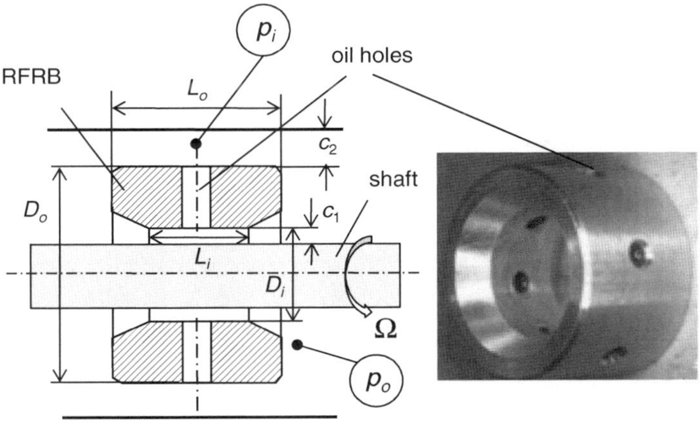

Fig. 6.26 Setup of a rotating floating ring bearing on the rotor shaft

The bearing friction power on the rotor shaft is written as

$$\begin{aligned} P_f &= F_f U_j = U_j \int_A \tau dA \\ &= \eta_i \frac{\partial U}{\partial h} A_j U_j \\ &\approx \eta_i \left(\frac{U_j}{h} \right) A_j U_j \end{aligned} \tag{6.85}$$

where

η_i	is the dynamic viscosity of the inner oil film;
$\partial U / \partial h$	is the velocity gradient in the oil film;
h	is the oil film thickness;
A_j	is the journal frictional area;
U_j	is the journal circumferential velocity.

In order to reduce the friction power, the velocity gradient is decreased when the bearing ring rotates with a circumferential velocity U_R at the RFRB. Hence, the reduced bearing friction power is smaller.

$$P_f^* \approx \eta_i \frac{(U_j - U_R)}{h} A_j U_j \tag{6.86}$$

The relative reduction of the friction power results from Eqs. (6.85 and 6.86) in

$$\begin{aligned}\frac{\Delta P_f}{P_f} &= \frac{P_f - P_f^*}{P_f} = 1 - \frac{P_f^*}{P_f} \\ &= \frac{U_R}{U_j} \approx \frac{\Omega_R}{\Omega} \equiv RSR\end{aligned} \tag{6.87}$$

where Ω_R is the ring angular velocity ($\Omega_R < \Omega$); Ω is the rotor angular velocity.

Equation (6.87) indicates that the friction power reduction is proportional to the ring speed ratio *RSR* that is defined as the ratio of the bearing ring speed to the rotor speed.

Torque acting upon the bearing ring is calculated using the Petroff's law that results from the Reynolds lubrication equation.

$$M = R\int_A \tau \mathrm{d}A = R\int_A \eta \frac{\partial U}{\partial h}\mathrm{d}A = \frac{2\pi\eta R^3 L\Omega}{c} \tag{6.88}$$

Applying Eq. (6.88) to the bearing ring, one obtains the driving torques acting on the inner side of the bearing

$$M_i = \frac{2\pi\eta_i R_i^3 L_i(\Omega - \Omega_R)}{c_1} \tag{6.89}$$

and the friction torque on the outer side of the bearing

$$M_o = -\frac{2\pi\eta_o R_o^3 L_o \Omega_R}{c_2} \tag{6.90}$$

The angular momentum law for the bearing ring is written as

$$\sum M = M_i + M_o = I_p\ddot{\theta} \tag{6.91}$$

In case of $\ddot{\theta} = 0$, the steady state *RSR* results using Eqs. (6.89, 6.90 and 6.91) as

$$RSR_{steady} \equiv \frac{\Omega_R}{\Omega} \approx \frac{1}{1 + \frac{\eta_o(T_o)}{\eta_i(T_i)}\left(\frac{L_o}{L_i}\right)\left(\frac{c_1}{c_2}\right)\left(\frac{D_o}{D_i}\right)^3} \tag{6.92a}$$

where

- η_i, η_o are the dynamic viscosity of the inner and outer oil films;
- L_i, L_o are the inner and outer bearing width;
- c_1, c_2 are the inner and outer radial bearing clearance;
- D_i, D_o are the inner and outer bearing diameter;
- I_p is the polar mass inertia moment of the bearing ring.

When the inner and outer relative eccentricities ε_1 and ε_2 considered, the modified steady state RSR^* results as [5]

$$RSR^*_{steady} \approx \frac{1}{1 + \frac{\eta_o(T_o)}{\eta_i(T_i)}\left(\frac{L_o}{L_i}\right)\left(\frac{c_1\sqrt{1-\varepsilon_1^2}}{c_2\sqrt{1-\varepsilon_2^2}}\right)\left(\frac{D_o}{D_i}\right)^3} \tag{6.92b}$$

In case of $\ddot{\theta} \neq 0$, one obtains the modified unsteady state ring speed ratio using Eqs. (6.89, 6.90 and 6.91) (see Fig. 6.27).

$$RSR_{unsteady} = RSR^*_{steady} - \frac{I_p\ddot{\theta}}{2\pi\Omega\left(\frac{\eta_i R_i^3 L_i}{c_1} + \frac{\eta_o R_o^3 L_o}{c_2}\right)} \tag{6.92c}$$

Since the bearing inner clearance is much smaller than the bearing outer clearance, the inner oil film temperature is higher than the outer oil film temperature due to the induced bearing friction, especially at increasing the rotor speeds. Hence, the oil viscosity ratio is larger than 1 (i.e., $\eta_o/\eta_i > 1$) leading to decreasing the ring speed ratio *RSR*. The computed behavior of the ring speed ratio versus rotor speed at various oil inlet temperatures is displayed in Fig. 6.28 [12].

The effective temperatures T_i and T_o of the inner and outer oil films are only induced by the bearing friction where the heat conduction of the bearing housing to the oil films is not taken into account. In fact, the heat conduction from the bearing housing increases the temperature of the outer oil film; therefore, the ring speed ratio increases. The ring speed ratio *RSR* varies between 15 and 30 % at low rotor speeds to about 90,000 rpm (1500 rps) at the oil inlet temperature of 90 °C. Hence, the friction power of the RFRB on the rotor shaft could be reduced by 20 %

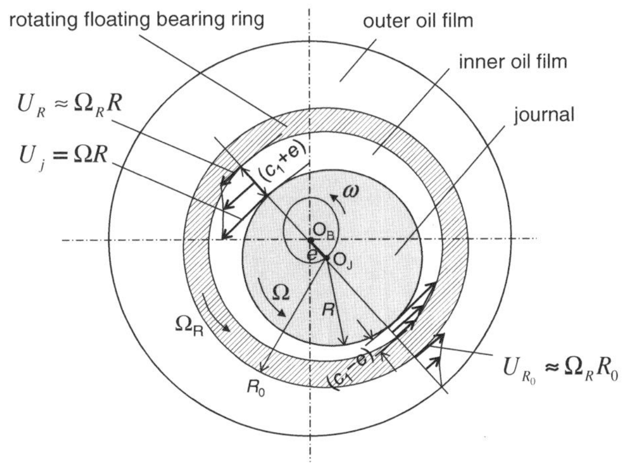

Fig. 6.27 Flow velocities in the oil films

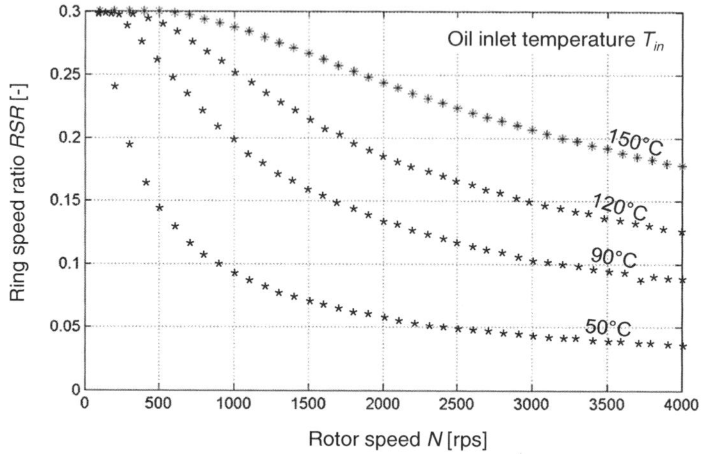

Fig. 6.28 Computed ring speed ratio RSR_{steady} versus rotor speed N at various T_{in}

compared to the semi-floating ring bearing in LET leading to improving the response behavior of the turbocharger.

The higher the rotor speed is, the higher the inner temperature of the oil film becomes due to bearing friction leading to reducing the ring speed. At high oil temperatures, the dynamic viscosity ratio of both outer and inner oil films does not change so much. Therefore, the ring speed ratio reduces slowly and tends to remain constant at the very high rotor speeds, as shown in Fig. 6.28. At decreasing the oil inlet temperature, a large bearing friction is generated in the inner oil film leading to a higher oil temperature in the inner bearing clearance. As a result, the ring speed of the bearing decreases. Generally, the ring speed ratio *RSR* decreases with the rotor speed at a constant oil inlet temperature and increases with the oil inlet temperature at a constant rotor speed.

The RFRB take advantages of more damping due to the outer oil film and reducing the bearing friction power, especially in LET. However, they need more oil volumetric flow rate, and the rotor deflection could be larger compared to the semi-floating ring bearings.

In fact, if we consider the unsteady state at run-up, the ring speed ratio is very low at small rotor speeds because the inertia term of the ring speed ratio (second term on the RHS) of Eq. (6.92c) is significantly large, especially at high angular accelerations. At increasing rotor speeds, this term becomes smaller; hence, the ring speed ratio increases. At further increasing the rotor speeds, the oil viscosity ratio between the outer and inner oil films is larger than 1 so that the ring speed ratio begins reducing.

6.5.4 Influential Parameters of RFRB

To optimize the bearing design in terms of bearing friction, effective temperature in the bearing, and rotor stability, some influential parameters are discussed. They are the typical ratios of D_o/D_i, L_o/L_i, and c_2/c_1 (see Fig. 6.26). They have mutual effects on the characteristics of the rotor response of turbochargers, as shown in Fig. 6.18. As a result, they must be carefully chosen according to the customer priorities.

In the following section, the influential parameters of the RFRB affecting the rotor responses are analyzed.

- **Ratio *Do/Di***

The larger the diameter ratio D_o/D_i at a given inner bearing diameter D_i is, the more the damping coefficient of the bearing acts upon the rotor due to the large outer surface of the bearing. It improves the rotor stability against the oil whirl. Moreover, the amplitude of the rotor response and induced airborne noises are reduced. However, the measure needs more oil volumetric flow rate in the bearings, and more blowby results from the corresponding large seal rings. Blowby is the volumetric flow rate of the charge air and exhaust gas leaking through the seal rings into oil in the bearing housing. Additionally, the large diameter ratio strongly reduces the ring speed ratio in Eqs. (6.92a, 6.92b and 6.92c). Hence, it can lessen the main advantage of low bearing friction of the RFRB. At an excessively large D_o, the stiffness and damping coefficients of the outer oil film reduce; therefore, the outward move of the bearing ring in the outer bearing clearance is larger. Hence, bearing contact wear could occur between the shaft and inner bearing; and possibly between the outer bearing ring and bearing housing.

- **Ratio *Lo/Li***

The larger the length ratio L_o/L_i at a small inner bearing width L_i is, the more the damping coefficient of the bearing acts upon the rotor due to the large outer surface of the bearing. It improves rotor stability against the oil whirl. Additionally, this measure reduces the rotor response and airborne noises. Moreover, it needs less oil volumetric flow rate through the bearings due to the increased axial flow resistance in the outer oil film. However, the large length ratio reduces the ring speed ratio in Eqs. (6.92a, 6.92b), and (6.92c). Hence, it could lessen the main advantage of the low bearing friction of the RFRB. The wider the outer width of the bearing L_o is, the shorter the bearing span is. That reduces rotor stability of the rotor. Similarly, at an excessively wide L_o, the bearing stiffness coefficient of the outer oil film increases. Therefore, the outward move of the bearing ring in the outer bearing clearance is more difficult and smaller. Hence, bearing contact wear could occur between the shaft and inner bearing.

- **Ratio c_2/c_1**

The larger the clearance ratio c_2/c_1 at a small inner bearing clearance c_1 is, the more the damping coefficient of the bearing is due to the large outer bearing clearance. It improves rotor stability and reduces the rotor response and airborne noises, especially the constant tone. Moreover, the large clearance ratio at the given inner bearing clearance increases the ring speed ratio because the bearing friction in the outer ring surface reduces in Eqs. (6.92a, 6.92b and 6.92c). However, it needs more oil flow rate due to the larger outer bearing clearance. At an excessively large outer bearing clearance c_2, the bearing stiffness coefficient of the outer oil film decreases. It causes a large rotor response with the conical vibration mode in the outer oil film leading to contact wear between the outer bearing ring and bearing housing.

Note that the typical ratios of the two oil film bearings should be considered under various operating conditions, such as the oil type, oil inlet temperature and pressure, rotor speed range, kind of the rotor balancing, etc.

6.6 Rolling Element Bearings

There are some common types of the rolling element bearings [7]:

- Angular contact ball bearings;
- Ball thrust bearings;
- Straight roller bearings;
- Tapered roller bearings;
- Needle roller bearings.

This section only focuses on two normal types of the rolling element bearings that could be applied to automotive turbochargers: angular contact ball and straight roller bearings, as shown in Fig. 6.29. The other types of rolling element bearings are either similar to the ball and roller bearings or have already been existing in [4, 7, 14].

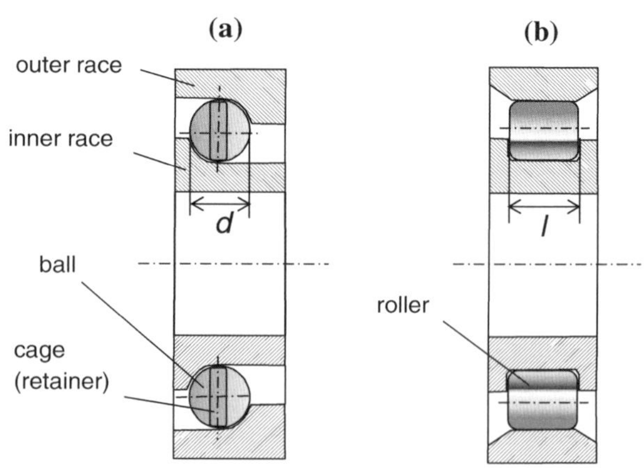

Fig. 6.29 **a** Angular contact ball bearing; **b** straight roller bearing

6.6.1 Characteristics of the Rolling Element Bearings

At first, some advantages and disadvantages of the angular contact ball bearings applied to automotive turbochargers are reviewed [2]. Ball bearings have no intrinsic damping due to very small clearances between the balls and races. As a result, they need an additional damping property to suppress the rotor amplitude at resonance and to reduce the airborne noises, such as unbalance whistle and constant tone (howling). In this case, a squeeze film damper is unconditionally necessary for ball bearings of the automotive turbochargers.

Contrary to the oil film bearings, they do not have the inner oil whirl; therefore, no constant tone occurs However, they usually have the multiple harmonic vibrations of 2X, 3X, 4X, or higher frequency orders and modulation sidebands frequencies due to unsuitable clearance of the oil damper between the outer race and bearing housing. Moreover, some asynchronous vibrations could occur in the ball bearings if defects of balls, cage, inner, and outer races take place. Such high-frequency vibration components induce high-frequency noises in the ball bearings that are uncomfortable and unwanted audible noises in the passenger vehicles.

Unbalance of the rolling elements, cage, and outer race cannot be directly balanced; however, it is very small. Sometimes unusual asynchronous responses could happen if the balls have slightly different diameters due to bearing wear and damage.

The bearing stiffness coefficient of the ball and roller bearings depends on the acting forces and other bearing characteristics.

Rolling element deflection

According to the Hertzian theory, the elastic deformation at the contact surface between the balls and races results as

$$\delta = \sqrt[3]{\frac{2.25(1-\nu^2)^2 F^2}{E^2 R}} \equiv c_K(E, \nu, r_1, r_2, \ldots)\ F^{2/3} \tag{6.93}$$

where

ν is the Poisson's ratio;
E is the elasticity module;
F is the acting force on the balls;
r_i is the radii of the ball bearing geometry;
R is the equivalent radius;
c_K is the deformation factor.

Using the elasticity module $E = 206$ GPa and Poisson's ratio $\nu = 0.3$, one obtains the deformation of the rolling elements empirically [8].

- *Ball bearings*:

$$\delta = 4.37 \times 10^{-4} F^{2/3} d^{-1/3} \tag{6.94}$$

where
δ is the ball deformation (mm);
d is the ball diameter (mm);
F is the acting force on the ball bearing (N).

- *Roller bearings*:

$$\delta = 0.77 \times 10^{-4} F^{0.9} l^{-0.8} \tag{6.95}$$

where
δ is the roller deformation (mm);
l is the roller effective length (mm);
F is the acting force on the roller bearing (N).

The equivalent radius R between the balls and inner race are given for various geometries of convex, plane, and concave contact surfaces in the ball bearings, as shown in Fig. 6.30 [8].

(a) Bearing force and its stiffness
The bearing force is empirically determined by the rolling element deformation, number of rolling elements, and some bearing parameters depending on the bearing types [8].

$$F = Sz \cdot \left(\frac{\delta}{C}\right)^n \tag{6.96}$$

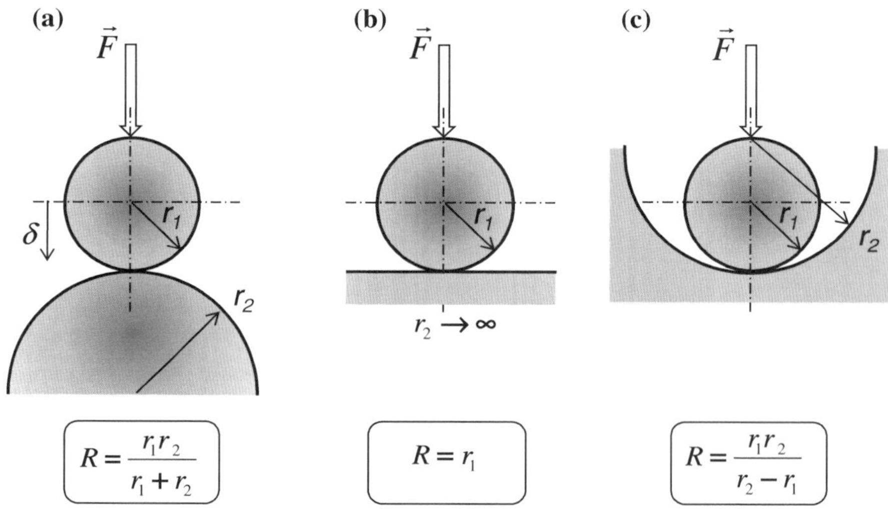

Fig. 6.30 Equivalent radius R of various contact surfaces **a** convex, **b** plane, **c** concave

where

z is the number of rolling elements (balls or rollers);
δ is the rolling element deformation (mm) in direction of the acting force F (N);
$S = 0.23$ (ball bearings); $= 0.24$ (roller bearings);
$n = 3/2$ (ball bearings); $= 10/9$ (roller bearings);
$C = 4.37 \times 10^{-4} d^{-1/3}$ (mm $N^{-2/3}$) for ball bearings with d (mm) of the ball diameter;
$C = 0.77 \times 10^{-4} l^{-0.8}$ (mm $N^{-0.9}$) for roller bearings with l (mm) of the roller effective length.

Combining the bearing force in Eq. (6.96) with the bearing parameter C, one obtains the bearing force in a function of rolling element deformation for the ball and roller bearings (see Fig. 6.31).

- *Ball bearings*:

$$F = 2.5 \times 10^4 z d^{0.5} \delta^{1.5} \text{ in N with } d \text{ (mm) and } \delta \text{ (mm).}$$

- *Roller bearings*:

$$F = 0.89 \times 10^4 z l^{0.9} \delta^{1.12} \text{ in N with } d \text{ (mm) and } \delta \text{ (mm).}$$

The rolling element deformation is derived from Eq. (6.96).

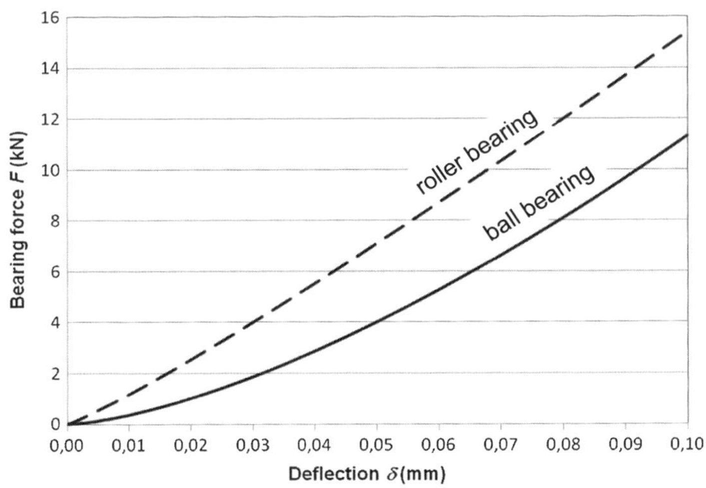

Fig. 6.31 Bearing forces versus deformation ($z = 8$, $d = l = 3.2$ mm)

$$\delta = C \cdot \left(\frac{F}{Sz}\right)^{\frac{1}{n}} \tag{6.97}$$

Inserting the bearing parameters C and S into Eq. (6.97), the rolling element deformation results in a function of the acting force for

- *Ball bearings*:

$$\delta = 1.2 \times 10^{-3} d^{-1/3} \left(\tfrac{F}{z}\right)^{2/3} \text{ in mm with } d \text{ (mm) and } F \text{ (N);}$$

- *Roller bearings*:

$$\delta = 0.28 \times 10^{-3} l^{-0.8} \left(\tfrac{F}{z}\right)^{0.9} \text{ in mm with } l \text{ (mm) and } F \text{ (N).}$$

The bearing stiffness coefficient results from the partial derivative of force with respect to deformation using Eqs. (6.96 and 6.97).

$$\begin{aligned} k &\equiv \left(\frac{\partial F}{\partial \delta}\right) \\ &= \frac{n}{C^n} S z \delta^{n-1} = \frac{n}{C} (S z)^{1/n} F^{\left(\frac{n-1}{n}\right)} \end{aligned} \tag{6.98}$$

Substituting the parameters C and S for each bearing type into Eq. (6.98), one obtains the bearing stiffness for

- *Ball bearings*:

$$k = 1.3 \times 10^3 z^{2/3} d^{1/3} F^{1/3} \text{ in N/mm; } d \text{ (mm); } F \text{ (N);}$$

- *Roller bearings*:

$$k = 4 \times 10^3 z^{0.9} l^{0.8} F^{0.1} \text{ in N/mm; } l \text{ (mm); } F \text{ (N).}$$

Figure 6.32 displays the bearing stiffness coefficients of the ball and roller bearings in a function of acting force in case of $z = 8$ rolling elements, the ball diameter d, and roller length l of 3.2 mm.

(b) Kinematics of the rolling element bearing

To calculate the angular velocities of the cage ω_c and rolling element ω_b (ball or roller), a simple bearing model is used in Fig. 6.33 [14, 15].

The inner race is fixed to the shaft, and it rotates with the rotor velocity ω_i; the outer race rotates with an angular velocity ω_o.

The rolling element contacts the inner and outer races at the radius r_i and r_o, respectively.

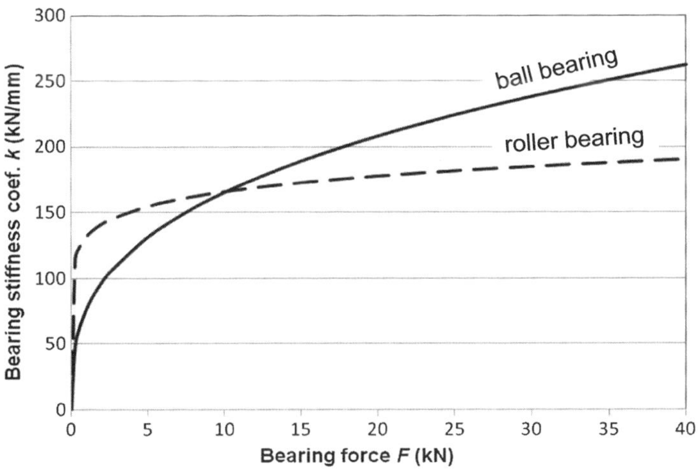

Fig. 6.32 Bearing stiffness coefficients versus acting force ($z = 8$, $d = l = 3.2$ mm)

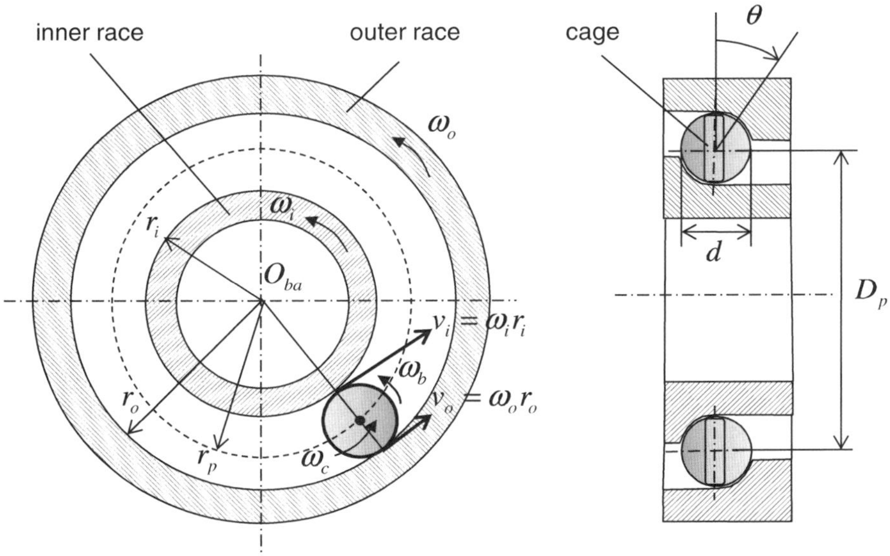

Fig. 6.33 Model of a ball bearing with the angular velocities

The pitch diameter D_p is defined as the maximum diametral distance of the centers between two rolling elements.

The angular velocity of the bearing cage ω_c about the bearing axis O_{ba} is written in

$$\omega_c = \frac{1}{2}\frac{(v_i + v_o)}{r_p} = \frac{v_i + v_o}{D_p} \tag{6.99}$$

The circumferential velocity v_i at the contact area between the inner race and the rolling element results as

$$v_i = \omega_i r_i = \frac{\omega_i}{2}(D_p - d\cos\theta) = \frac{D_p}{2}\omega_i\left(1 - \frac{d}{D_p}\cos\theta\right) \tag{6.100}$$

Similarly, the circumferential velocity v_o at the contact area between the outer race and the rolling element is written as

$$v_o = \omega_o r_o = \frac{\omega_o}{2}(D_p + d\cos\theta) = \frac{D_p}{2}\omega_o\left(1 + \frac{d}{D_p}\cos\theta\right) \tag{6.101}$$

Substituting Eqs. (6.99, 6.100, and 6.101), one obtains the angular velocity of the cage.

$$\omega_c = \frac{1}{2}\left[\omega_i\left(1 - \frac{d}{D_p}\cos\theta\right) + \omega_o\left(1 + \frac{d}{D_p}\cos\theta\right)\right] \tag{6.102}$$

Similarly, one calculates the angular velocity ω_b of the rolling element about its own axis without spinning and slipping at the races.

$$\omega_b = \frac{v_i - v_o}{d} \tag{6.103}$$

where the velocity direction of ω_b is the same direction of $(v_i - v_o)$.

Substituting Eqs. (6.100 and 6.101) into Eq. (6.103), one obtains the angular velocity of the rolling element.

$$\omega_b = \frac{1}{2}\frac{D_p}{d}\left[\omega_i\left(1 - \frac{d}{D_p}\cos\theta\right) - \omega_o\left(1 + \frac{d}{D_p}\cos\theta\right)\right] \tag{6.104}$$

In a cylinder roller bearing with $\theta = 0$, the angular velocities of the cage and rolling element become obviously at $\cos\theta = 1$ in Eqs. (6.102 and 6.104).

$$\omega_c = \frac{1}{2}\left[\omega_i\left(1 - \frac{d}{D_p}\right) + \omega_o\left(1 + \frac{d}{D_p}\right)\right] \tag{6.105}$$

and

$$\omega_b = \frac{1}{2}\frac{D_p}{d}\left[\omega_i\left(1 - \frac{d}{D_p}\right) - \omega_o\left(1 + \frac{d}{D_p}\right)\right] \tag{6.106}$$

Additionally, at a nonrotating outer race of the bearing ($\omega_o = 0$), the angular velocities of the cage results from Eqs. (6.105).

$$\omega_c = \frac{\omega_i}{2}\left(1 - \frac{d}{D_p}\right) < \frac{\omega_i}{2} \tag{6.107}$$

It indicates that the angular velocity of the cage of the bearing is always less than a half of the rotor velocity.

The angular velocity of the rolling element becomes at $\omega_o = 0$ for a nonrotating outer race.

$$\omega_b = \frac{\omega_i}{2}\left(\frac{D_p}{d} - 1\right) \tag{6.108}$$

In case of $D_p/d \geq 2$, the angular velocity of the rolling element about its own axis is larger than or equal to a half of the rotor velocity.

$$\omega_b = \frac{\omega_i}{2}\left(\frac{D_p}{d} - 1\right) \geq \frac{\omega_i}{2} \tag{6.109}$$

6.6.2 Squeeze Film Damper with Ball Bearings

As discussed earlier, the rolling element bearings have no intrinsic damping; therefore, they need an additionally damping measure in turbochargers. In fact, they are usually integrated in the squeeze film damper (diameter D, and length L) where the rotor shaft is built into the bearings at the inner races. The squeeze film damper is floating, but nonrotating in the bearing housing with the squeeze oil film of a radial clearance c, as shown in Fig. 6.34. In case of a nonrotating damper, its whirl velocity ω equals zero.

Figure 6.34 shows a setup of ball bearings in application for the automotive turbochargers. Both ball bearings are built at the ends of a squeeze film damper

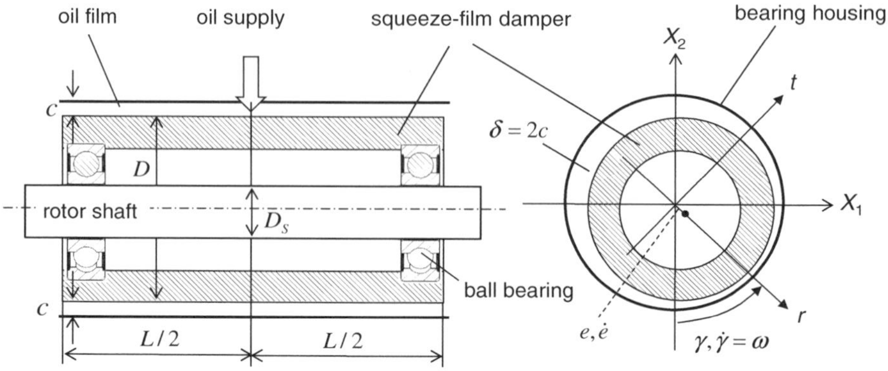

Fig. 6.34 Squeeze film damper with ball bearings for turbochargers

(called *bearing cartridge*) in which the rotor shaft is supported on the inner races of the bearings. The oil film between the bearing housing and bearing cartridge provides the bearing system with an external damping using the squeeze effect of the oil film to keep the rotor stable and to reduce the induced noises of the turbochargers.

In case of *no-end oil damper* where the damping oil continually flows outside through the squeeze film, the theory of short bearings is generally used to calculate the damping forces. There are four cases: non-cavitating and cavitating squeeze oil films, each with *no-end* and *end oil damper*. The 2π-theory is applied to the first case (non-cavitating); the π-theory is for the second case (cavitating).

Case 1:
No-end oil damper with non-cavitating squeeze oil film (2π-Theory)
In this case, the cross-coupled damping coefficients d_{rt} and d_{tr} equal zero. The damping force results from the short bearing theory [3].

$$F_d = \begin{pmatrix} F_r \\ F_t \end{pmatrix} = -\begin{bmatrix} d_{rr} & 0 \\ 0 & d_{tt} \end{bmatrix} \begin{pmatrix} \dot{e} \\ e\dot{\gamma} \end{pmatrix} \tag{6.110}$$

where

d_{rr} is the damping coefficient in the normal direction r;
d_{tt} is the damping coefficient in the tangential direction t;
$d_{rt} = d_{tr} = 0$, the cross-coupled damping coefficients;
$\dot{e}$ is the eccentricity velocity of the damper;
$\dot{\gamma}$ is the whirl velocity ω of the damper ($\omega = 0$ for nonrotating).

The damping coefficients of the oil damper are written as

$$d_{rr} = 4\pi\eta D\left(\frac{L}{\delta}\right)^3\left[\frac{1+2\varepsilon^2}{(1-\varepsilon^2)^{5/2}}\right] \tag{6.111}$$

$$d_{tt} = 4\pi\eta D\left(\frac{L}{\delta}\right)^3\left[\frac{1}{(1-\varepsilon^2)^{3/2}}\right] \tag{6.112}$$

within

η is the oil dynamic viscosity;
δ is the diametral damper clearance;
L is the damper length;
D is the damper diameter;
ε is the relative eccentricity of the damper ($\varepsilon = e/c$).

Thus, the damping forces of the oil damper result as

$$F_r = -d_{rr}\dot{e} = -4\pi\eta D\left(\frac{L}{\delta}\right)^3\left[\frac{1+2\varepsilon^2}{(1-\varepsilon^2)^{5/2}}\right]c\dot{\varepsilon} \tag{6.113}$$

$$F_t = -d_{tt}e\dot{\gamma} = -4\pi\eta D\left(\frac{L}{\delta}\right)^3\left[\frac{1}{(1-\varepsilon^2)^{3/2}}\right]c\varepsilon\dot{\gamma} \tag{6.114}$$

Case 2:
No-end oil damper with cavitating squeeze oil film (π-Theory)
In this case, the cross-coupled damping coefficients d_{rt} and d_{tr} are not equal to zero. The damping force results from the short bearing theory [3].

$$F_d = \begin{pmatrix} F_r \\ F_t \end{pmatrix} = -\begin{bmatrix} d_{rr} & d_{rt} \\ d_{tr} & d_{tt} \end{bmatrix}\begin{pmatrix} \dot{e} \\ e\dot{\gamma} \end{pmatrix} \tag{6.115}$$

where

d_{rr} is the damping coefficient in the normal direction r;
d_{tt} is the damping coefficient in the tangential direction t;
d_{rt}, $d_{tr} \neq 0$, the cross-coupled damping coefficients;
$\dot{e}$ is the eccentricity velocity of the damper;
$\dot{\gamma}$ is the whirl velocity ω of the damper ($\omega = 0$ for nonrotating).

In case of a very small eccentricity velocity, the damping coefficients of the oil damper are written as

$$d_{rr} = 2\pi\eta D\left(\frac{L}{\delta}\right)^3\left[\frac{1+2\varepsilon^2}{(1-\varepsilon^2)^{5/2}}\right] \tag{6.116}$$

$$d_{tt} = 2\pi\eta D\left(\frac{L}{\delta}\right)^3\left[\frac{1}{(1-\varepsilon^2)^{3/2}}\right] \tag{6.117}$$

$$d_{rt} = d_{tr} = 4\eta D\left(\frac{L}{\delta}\right)^3\left[\frac{2\varepsilon}{(1-\varepsilon^2)^2}\right] \tag{6.118}$$

within

η is the oil dynamic viscosity;
δ is the diametral damper clearance;
L is the damper length;
D is the damper diameter;
ε is the relative eccentricity of the damper ($\varepsilon = e/c$).

Thus, the damping forces of the oil damper result as

$$\begin{aligned} F_r &= -d_{rr}\dot{e} - d_{rt}e\dot{\gamma} \\ &= -2\pi\eta D\left(\frac{L}{\delta}\right)^3\left[\frac{1+2\varepsilon^2}{(1-\varepsilon^2)^{5/2}}\right]c\dot{\varepsilon} - 4\eta D\left(\frac{L}{\delta}\right)^3\left[\frac{2\varepsilon}{(1-\varepsilon^2)^2}\right]c\varepsilon\dot{\gamma} \end{aligned} \tag{6.119}$$

$$\begin{aligned} F_t &= -d_{tr}\dot{e} - d_{tt}e\dot{\gamma} \\ &= -4\eta D\left(\frac{L}{\delta}\right)^3\left[\frac{2\varepsilon}{(1-\varepsilon^2)^2}\right]c\dot{\varepsilon} - 2\pi\eta D\left(\frac{L}{\delta}\right)^3\left[\frac{1}{(1-\varepsilon^2)^{3/2}}\right]c\varepsilon\dot{\gamma} \end{aligned} \tag{6.120}$$

In case of end oil damper where the damping oil remains in the squeeze oil film closed by an O seal ring, the theory of long bearings is generally used to calculate the damping forces. There are two cases: non-cavitating and cavitating squeeze oil films. The 2π-theory is applied to the first case; the π-theory is for the second case.

Case 3:
End oil damper with non-cavitating squeeze oil film (2π-Theory)
In this case, the cross-coupled damping coefficients d_{rt} and d_{tr} equal zero. The damping force results from the long bearing theory [3].

$$F_d = \begin{pmatrix} F_r \\ F_t \end{pmatrix} = -\begin{bmatrix} d_{rr} & 0 \\ 0 & d_{tt} \end{bmatrix}\begin{pmatrix} \dot{e} \\ e\dot{\gamma} \end{pmatrix} \tag{6.121}$$

where

d_{rr} is the damping coefficient in the normal direction r;
d_{tt} is the damping coefficient in the tangential direction t;
$d_{rt} = d_{tr} = 0$, the cross-coupled damping coefficients;
$\dot{e}$ is the eccentricity velocity of the damper;
$\dot{\gamma}$ is the whirl velocity ω of the damper ($\omega = 0$ for nonrotating).

The damping coefficients of the oil damper are written as

$$d_{rr} = 12\pi\eta L\left(\frac{D}{\delta}\right)^3\left[\frac{1}{(1-\varepsilon^2)^{3/2}}\right] \quad (6.122)$$

$$d_{tt} = 24\pi\eta L\left(\frac{D}{\delta}\right)^3\left[\frac{1}{(2+\varepsilon^2)(1-\varepsilon^2)^{1/2}}\right] \quad (6.123)$$

within

η is the oil dynamic viscosity;
δ is the diametral damper clearance;
L is the damper length;
D is the damper diameter;
ε is the relative eccentricity of the damper ($\varepsilon = e/c$).

Thus, the damping forces of the oil damper result as

$$F_r = -d_{rr}\dot{e} = -\pi\eta L\left(\frac{D}{\delta}\right)^3\left[\frac{12}{(1-\varepsilon^2)^{3/2}}\right]c\dot{\varepsilon} \quad (6.124)$$

$$F_t = -d_{tt}e\dot{\gamma} = -\pi\eta L\left(\frac{D}{\delta}\right)^3\left[\frac{24}{(2+\varepsilon^2)(1-\varepsilon^2)^{1/2}}\right]c\varepsilon\dot{\gamma} \quad (6.125)$$

Case 4:
End oil damper with cavitating squeeze oil film (π-Theory)
In this case, the cross-coupled damping coefficients d_{rt} and d_{tr} are not equal to zero. The damping force results from the short bearing theory [3].

$$F_d = \begin{pmatrix} F_r \\ F_t \end{pmatrix} = -\begin{bmatrix} d_{rr} & d_{rt} \\ d_{tr} & d_{tt} \end{bmatrix}\begin{pmatrix} \dot{e} \\ e\dot{\gamma} \end{pmatrix} \quad (6.126)$$

where

d_{rr} is the damping coefficient in the normal direction r;
d_{tt} is the damping coefficient in the tangential direction t;
$d_{rt}, d_{tr} \neq 0$ are the cross-coupled damping coefficients;
$\dot{e}$ is the eccentricity velocity of the damper;
$\dot{\gamma}$ is the whirl velocity ω of the damper ($\omega = 0$ for nonrotating).

In case of a zero eccentricity velocity, the damping coefficients of the oil damper are written as

$$d_{tt} = 12\pi\eta L\left(\frac{D}{\delta}\right)^3\left[\frac{1}{(2+\varepsilon^2)(1-\varepsilon^2)^{1/2}}\right] \tag{6.127}$$

$$d_{rt} = d_{tr} = 24\eta L\left(\frac{D}{\delta}\right)^3\left[\frac{\varepsilon}{(2+\varepsilon^2)(1-\varepsilon^2)}\right] \tag{6.128}$$

within

η is the oil dynamic viscosity;
δ is the diametral damper clearance;
L is the damper length;
D is the damper diameter;
ε is the relative eccentricity of the damper ($\varepsilon = e/c$).

Thus, the damping forces of the oil damper result as

$$F_r = -d_{rt}e\dot{\gamma} = -24\eta L\left(\frac{D}{\delta}\right)^3\left[\frac{\varepsilon}{(2+\varepsilon^2)(1-\varepsilon^2)}\right]c\varepsilon\dot{\gamma} \tag{6.129}$$

$$F_t = -d_{tt}e\dot{\gamma} = -12\pi\eta L\left(\frac{D}{\delta}\right)^3\left[\frac{1}{(2+\varepsilon^2)(1-\varepsilon^2)^{1/2}}\right]c\varepsilon\dot{\gamma} \tag{6.130}$$

The damping forces F_r and F_t in the rotating coordinate system (r, t) are transformed into F_1 and F_2 in the inertial coordinate system (X_1, X_2) using Eq. (6.58), cf. Fig. 6.34. The transformed damping forces F_1 and F_2 in the inertial coordinate system (X_1, $X2$) are required for the computation of nonlinear rotordynamics in Chap. 7.

The damping forces in the inertial coordinate system (X_1, X_2) are written as

$$\begin{pmatrix} F_1 \\ F_2 \end{pmatrix} = \begin{pmatrix} \sin\gamma & \cos\gamma \\ -\cos\gamma & \sin\gamma \end{pmatrix}\begin{pmatrix} F_r \\ F_t \end{pmatrix}.$$

6.6.3 Bearing Defect-Related Frequencies

Defects in the rolling element bearings, such as bearing defects in the inner and outer races, cage (retainer), and rolling elements (balls or rollers) generate the asynchronous vibration components with high-frequency orders and sidebands frequencies due to frequency modulations. They are called the bearing defect-related frequencies.

Figure 6.35 shows the key elements and their geometries of an angular ball bearing. The balls with diameter d are held in the cage (retainer); the pitch diameter

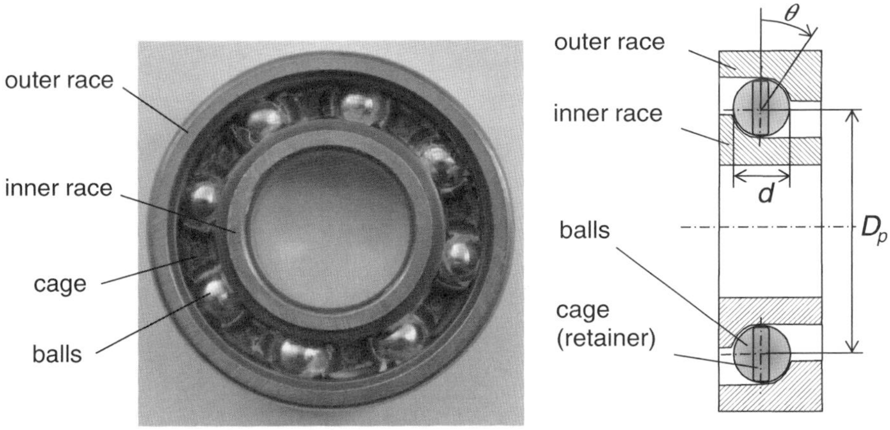

Fig. 6.35 Key elements of an angular contact ball bearing

D_p is defined as the maximum diametral distance between the centers of two balls. The contact angle θ is the angle between the line perpendicular to the shaft and the radius of the ball at the contact point at the outer race. The inner race is fixed to the shaft and it rotates with the rotor speed N (rps); the outer race is mounted in the bearing cartridge supported by the squeeze oil film (see Fig. 6.34).

The bearing defect-related frequencies are computed from the bearing geometries, number of balls, and rotor speed [2, 9, 15].

The bearing defect-related frequencies are displayed in Fig. 6.36 for a ball bearing containing $Z = 8$ balls with $d = 3.2$ mm, $D_p = 12.2$ mm, and $\theta = 45°$. The frequencies of the bearing defects *bpfi*, 2*bsf*, and *bpfo* are supersynchronous ($f > 1X$) and the frequency *ftf* is subsynchronous ($f < 1X$).

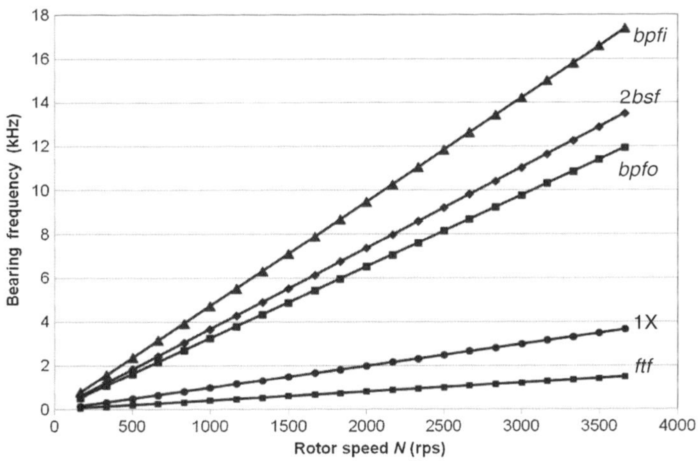

Fig. 6.36 Bearing defect-related frequencies

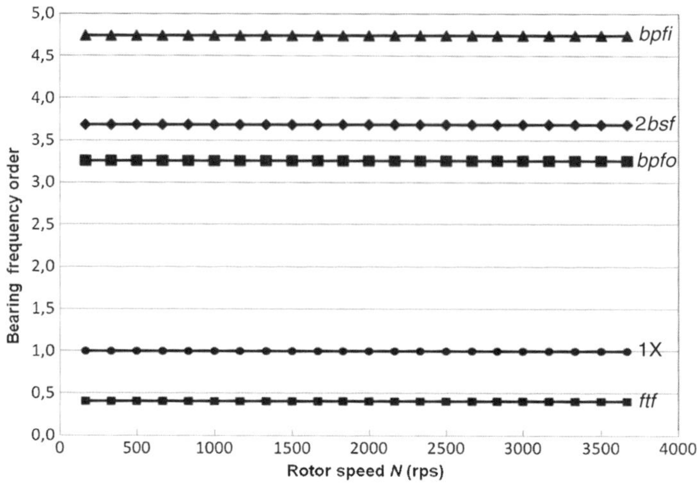

Fig. 6.37 Bearing defect-related frequency orders

- **Fundamental train frequency (*ftf*)**

It is caused by the defects of cage depending on the rotor speed N (rps).

$$ftf = \frac{N}{2}\left(1 - \frac{d}{D_p}\cos\theta\right) \tag{6.131}$$

- **Ball passing frequency over defective inner race (*bpfi*)**

It is generated when the balls or rollers pass over the defective inner race.

$$bpfi = Z\frac{N}{2}\left(1 + \frac{d}{D_p}\cos\theta\right) \tag{6.132}$$

- **Ball passing frequency over defective outer race (*bpfo*)**

It is generated when the balls or rollers pass over the defective outer race.

$$bpfo = Z\frac{N}{2}\left(1 - \frac{d}{D_p}\cos\theta\right) \tag{6.133}$$

- **Two times ball spin frequency (2*bsf*)**

It is induced by the defective balls or rollers spinning over the inner and outer races.

$$2bsf = N\frac{D_p}{d}\left[1 - \left(\frac{d}{D_p}\right)^2 \cos^2\theta\right] \quad (6.134)$$

where Z is the number of rolling elements.

The bearing defect-related frequency orders are displayed in Fig. 6.37 for a ball bearing containing $Z = 8$ balls with $d = 3.2$ mm, $D_p = 12.2$ mm, and $\theta = 45°$. The frequency orders of the bearing defects *bpfi*, *2bsf*, and *bpfo* are larger than 1 (supersynchronous) and the frequency *ftf* is smaller than 1 (subsynchronous).

References

1. DIN 31653—Parts 1, 2, and 3: German industry norm—hydraulic plan thrust bearings under steady-state conditions (1991)
2. Ehrich, F.: Handbook of Rotordynamics. Krieger Publishing Company, Malabar (2004)
3. Gasch, R., Nordmann, R., Pfuetzner, H.: Rotordynamik, 2. Springer, Auflage (2006)
4. Hamrock, B.; Schmid, S.R.; Jacobson, B.O.: Fundamentals of Fluid Film Lubrication, 2nd edn. Marcel Dekker Inc., New York (2004)
5. Hori, Y.: Hydrodynamic Lubrication. Springer, Heidelberg (2006)
6. Kennedy et al.: Tribology, Lubrication, and Bearing Design—The CRC Handbook of Mechanical Engineers. CRC Press, Boca Raton (1988)
7. Khonsari, M., Booser, E. (2008): Applied Tribology and Bearing Design and Lubrication, 2nd edn. Wiley, New Jersey (2008)
8. Kraemer, E.: Rotordynamics of Rotors and Foundations. Springer, Heidelberg (1993)
9. Mate, C.M.: Tribology on the Small Scale. Oxford University Press, Oxford (2008)
10. Nguyen-Schäfer, H.: Internal program CAF (Computing Axial Forces). BMTS (2008)
11. Nguyen-Schäfer, H., Di Giandomenico, G.: Internal program TBD (Thrust Bearing Design). BMTS (2009)
12. Nguyen-Schäfer, H., Di Giandomenico, G.: Internal program RFRB (Rotating Floating Ring Bearings). BMTS (2009)
13. Nguyen-Schäfer, H., Sprafke, P.: Numerical study on interaction effects of the bubbles induced by air-release and cavitation in hydraulic systems. In: 10th Bath International Fluid Power Workshop, Bath, UK. Research Studies Press (1997)
14. Szeri, A.Z.: Fluid Film Lubrication. Cambridge University Press, Cambridge (1998)
15. Taylor, J., Kirkland, D.W.: The Bearing Analysis Handbook. Vibration Consultants Inc., Tampa (2004)
16. Vance, J.: Rotordynamics of Turbomachinery. Wiley, New York (1988)

Chapter 7
Nonlinear Rotordynamics of Turbochargers

7.1 Boundary Conditions of Rotordynamics

Some uncertain boundary conditions play a key role for the stability and functionality of the rotor, as shown in Fig. 7.1. Unfortunately, it is very difficult or impossible to take all of them into account in the rotordynamic computation where only the rotor containing the compressor, turbine wheels, rotor shaft, radial bearings, and seal rings is considered. The boundary conditions are generally assumed ideal in the computation, such as sufficient supply oil, good parallelism of the thrust bearing, all wedges of the thrust bearing having the same slope, good quality of the radial bearing in terms of the bearing non-coaxiality, surface roundness, and characteristics of surface roughness. Note that the nonlinear rotordynamic responses of the rotor strongly depend on the initial and boundary conditions. For this reason, a little change of these conditions could bring the rotor response from a stable to an unstable state and vice versa. Therefore, the rotordynamic computation cannot exactly predict the instability and malfunctions of the rotor, or prevent it from damage at inappropriate boundary conditions, like oil insufficiency, contaminated oil with hard particles, foamy oil, or oil coking in the bearing during operation. Nevertheless, the computational results help us better understand the rotor response. Further improvements could be done for the rotor stability and reduction of the bearing friction. Additionally, experimental measurements also provide us with the real rotor response at such uncertain boundary conditions that could not be considered in the computation.

In the following section, some typical uncertain boundary conditions and their consequences in the automotive turbochargers are discussed.

If the compressor wheel speed takes turns changing from the high circumferential speeds ($U > 560$ m/s) to low speeds and vice versa during operation with a driving cycle for a long-time interval. The ultimate tensile stress of the compressor reduces with the operating cycles (about 10^5 cycles); it is called low-cycle fatigue (LCF) leading to damage of the compressor wheel. The partly broken compressor

H. Nguyen-Schäfer, *Rotordynamics of Automotive Turbochargers*,
Springer Tracts in Mechanical Engineering, DOI 10.1007/978-3-319-17644-4_7

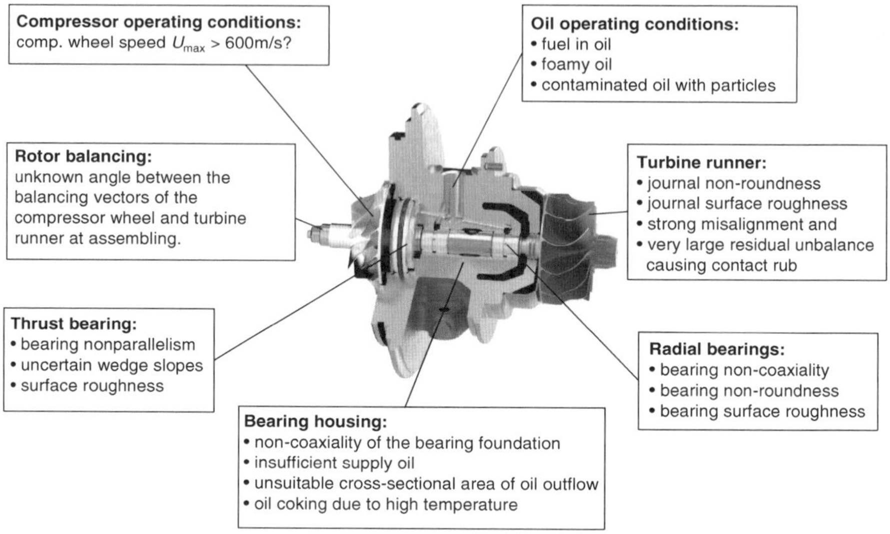

Fig. 7.1 Uncertain boundary conditions in the rotordynamic computation

wheel induces extremely large unbalances that cause the rotor instability and finally damage the bearings.

In case of oil insufficiency, the bearings operate in a poorly lubricated oil condition; hence, that causes the rotor instability and damages the bearings due to wear and seizure. Moreover, contaminated oil with hard particles or diluted oil with fuel and water causes wear in the bearing and reduces the rotor stability of the turbocharger.

Improper balancing of turbochargers induces excessively large unbalances that exceed the load capability of the bearings leading to bearing wear and seizure of the shaft in the bearing. Finally, the bearings are damaged and there is failure. In this case, asynchronous noises induced by contact between the shaft and bearings are strongly audible from the turbocharger.

Misaligned turbine shaft due to production mistake induces the dynamic unbalance of the rotor. It causes high rotor amplitudes due to the conical mode of the vibration response.

The non-roundness of the journal in bearings and its surface roughness cause unsuitable pressure distributions in the bearing that induce the rotor instability and bearing wear. Furthermore, the non-coaxiality of the bearing foundation is to blame for the rotor misalignment leading to the rotor instability and inducing high-order frequencies.

Finally, the oil outflow in the bearing housing is very important for the bearing functionality and stability. If the foaming oil blocks the bearing outlet and does not leave the bearing outlet quickly, the effective oil temperature increases inside the bearing causing the decrease of oil viscosity. Hence, the bearing stiffness and damping coefficients drop leading to the rotor instability and oil coking in the bearings as well.

7.2 Vibration Equations of the Rotor with RFRBs

Figure 7.2 displays the rotor of an automotive turbocharger including the rotor shaft, compressor wheel, turbine wheel, rotating floating ring bearings (RFRBs), seal rings, thrust rings. All components must be taken into account in the rotor-dynamic computation to study the vibration response of the rotor, such as the frequency components in Waterfall diagram, shaft orbit in the phase plane, and rotor response in the time diagram.

The continuum vibration equations of the rotor are discretized by two methods, the finite element and transfer matrix methods. Taking discretization of the vibration equations, the rotor is divided into a finite number of elements in the computational model including concentrated masses, cylindrical elements, disks, and interfaces between the rotor, bearings, and seal rings.

First, the finite element method, a discretization method of the continuum structure of the rotor, is based on the principle of D'Alembert in which the sum of virtual work of all rotor elements equals zero, as shown in Sect. 8.4.1. Using the Lagrange's equations (see Chap. 5), the vibration equations of the rotor are derived and expressed in a discrete matrix equation.

Second, another alternative method is the transfer matrix method developed by Myklestad and Prohl. The first finite element of the rotor is connected with the last one by the overall transfer matrix that consists of the transfer matrices of the corresponding rotor elements. The main advantage of this method is to spare the computational working memory compared to the finite element method using the Lagrange's equations. Both methods are not discussed further in detail here because one can find them in other literatures, e.g., in [9, 21].

The vibration equations of the rotor with N degrees of freedom are written in the discrete matrix equation (see Fig. 7.2) [4–7, 12].

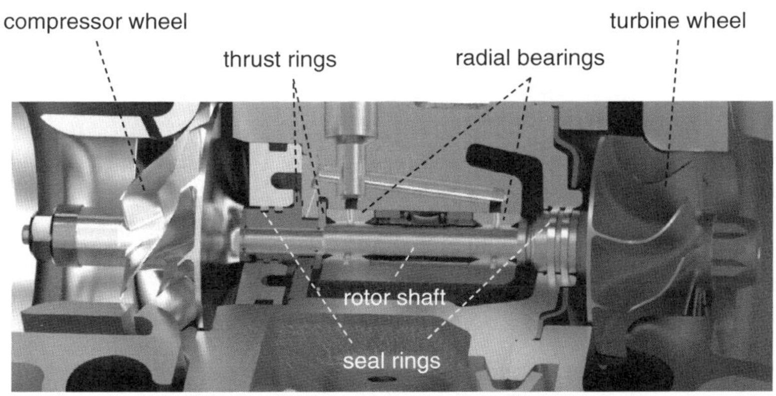

Fig. 7.2 Setup of the rotor of an automotive turbocharger (Courtesy BMTS)

$$\mathbf{M}\ddot{\mathbf{x}} + \mathbf{C}_{SG}\dot{\mathbf{x}} + \mathbf{K}_S\mathbf{x} = \mathbf{f}(t) \tag{7.1}$$

where

N	is the degrees of freedom (DOF) of the whole rotor;
$\mathbf{M} = (N \times N)$	mass matrix containing masses and mass inertia moments of the rotor with m finite elements, cf. Eq. (5.53). The sparse matrix $\mathbf{M}$ consists of m submatrices that are in the band diagonal of $\mathbf{M}$ with almost nonzero elements [9, 21].
$\mathbf{C}_{SG} = (N \times N)$	damping coefficient and gyroscopic sparse matrix of the rotor;
$\mathbf{K}_S = (N \times N)$	stiffness coefficient sparse matrix containing diagonal and cross-coupled stiffness coefficients of the shaft;
$\mathbf{x} = (N \times 1)$	rotor response vector including all DOF of the rotor with two translational and two rotational displacements at each station of the element;
$\mathbf{f}(t) = (N \times 1)$	vector of the unbalance and nonlinear bearing forces.

To reduce the second-order to first-order equation system, the $(2N \times 1)$ vector $\mathbf{z}$ of the vibration response is defined as

$$\begin{aligned} \mathbf{z} &\equiv \begin{bmatrix} \mathbf{x} \\ \mathbf{y} \end{bmatrix} = \begin{bmatrix} \mathbf{x} \\ \dot{\mathbf{x}} \end{bmatrix} \\ \Rightarrow \dot{\mathbf{z}} &= \begin{bmatrix} \dot{\mathbf{x}} \\ \dot{\mathbf{y}} \end{bmatrix} = \begin{bmatrix} \dot{\mathbf{x}} \\ \ddot{\mathbf{x}} \end{bmatrix} = \begin{bmatrix} \mathbf{y} \\ \dot{\mathbf{y}} \end{bmatrix} \end{aligned} \tag{7.2}$$

Substituting the vector $\mathbf{z}(t)$ into Eq. (7.1) and using Eq. (7.2), the $2N$ first-order vibration equations of the rotor result as

$$\begin{aligned} \begin{bmatrix} \dot{\mathbf{x}} \\ \ddot{\mathbf{x}} \end{bmatrix} &= \begin{bmatrix} \mathbf{0} & \mathbf{I} \\ -\mathbf{M}^{-1}\mathbf{K}_S & -\mathbf{M}^{-1}\mathbf{C}_{SG} \end{bmatrix} \begin{bmatrix} \mathbf{x} \\ \dot{\mathbf{x}} \end{bmatrix} + \begin{bmatrix} \mathbf{0} \\ \mathbf{M}^{-1}[\mathbf{U}(\Omega) + \mathbf{F}_B(\mathbf{x}, \dot{\mathbf{x}}, t)] \end{bmatrix} \\ \Leftrightarrow \dot{\mathbf{z}} &= \begin{bmatrix} \mathbf{0} & \mathbf{I} \\ -\mathbf{M}^{-1}\mathbf{K}_S & -\mathbf{M}^{-1}\mathbf{C}_{SG} \end{bmatrix} \mathbf{z} + \begin{bmatrix} \mathbf{0} \\ \mathbf{M}^{-1}[\mathbf{U}(\Omega) + \mathbf{F}_B(\mathbf{x}, \dot{\mathbf{x}}, t)] \end{bmatrix} \end{aligned} \tag{7.3}$$

where

$\mathbf{U}$	is the unbalance force and moment;
$\mathbf{F}_B$	is the nonlinear bearing force computed by Eqs. (6.83 and 6.84) or results from the impedance table.

The coupled nonlinear vibration equation system (7.3) is written in the simple equation matrix of $\mathbf{z}$.

$$\dot{\mathbf{z}} = \mathbf{A}\mathbf{z} + \mathbf{b}(\mathbf{z}, \Omega, t) \tag{7.4}$$

where

A = (2*N* × 2*N*) sparse matrix of mass, mass inertia moments, stiffness, and gyroscopic effect of the rotor;

b = (2*N* × 1) vector of external forces including unbalance and nonlinear bearing forces within

$$\mathbf{A} = \begin{bmatrix} \mathbf{0} & \mathbf{I} \\ -\mathbf{M}^{-1}\mathbf{K}_S & -\mathbf{M}^{-1}\mathbf{C}_{SG} \end{bmatrix};$$

$$\mathbf{b}(\mathbf{z}, \Omega, t) = \begin{bmatrix} \mathbf{0} \\ \mathbf{M}^{-1}[\mathbf{U}(\Omega) + \mathbf{F}_B(\mathbf{z}, t)] \end{bmatrix}$$

The sparse matrix **A** containing the matrices **M**, $\mathbf{K}_S$, and $\mathbf{C}_{SG}$ is obviously non-symmetric; i.e., $\mathbf{A} \neq \mathbf{A}^\mathbf{T}$ where $\mathbf{A}^\mathbf{T}$ is the transposed matrix of **A**. As a result, the bimodal method [16] is applied to solve the first-order nonlinear equation matrix of **z** with a (2*N* × 2*N*) large sparse matrix **A** of *N* DOF of the whole rotor.

The eigenvalue λ of the vibration system is found solving the characteristic equation of the homogeneous vibration Eq. (7.4).

$$(\mathbf{A} - \lambda\mathbf{I})\mathbf{r} = \mathbf{0} \tag{7.5}$$

where **r** is defined by the right eigenvector.

Changing the matrix **A** into $\mathbf{A}^\mathbf{T}$ in the homogeneous Eq. (7.4), one obtains

$$\dot{\mathbf{z}} - \mathbf{A}^T\mathbf{z} = \mathbf{0} \tag{7.6}$$

With $\mathbf{z} = \mathbf{l}e^{\lambda t}$ one gets the characteristic equation of Eq. (7.6).

$$\left(\mathbf{A}^T - \lambda\,\mathbf{I}\right)\mathbf{l} = \mathbf{0} \tag{7.7}$$

where **l** is defined by the left eigenvector.

In the following section, it proves that the right and left eigenvectors of Eqs. (7.5 and 7.7) are orthogonal; that means

$$(\mathbf{r}) \cdot (\mathbf{l}^T) = (\mathbf{l}) \cdot (\mathbf{r}^T) = \mathbf{I}$$

with the unit matrix **I**.

The left eigenvector $\mathbf{l_i}$ corresponding to the eigenvalue λ_i results from Eq. (7.7).

$$\mathbf{A}^T\mathbf{l}_i = \lambda_i\mathbf{l_i} \tag{7.8}$$

Multiplying both sides of Eq. (7.8) by $\mathbf{r}_j^T$, one obtains

$$\mathbf{r}_j^T\mathbf{A}^T\mathbf{l}_i = \lambda_i\mathbf{r}_j^T\mathbf{l_i} \tag{7.9}$$

Having transposed both sides of Eq. (7.9), it gives

$$\mathbf{l}_i^T \mathbf{A} \mathbf{r}_j = \lambda_i \mathbf{l}_i^T \mathbf{r}_j \tag{7.10}$$

Similar to the left eigenvector, one obtains the corresponding relation for the right eigenvector.

$$\mathbf{l}_i^T \mathbf{A} \mathbf{r}_j = \lambda_j \mathbf{l}_i^T \mathbf{r}_j \tag{7.11}$$

The orthogonality of the eigenvectors is given by substituting Eqs. (7.10 and 7.11).

$$(\lambda_i - \lambda_j) \mathbf{l}_i^T \mathbf{r}_j = 0 \tag{7.12}$$

since

$$\begin{aligned} i \neq j &\Rightarrow \mathbf{l}_i^T \mathbf{r}_j = 0; \\ i = j &\Rightarrow \mathbf{l}_i^T \mathbf{r}_i = a_i^2 \neq 0 \end{aligned}$$

Normalizing the eigenvectors by a_i given in Eq. (7.12), the orthonormality of the normalized right and left eigenvectors is written using Kronecker delta as

$$\mathbf{l}_i^{*T} \mathbf{r}_j^* = \delta_{ij} = \begin{cases} 0, & i \neq j \\ 1, & i = j \end{cases} \tag{7.13}$$

where

$$\mathbf{l}_i^{*T} = \frac{\mathbf{l}_i^T}{a_i}; \ \mathbf{r}_j^* = \frac{\mathbf{r}_j}{a_i}.$$

The modal right and left matrices **R** and **L** result from the corresponding normalized eigenvectors **l*** and **r***.

$$\mathbf{R} = \left(\mathbf{r}_j^*\right); \ \mathbf{L} = \left(\mathbf{l}_i^*\right); \quad i, j = 1, \ldots, 2N \tag{7.14}$$

The new response vector $\mathbf{w}(t)$ is defined so that

$$\mathbf{z}(t) = \mathbf{R}\mathbf{w}(t) \tag{7.15}$$

Inserting Eq. (7.15) into Eq. (7.4) and multiplying both sides by the transposed left modal matrix $\mathbf{L}^\mathbf{T}$, one obtains the vibration equation of the rotor.

$$(\mathbf{L}^\mathbf{T}\mathbf{R})\dot{\mathbf{w}}(t) = (\mathbf{L}^\mathbf{T}\mathbf{A}\mathbf{R})\mathbf{w}(t) + \mathbf{L}^\mathbf{T}\mathbf{b}(\mathbf{w}, t) \tag{7.16}$$

Due to the orthonormality of the right and left normalized eigenvectors, as proved in Eq. (7.13), one obtains two relations between the matrices **L** and **R** as

$$\mathbf{L}^{\mathbf{T}}\mathbf{R} = \mathbf{I} \tag{7.17a}$$

and

$$\begin{aligned}\mathbf{L}^{\mathbf{T}}\mathbf{A}\mathbf{R} &= (\mathbf{l}_i^{*T}\mathbf{A}\mathbf{r}_j^*) = \lambda_j \mathbf{l}_i^{*T}\mathbf{r}_j^* \\ &= \begin{cases} 0, & i \neq j \\ \lambda_j, & i = j \end{cases}\end{aligned} \tag{7.17b}$$

Therefore,

$$\mathbf{L}^{\mathbf{T}}\mathbf{A}\mathbf{R} \equiv \Lambda = \begin{bmatrix} \lambda_1 & 0 & 0 & 0 & 0 \\ 0 & \cdot & 0 & 0 & 0 \\ 0 & 0 & \lambda_k & 0 & 0 \\ 0 & 0 & 0 & \cdot & 0 \\ 0 & 0 & 0 & 0 & \lambda_{2N} \end{bmatrix} \tag{7.18}$$

Obviously, the matrix $\mathbf{\Lambda}$ is a diagonal matrix of the eigenvalues λ.

Substituting Eqs. (7.17a, 7.17b, and 7.18) into Eq. (7.16), one obtains

$$\dot{\mathbf{w}}(t) = \Lambda \mathbf{w}(t) + \mathbf{c}(\mathbf{w}, t) \tag{7.19}$$

where the vector $\mathbf{c}$ on the RHS is defined as

$$\mathbf{c}(\mathbf{w}, t) \equiv \mathbf{L}^{\mathbf{T}}\mathbf{b}(\mathbf{w}, t).$$

Equation (7.19) can be written in $2N$ first-order decoupled nonlinear equations as follows

$$\dot{w}_k(t) = \lambda_k w_k(t) + c_k(w_k, t); \quad k = 1, \ldots, 2N. \tag{7.20}$$

within

$$c_k(w_k, t) = \mathbf{l}_k^{*T}\mathbf{b}(w_k, t) = \sum_{i=1}^{2N} l_{i,k}^* b_i(w_k, t) \tag{7.21}$$

Using Runge–Kutta scheme with high order for solving Eq. (7.20), one obtains the solution $w_k(t)$ for the k degree of freedom and the response vector $\mathbf{w}(t)$.

Then inserting $\mathbf{w}(t)$ into Eq. (7.15), one computes backward the vector $\mathbf{z}(t)$.

$$z_i(t) = \sum_{k=1}^{2N} r_{i,k}^* w_k(t); \quad i = 1, \ldots, 2N \tag{7.22}$$

The unknown vibration response vector $\mathbf{x}(t)$ results from Eq. (7.22) as

$$x_i(t) = z_i(t); \quad i = 1, \ldots, N \tag{7.23}$$

where N is the DOF of the whole rotor.

7.3 Synchronous and Asynchronous Vibrations

The nonlinear bearing forces, as discussed in Chap. 6, are mainly to blame for the rotor responses in nonlinear rotordynamics. The vibration responses of the nonlinear system result from superimposing the asynchronous frequencies of the malfunctions (unbalance, oil whirls, rotor misalignment, excessively large unbalance, contact rub, wear, and sidebands of the frequency modulations) on the unbalance harmonics, as shown in Fig. 7.3. Additionally, the operating conditions of the turbochargers at oil shortage, contaminated oil, or high oil temperatures affect the rotor responses. Depending on the malfunctions, the rotor responses consist of at least two or many frequency components of the unbalance excitation, oil whirls, and sidebands. Therefore, they are periodic or quasiperiodic vibrations that are superimposed by the synchronous and asynchronous frequency components (cf. Chap. 3) [1, 3, 5, 17].

The unbalance force excites the rotor with a harmonic vibration that is synchronous with the rotor speed at the frequency order of 1X. The rotor amplitude reaches a resonance when the eigenfrequency of the rotor equals the rotor frequency. The resonances occur at the intersections of the rotor eigenfrequencies and the synchronous excitation line 1X in Campbell diagram.

Oil whirl is a kind of a self-excited vibration with the frequency order less than 1 (i.e., $\omega/\Omega < 1X$). Oil whirling in the bearing clearance induces the oil whirl. The rotational kinetic energy of the oil whirling excites the rotor and induces the

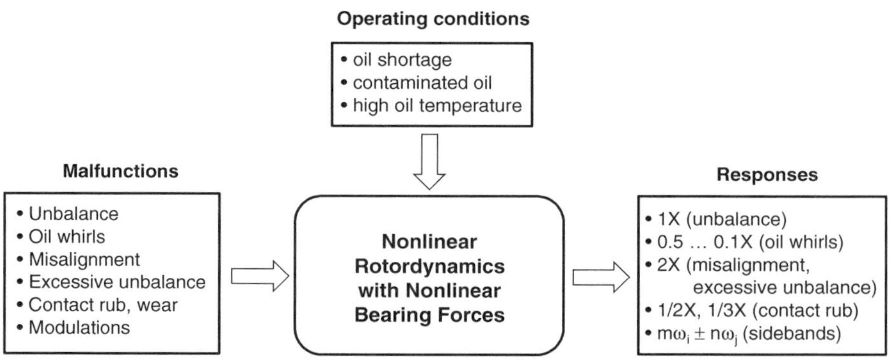

Fig. 7.3 Rotor responses in nonlinear rotordynamics

subsynchronous self-excited vibration. When the whirl frequency ω equals the first critical bending frequency of the rotor, oil whip takes place in which the rotor amplitude continuously increases with time. The oil whip is very dangerous and destructive because the rotor deflection rises with time in the bearing clearance and touches the bearing. It leads to bearing damage due to wear or seizure of the journal in the bearing. The rotor speed where the oil whip begins occurring is called the threshold speed Ω_{th}. In nonlinear rotordynamics, the rotor amplitude is mostly limited on the limit cycle, as discussed in Chap. 4. As long as the limit cycle is smaller the bearing clearance, the bearing contact and wear do not happen.

The oil whirling frequency of the inner oil film (inner oil whirl frequency) results from the rotor speed Ω and bearing ring speed Ω_R (see Sect. 7.5) [8], cf. Fig. 7.4.

$$\omega_i = \lambda_i(\Omega + \Omega_R) = \lambda_i\Omega\,(1 + RSR) \tag{7.24}$$

where

λ_i is the fluid circumferential average velocity ratio of the inner oil film [12];

$RSR \equiv \frac{\Omega_R}{\Omega}$ is the ring speed ratio.

The ring speed ratio of the bearing has been derived in Eq. (6.92).

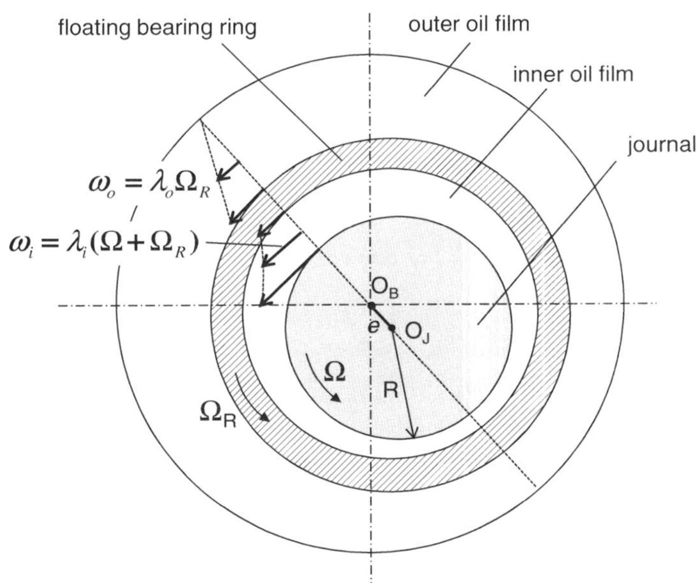

Fig. 7.4 Oil whirl frequencies in the rotating floating radial bearing

$$RSR \approx \frac{1}{1 + \frac{\eta_o(T_o)}{\eta_i(T_i)}\left(\frac{L_o}{L_i}\right)\left(\frac{c_1}{c_2}\right)\left(\frac{D_o}{D_i}\right)^3} \tag{7.25}$$

The frequency order of the inner oil whirl results from Eq. (7.24) in

$$\frac{\omega_i}{\Omega} = \lambda_i(1 + RSR) \tag{7.26}$$

The threshold speed Ω_{th} of the rotor is found setting the inner oil whirl frequency ω_i and equals the critical frequency $\Omega_{cr,1}$.

$$\begin{aligned} \omega_i &= \lambda_i(1 + RSR)\Omega_{th} \\ &= \Omega_{cr,1} = \sqrt{\frac{K_{sys}}{m}} \end{aligned} \tag{7.27}$$

Thus, the threshold speed Ω_{th} of the rotor is written as

$$\Omega_{th} = \frac{1}{\lambda_i(1 + RSR)}\sqrt{\frac{K_{sys}}{m}} \tag{7.28}$$

where K_{sys} is the rotor stiffness coefficient including the shaft and bearing stiffness coefficients.

According to the computation of the ring speed ratio *RSR* in Fig. 6.28, it varies with the oil inlet temperature from 0.30 to 0.05. Therefore, at low rotor speeds with $\lambda_i \approx 0.5$ and $RSR \approx 0.30$, the frequency order of the inner oil whirl results in

$$\frac{\omega_i}{\Omega} \approx 0.7 \text{ at low rotor speeds } \Omega. \tag{7.29a}$$

However, the oil average velocity ratio λ_i reduces at increasing journal eccentricities due to the oil friction in the narrow oil film (see Fig. 7.11). Hence, it can be reduced from 0.5 to zero at high rotor speeds. In this case, the frequency order of the inner oil whirl is calculated with $\lambda_i \approx 0.3$ and $RSR \approx 0.1$.

$$\frac{\omega_i}{\Omega} \approx 0.25\ldots0.30 \text{ at high rotor speeds } \Omega. \tag{7.29b}$$

The oil whirl frequency of the outer oil film (outer oil whirl frequency) results only from the ring speed Ω_R.

$$\omega_o = \lambda_o \Omega_R \tag{7.30}$$

where

λ_o is the oil average velocity ratio of the outer oil film;

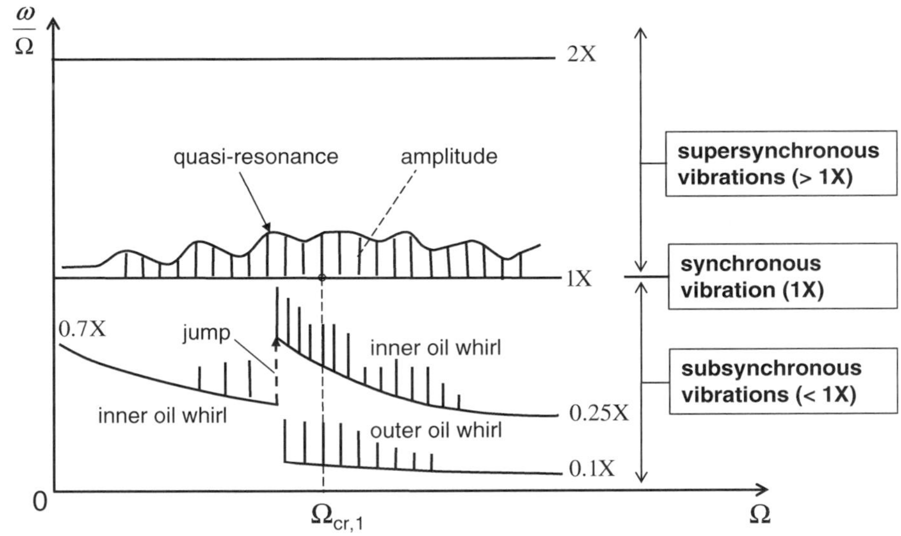

Fig. 7.5 Modified Waterfall diagram of a typical turbocharger

RSR ~ (0.30…0.15) at 120 °C oil inlet temperature, as displayed in Fig. 6.28.

The ratio λ_o of the oil velocity decreases with the bearing ring eccentricity; therefore, the outer oil whirl frequency order results in

$$\frac{\omega_o}{\Omega} \approx 0.10\ldots0.15 \tag{7.31}$$

The response frequency orders of a typical turbocharger are found in Waterfall diagram (see Fig. 7.5). The response orbit is a forward whirl, as displayed in Case History #3 (Sect. 3.6). In nonlinear rotordynamics, there is no apparent resonance but only the quasi-resonance occurs, unlike the typical resonance in linear rotordynamics. The reason for the quasi-resonance is explained that the bearing stiffness and damping coefficients increase with the rotor deflection. Therefore, the eigenfrequency of the rotor continuously increases with the rotor deflection at raising the rotor speed so that the typical resonance does not happen obviously, instead a slight increase of the rotor amplitude. However, the resonance amplitude is relatively small compared to the oil whirl amplitudes. The orbit of the rotor response is limited on the limit cycle in nonlinear rotordynamics, as discussed in Chap. 4.

In case of a contact rub at the normal tight condition in the bearings, the rotor response results from the superimposition of the subsynchronous frequency components due to contact-rub-related nonlinearity on the unbalance vibration. In fact, the unbalance forces excite the subsynchronous frequency components of the nonlinear rotor to induce the rotor response. At the heavy rub in the bearings, the rub subsynchronous frequency components have fractional frequency orders of (1/2)X, (1/3)X, or higher orders. The fractional frequency orders (1/*N*)X indicate that one convolution of the whirl cycle is done in every N revolutions of the rotor; i.e., one rub contact takes place in the bearing in every N cycles of the rotor. In fact, the rotor bounces after it touched the bearing. Until the next contact rub occurs between the rotor and bearing, the rotor has made N revolutions during this time. Hence, the contact rub has a subsynchronous frequency order of $(1/N)$X. The rotor orbit corresponding to the unbalance and contact rub is a forward whirl, as shown in Case History #4 (Sect. 3.6).

In case of the misaligned or excessively large unbalanced rotor, such super-synchronous vibrations with multiple harmonic frequency orders of 2X, 3X occur in the nonlinear rotor. An excessively large unbalance of the rotor generally causes a large rotor deflection leading to the contact rub at normally tight condition. Therefore, the subsynchronous frequency components with fractional frequency orders (1/2)X or (1/3)X occur as the consequence of the excessively large unbalance. The response orbit of the misaligned rotor is a forward whirl with a "banana" shape. At increasing radial bearing load, the response orbit changes from the "banana" to "lying-eight" shape. The banana shape orbit bounces as it comes close to the bearing wall. In case of the very high radial loads in the bearing, the response orbit turns into the "lying-eight" shape. On the contrary, the "lying-eight" shape has the forward and backward loops. At the very high radial load at the bearing wall, the rotor bounces in the backward loop. As far away from the bearing wall, the radial load is decreased, the rotor moves in the forward loop, as displayed in Case History #5 (Sect. 3.6).

In the strongly nonlinear rotor, the sideband frequencies occur due to frequency modulations between the harmonic and asynchronous frequencies. The sideband frequencies consist of the sum of the multiple integers of all frequencies acting on the rotor. The frequency modulation will be discussed later in Sect. 7.6.

In fact, it is very difficult to compute exactly such nonlinear effects, like the contact rub and wear due to the misalignment and excessively large unbalance of the rotor in the computation of nonlinear rotordynamics. The rotor response induced by the unbalance, inner, and outer oil whirls can be simulated in nonlinear rotordynamics. However, the rotor responses caused by misalignment, contact rub, and wear could be recognized by the sideband frequencies in the measured Waterfall diagram.

7.4 Frequency Analysis in Waterfall Diagram

In nonlinear rotordynamics, Waterfall diagram (frequency spectrum diagram) is used to analyze the frequency components over the rotor speed and whirl frequency. In fact, Waterfall diagram is the FFT (Fast Fourier Transform) of the rotor response; it is a three-dimensional diagram of spectra of the rotor deflection δ varying with the whirl frequency ω and rotor speed Ω. The rotor deflection is displayed with the color scale in the vertical direction (see Fig. 7.6).

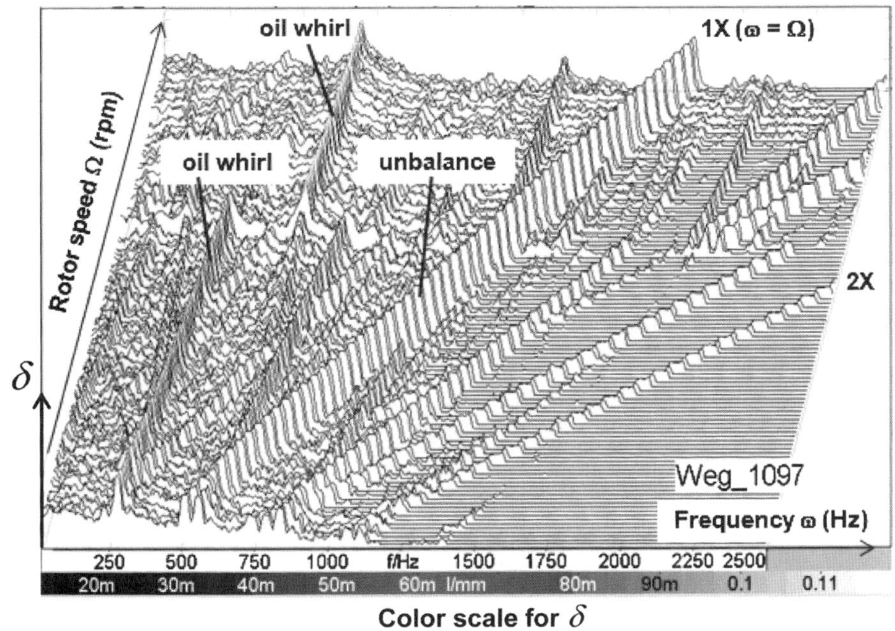

Fig. 7.6 Typical three-dimensional Waterfall diagram

To simplify the spectrum analysis, one plots the 3-D Waterfall diagram in a spectrum diagram of frequency Ω-(ω/Ω), as shown in Figs. 7.5 and 7.8. The color displayed in the frequency lines indicates the rotor amplitude; e.g., red for the maximum amplitude, blue for the minimum amplitude, and other colors for the amplitudes between the maximum and minimum values. This 2-D Waterfall diagram is called the modified Waterfall diagram, which is generally applicable to the turbocharger industry instead of the 3-D Waterfall diagram.

The quasi-resonance of the rotor occurs at the first critical speed $\Omega_{cr,1}$ at which the rotor speed Ω equals the eigenfrequency ω in the excitation line 1X, as shown in Fig. 7.7. By further increasing the rotor speed, the inner oil film temperature is higher than the outer oil film temperature due to larger bearing friction in the inner oil film. As a result, the damping coefficient of the inner oil film decreases. As soon as the destabilizing force induced by the cross-coupled stiffness coefficient exceeds the damping force of the bearing, the inner oil whirl occurs with a frequency order of nearly 0.7X at low rotor speeds. The frequency order of the inner oil whirl is continuously reducing because the inner oil film temperature further increases with the rotor speed, as discussed in Sect. 7.3.

At increasing rotor speeds, the rotor unbalance becomes larger; hence, the vibration mode of the inner oil whirl changes from the conical mode into cylindrical mode in the inner bearing clearance leading to the jump of the inner oil whirl frequency to a higher whirl frequency, as shown in Fig. 7.8. The higher the rotor

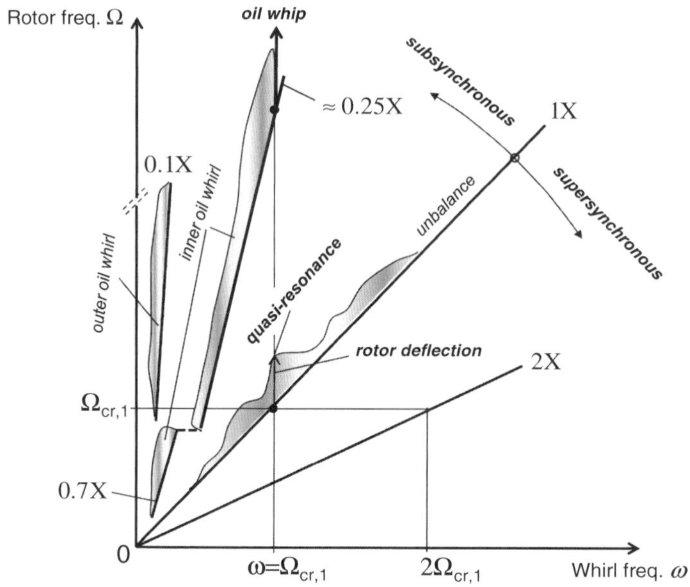

Fig. 7.7 Modified two-dimensional Waterfall diagram

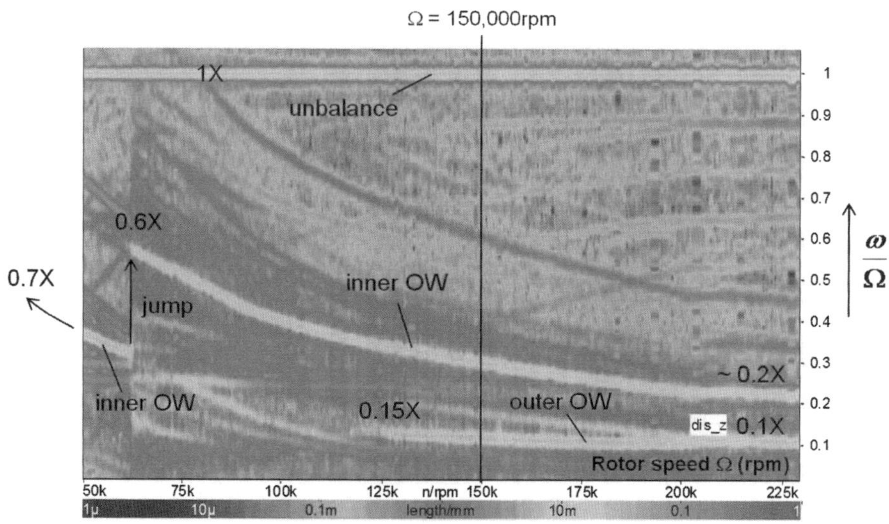

Fig. 7.8 Measured Waterfall diagram of a turbocharger at 120 °C oil [10]

unbalance is, the earlier the jump occurs; and vice versa. Therefore, the ring speed ratio *RSR* suddenly rises; the frequency order of the inner oil whirl jumps into the higher frequency order before the quasi-resonance occurs (see Figs. 7.5 and 7.7).

After that, the inner oil whirl frequency order decreases with the rotor speed due to the increase of the inner oil film temperature and finally reaches a frequency order of 0.20X to 0.25X.

Oil whip occurs at the threshold rotor speed at which the whirl frequency at 0.25X equals the first bending critical speed, as shown in Fig. 7.7. In this case, the oil whip occurs at the rotor speed of about four times of the first critical speed $\Omega_{cr,1}$. Therefore, the first critical frequency of the rotor should be designed as high as possible so that the oil whip does not happen in the operating speed range of the turbocharger. The outer oil whirl has the conical or cylindrical vibration mode; its frequency order is relatively small, reducing from 0.15X to 0.10X with the rotor speed according to Eq. (7.31). In case of the conical mode, the rotor amplitude induced by the outer oil whirl is much larger than the inner oil whirl amplitude because the outer bearing clearance is larger. Note that the outer oil whirl is less noisy because its whirl frequency is very small. The intensity of airborne noises is proportional to squared whirl frequency ($I_{noise} \sim \omega^2$). On the contrary, the amplitude of the inner oil whirl is much smaller but much noisy because the inner oil whirl frequency is higher. Hence, the constant tone is induced by the inner oil whirl, not the outer oil whirl in the bearing.

Figure 7.8 shows that the inner oil whirl frequency jumps from 0.3X to about 0.6X at the rotor speed of 65,000 rpm. The frequency jump occurs at the change of the vibrational mode of the rotor response, from the conical vibration mode into the cylindrical mode. At the rotor speed of 150,000 rpm in Waterfall diagram, the rotor consists of three frequency components of the unbalance (1X), inner oil whirl (ca. 0.30X), and outer oil whirl (nearly 0.12X). The amplitudes of each frequency component are given in the curve of the frequency order ω/Ω by the deflection color scale. The computed journal locus in the rotating floating ring bearings are displayed in Figs. 3.9c and 7.32.

7.5 Oil Whirl and Oil Whip in the Turbochargers

Oil whirl is a kind of self-excited instability with a subsynchronous frequency that usually occurs in automotive turbochargers. The occurrence of the oil whirl is caused by the oil whirling in the bearing clearance; therefore, it has nothing to do with the rotor unbalance. That means, it happens in the bearing without condition whether the rotor is well balanced or unbalanced. However, the strongly unbalanced rotor could reduce the oil whirl amplitude when the rotor is close to the bearing wall. Instability of the oil whirl at the threshold rotor speed leads to oil whip, which damages the bearing due to wear and seizure of the rotor in the bearing. Moreover, the inner oil whirl causes the airborne noise, called constant tone that is audible in the middle working range of the rotor speeds, particularly at the low oil temperature of 50 °C. Therefore, the inner oil whirl is unwanted in automotive turbochargers. In the following section, the root cause of the oil whirl is investigated and discussed, how to come up with the measure in order to reduce or avoid the self-excitation instability of the oil whirl, and the induced airborne noise in the turbochargers.

The computing model of the oil whirling in the inner bearing clearance shown in Figs. 7.9 and 7.10 is used to study the cause of the oil whirl. At first, the rotating journal moves outward from the bearing center due to the rotor unbalance causing

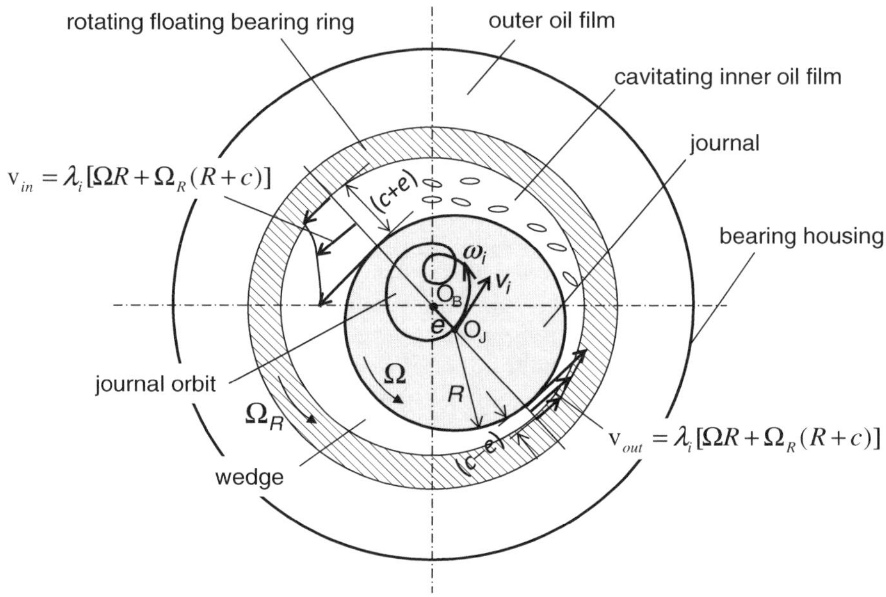

Fig. 7.9 Oil whirling model in the inner bearing clearance

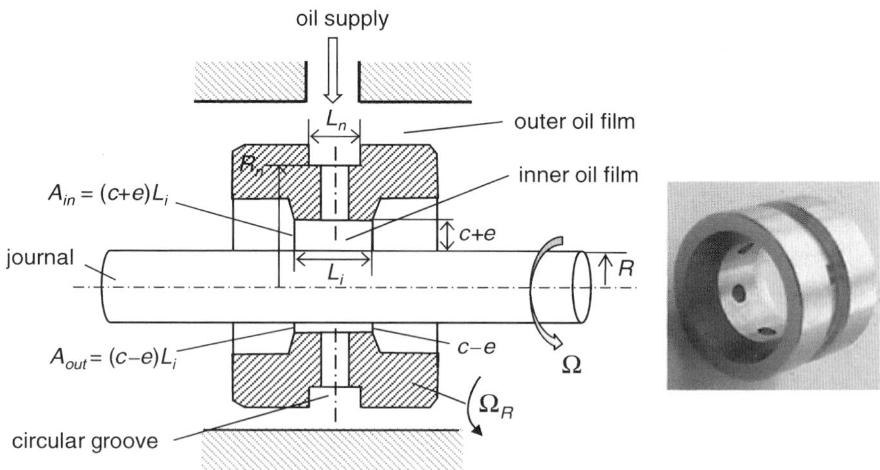

Fig. 7.10 A radial bearing with outer circular groove

an eccentricity of the journal in the bearing clearance. Due to the rotational kinetic energy of the journal, the pressure increase of the oil film is generated in the convergent wedge of the inner bearing clearance. On the contrary, the low oil pressure occurs in the divergent wedge of the inner bearing clearance. As a result, the resulting oil pressure acting on the journal creates the bearing force that consists of the stiffness and damping forces.

7.5.1 Root Cause of the Oil Whirl

Considering the convergent bearing clearance, the volumetric flow rate of oil at the inlet of the wedge is calculated as

$$\begin{aligned}\dot{Q}_{in} &= v_{in}A_{in} \\ &= \lambda_i \cdot [\Omega R + \Omega_R(R + \mathrm{c})] \cdot (c + e)L_i\end{aligned} \tag{7.32}$$

where

λ_i is the fluid circumferential average velocity ratio [12];
c is the radial inner bearing clearance;
e is the journal eccentricity;
L_i is the inner bearing width;
R is the journal radius.

Analogous to Eq. (7.32), the volumetric flow rate of oil at the outlet of the wedge results as

$$\begin{aligned}\dot{Q}_{out} &= v_{out}A_{out} \\ &= \lambda_i \cdot [\Omega R + \Omega_R(\mathrm{R} + \mathrm{c})] \cdot (c - e)L_i\end{aligned} \tag{7.33}$$

Obviously, the oil volumetric inflow in the convergent wedge is larger than the oil outflow from the wedge. To comply with the mass balance of incompressible oil in the wedge, the journal must move away from the wedge with a whirling velocity v_i perpendicular to the journal eccentricity e. The oil can only flow in the axial direction to the ends of the bearing; however, its axial outflow $\dot{Q}_{ax}$ is very small due to the small inner bearing clearance. Therefore, the surplus oil flow rate is used for moving the journal outward. As a result, the journal whirls with a forward whirl speed ω_i corresponding to v_i, as shown in Fig. 7.9.

The surplus of oil flow rate in the wedge results from Eqs. (7.32 and 7.33).

$$\begin{aligned}\Delta\dot{Q} &= \dot{Q}_{in} - \dot{Q}_{out} \\ &\approx 2RL_i e\lambda_i(\Omega + \Omega_R) \\ &= 2RL_i e\lambda_i\Omega(1 + RSR)\end{aligned} \tag{7.34}$$

where $RSR = \Omega_R/\Omega$ is the ring speed ratio of the bearing.

The oil volumetric flow rate is required for the journal whirling.

$$\begin{aligned}\dot{Q}_w &= v_i A_i \\ &= (e\omega_i)\cdot(2RL_i) = 2RL_i e\omega_i\end{aligned} \tag{7.35}$$

where

A_i is the whirling area of the journal ($A_i = 2RL_i$);

ω_i is the whirl speed of the journal.

Using the continuity equation for the incompressible oil flow in the converged wedge, one obtains the relation of the oil flow rates.

$$\sum_i \dot{Q}_i = \Delta\dot{Q} - \dot{Q}_w - \dot{Q}_{ax} = 0 \tag{7.36}$$

Thus,

$$2RL_i e\lambda_i(\Omega + \Omega_R) = 2RL_i e\omega_i + \dot{Q}_{ax} \tag{7.37}$$

The forward whirl speed of the journal results as

$$\omega_i = \lambda_i(\Omega + \Omega_R) - \frac{\dot{Q}_{ax}}{2RL_i e} \tag{7.38}$$

where $\lambda_i \approx 0.5$ near the bearing center (see Fig. 7.11).

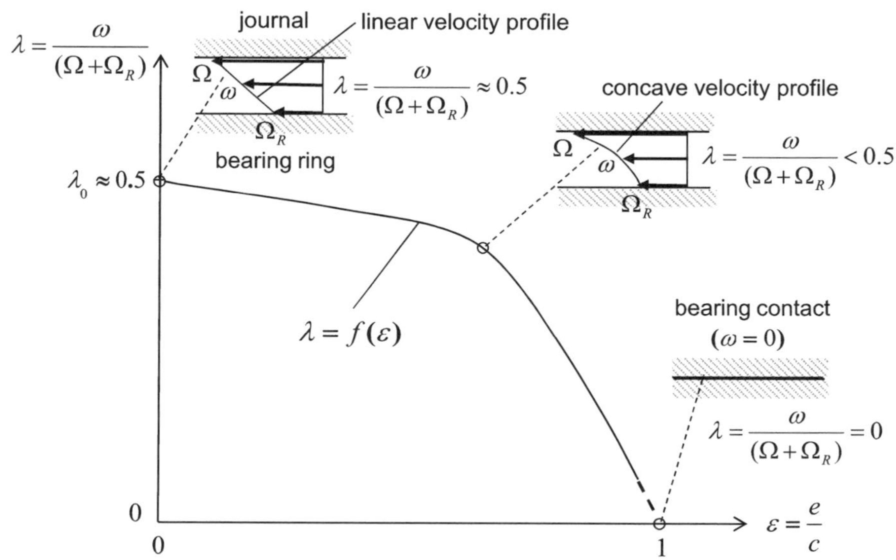

Fig. 7.11 Velocity ratio λ versus journal relative eccentricity ε

Since the axial flow rate is negligibly small, the whirl speed of the journal is written in terms of the rotor speed and ring speed ratio.

$$\begin{aligned}\omega_i &= \lambda_i(\Omega + \Omega_R) \approx \frac{1}{2}(\Omega + \Omega_R)\\ &= \frac{\Omega}{2}(1 + \frac{\Omega_R}{\Omega}) = \left(\frac{1 + RSR}{2}\right)\Omega\end{aligned} \tag{7.39a}$$

Equation (7.39a) proves that the assumption of the whirling speed of the inner oil whirl used in Eq. (7.24) is correct. In case semi-floating ring or fixed bearings at the bearing ring speed Ω_R equals zero ($RSR = 0$), the oil whirl is usually called the half-frequency whirl ($\omega_i = \Omega/2$). This forward whirl speed ω_i of the journal induces the airborne noise called constant tone that will be discussed later in Sect. 7.7.

The inner oil whirl is caused by the journal whirling with the whirl speed ω_i that depends on the velocity ratio λ_i. In fact, this velocity ratio varies with the journal relative eccentricity ε. Near the bearing center ($\varepsilon \approx 0$), the velocity ratio λ_i is approximately 0.475 ($\approx$ 0.5) by measurements in [12]. At increasing journal eccentricity ε, the flow resistance in the bearing clearance rises due to the viscous friction in the oil boundary layer. Hence, the velocity profile becomes much concave leading to decrease of the velocity ratio. Due to bearing contact with the journal, the velocity ratio λ goes to zero as the journal moves close to the bearing wall ($\varepsilon \to 1$). Figure 7.11 displays the qualitative behavior of the average velocity ratio versus the journal relative eccentricity.

In case of the radial bearing with a circular groove in the outer bearing surface (see Fig. 7.10), two outer oil whirls (twofold outer oil whirl) mostly occur in Waterfall diagram. According to Eq. (7.30), the first outer oil whirl speed in the area outside the circular groove is calculated at a negligibly small axial flow rate due to the small outer bearing clearance.

$$\omega_{o,1} \approx \lambda_o \Omega_R \tag{7.39b}$$

where Ω_R is the bearing ring speed.

The second outer oil whirl speed in the area inside the circular groove is given according to Eq. (7.38) as follows:

$$\omega_{o,2} = \lambda_o \Omega_R - \frac{\dot{Q}_{ax}}{2R_n L_n e_n} < \omega_{o,1} \tag{7.39c}$$

The axial flow rate in Eq. (7.39c) is no longer negligible since the oil flows into the inner oil film through all radial oil holes. The difference between both outer whirl frequencies is proportional to the second term on the RHS of Eq. (7.39c). Similarly, the twofold inner oil whirl is induced in the inner oil film.

7.5.2 Threshold of Instability

To study the threshold of instability of the oil whirl, the force balance of the journal in the bearing is applied to the computational journal model, as displayed in Fig. 7.12. In the radial direction r, the journal stiffness force$-Ke$, bearing cross-coupled damping force $jc_c\dot{e}$, inertial force $-m\ddot{e}$, and centripetal force $mr_\varepsilon\omega^2 e^{j\omega t}$ act on the journal. In the tangential direction t, the destabilizing force jk_ce is caused by the cross-coupled stiffness coefficient k_c of the oil film and the damping force $-c\dot{e}$ of the bearing.

Using the impulse momentum theorem, the vibration equation of the journal is written in the rotating coordinate system (r, t) [12, 20, 21].

$$m\ddot{e} + (c - jc_c)\dot{e} + (K - jk_c)e = mr_\varepsilon\omega^2 e^{j\omega t} \tag{7.40}$$

where

m is the rotor mass;
j is the imaginary unit ($j^2 = -1$);
e is the eccentricity of the journal;
c is the diagonal damping coefficient of the oil film bearing;
c_c is the cross-coupled damping coefficient of the oil film bearing;
K is the effective stiffness coefficient of the rotor-bearing system;
k_c is the cross-coupled stiffness coefficient of the oil film bearing;
ω is the whirl speed of the journal in the bearing;
r_ε is the unbalance radius of the rotor.

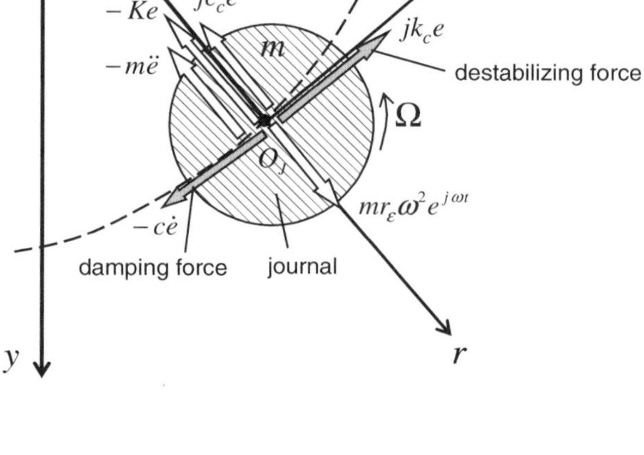

Fig. 7.12 Forces acting upon the journal in an oil film radial bearing

The journal response can be expressed as an exponential function of the eigenvalue s and time t.

$$e(t) = Ae^{st} \tag{7.41}$$

where s is the complex eigenvalue including the real part α and imaginary part ω_n.

$$s = \alpha \pm j\omega_n \tag{7.42}$$

Differentiating Eq. (7.41) with respect to time, one obtains

$$\begin{aligned} \dot{e} &= sAe^{st}; \\ \ddot{e} &= s^2 Ae^{st} \end{aligned} \tag{7.43}$$

Substituting Eqs. (7.41 and 7.43) into the homogeneous Eq. (7.40), the characteristic equation results in

$$D(s) \equiv ms^2 + (c - jc_c)s + (K - jk_c) = 0 \tag{7.44}$$

The cross-coupled stiffness coefficient k_c of the bearing is generated by the whirl speed ω. Therefore, it is proportional to $\lambda\Omega$ [12]

$$k_c \equiv c\lambda\Omega \tag{7.45}$$

Using Eq. (7.45), the characteristic Eq. (7.44) is written as

$$D(s) \equiv ms^2 + (c - jc_c)s + (K - jc\lambda\Omega) = 0 \tag{7.46}$$

For satisfying the stability of the rotor, the real part of the complex eigenvalue s must be negative.

Therefore, the real part α calculated in Appendix C is written as

$$\alpha = -\frac{c}{2m} \pm \frac{1}{\sqrt{2}} \left[-E + \sqrt{E^2 + F^2} \right]^{\frac{1}{2}} < 0 \tag{7.47}$$

where

$$E \equiv \frac{K}{m} - \left(\frac{c^2 - c_c^2}{4m^2} \right); \quad F \equiv \frac{c}{m} \left(\lambda\Omega - \frac{c_c}{2m} \right)$$

Solving Eq. (7.47), one obtains the rotor stability condition of the oil whirl

$$0 < \lambda\Omega < \frac{c_c}{2m} + \sqrt{\frac{K}{m} + \left(\frac{c_c}{2m} \right)^2} \tag{7.48}$$

The frequency of the threshold of instability results at $\alpha = 0$, where the oil whirl instability begins to take place.

$$\Omega_{th} = \frac{1}{\lambda}\left(\frac{c_c}{2m} + \sqrt{\frac{K}{m} + \left(\frac{c_c}{2m}\right)^2}\right) \tag{7.49}$$

where λ is the average fluid velocity ratio (see Fig. 7.11).

The rotor response of Eq. (7.40) can be generally written as

$$e(t) = Ae^{j(\omega t + \beta)} \tag{7.50}$$

Substituting Eqs. (7.43) and (7.50) into Eq. (7.40) with $k_c = c\lambda\Omega$, one obtains

$$\left((K - m\omega^2 + c_c\omega) + jc(\omega - \lambda\Omega)\right)Ae^{j(\omega t+\beta)} = mr_\varepsilon\omega^2 e^{j\omega t} \tag{7.51}$$

The response amplitude of the journal results as

$$A = \frac{\left|mr_\varepsilon\omega^2 e^{-j\beta}\right|}{\left|(K - m\omega^2 + c_c\omega) + jc(\omega - \lambda\Omega)\right|} \tag{7.52}$$

Calculating the norms, the rotor amplitude becomes

$$A = \frac{mr_\varepsilon\omega^2}{\sqrt{(K - m\omega^2 + c_c\omega)^2 + c^2(\omega - \lambda\Omega)^2}} \tag{7.53}$$

The frequency of the threshold of instability happens when the amplitude A goes to infinity. Therefore, both operating conditions are necessary at the same time:

$$K - m\omega^2 + c_c\omega = 0 \tag{7.54a}$$

and

$$c(\omega - \lambda\Omega) = 0 \tag{7.54b}$$

The first and second conditions in Eqs. (7.54a and 7.54b) give

$$\omega_{1,2} = \frac{c_c}{2m} \pm \sqrt{\frac{K}{m} + \left(\frac{c_c}{2m}\right)^2} \tag{7.55a}$$

and

$$\omega = \lambda\Omega \Rightarrow \Omega = \frac{\omega}{\lambda} \tag{7.55b}$$

Thus, the frequency of the threshold of instability of the rotor fulfills both necessary conditions.

$$\begin{aligned}\Omega_{th} &= \frac{\omega_i}{\lambda} \\ &= \frac{1}{\lambda}\left(\frac{c_c}{2m} + \sqrt{\frac{K}{m} + \left(\frac{c_c}{2m}\right)^2}\right)\end{aligned} \tag{7.56}$$

Note that the threshold frequency Ω_{th} derived in Eq. (7.56) is the same threshold frequency in Eq. (7.49) that results from solving the characteristic Eq. (7.46).

Oil whip occurs at the rotor speed at which the oil whirl frequency equals the first bending critical frequency. The oil whip, a kind of self-excited instability, is very dangerous and destructive since the rotor defection continuously increases with time in the bearing clearance even at a constant rotor speed leading to bearing damage due to wear or seizure of the journal in the bearing.

Therefore, the threshold speed of the rotor is calculated as

$$\begin{aligned}\omega = \lambda\Omega_{th} &= \Omega_{cr,1} \\ \Rightarrow \Omega_{th} &= \frac{\Omega_{cr,1}}{\lambda} = \frac{1}{\lambda}\sqrt{\frac{K}{m}}\end{aligned} \tag{7.57}$$

where $\Omega_{cr,1}$ is the first bending critical frequency of the rotor.

In case the term $c_c/(2m)$ is negligibly small, the stability condition of Eq. (7.48) for the rotor response results using Eq. (7.57) as

$$0 < \Omega < \frac{1}{\lambda}\sqrt{\frac{K}{m}} = \Omega_{th} \tag{7.58}$$

where K is the rotor stiffness coefficient containing the shaft and bearing stiffness coefficients.

7.6 Modulations of Vibrations

In the nonlinear rotordynamic system, some malfunctions, such as, rotor unbalance, oil whirl, misalignment, heavy rub, and excessively large unbalance cause truncations of the vibrations. Therefore, the sidebands are additionally generated. The sideband frequencies result from adding and subtracting multiple integers of the frequency components of the malfunctions. This is called frequency modulation (FM) of the frequency components in the nonlinear system.

Most of the sidebands occur at high rotor speeds and amplify the amplitude of the rotor response. The sidebands happen at the sub- and supersynchronous frequencies that are generated by frequency components of the malfunctions. Note that the more nonlinear the rotor system is, the more visibly the sidebands appear in Waterfall diagram. The response amplitude of the strongly nonlinear system results from all component amplitudes of the malfunctions and sidebands. As a result, the rotor response amplitude is intensified; the rotor behavior becomes much more nonlinear at the large rotor amplitudes.

7.6.1 Responses of Nonlinear Vibration Systems

Considering a nonlinear rotordynamic system, the system response $y(x)$ results from the input signal $x(t)$ with a large amplitude and the system transfer impedance $Z(\varepsilon)$ (see Fig. 7.13).

The nonlinear response $y(x)$ of the system is written in the second order of the input function $x(t)$.

$$\begin{aligned} y(x) &= f(Z, x, \varepsilon) = Z \cdot \left(x + \varepsilon\,\mathrm{sgn}(x)x^2 + \cdots\right) \\ &= Z \cdot (x + \varepsilon^* x^2 + \cdots) \end{aligned} \tag{7.59}$$

where of the system;

x	is the input signal with large amplitudes;
Z	is the complex transfer impedance;
Zx	is the linear term of the response;
E	is the nonlinear system factor ($\varepsilon \ll 1$);
$Z\varepsilon^* x^2$	is the nonlinear term of the response;
$\varepsilon^* = \varepsilon$ sgn(x) within sgn(x)	is the signum function of x, which is defined as

$$\mathrm{sgn}(x) = \begin{cases} -1 \text{ if } x < 0 \\ 0 \text{ if } x = 0 \\ +1 \text{ if } x > 0 \end{cases}$$

Therefore,

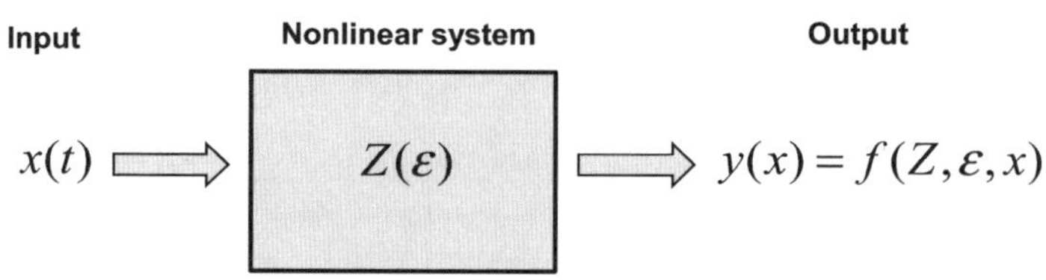

Fig. 7.13 Response of a nonlinear vibration system

$$\varepsilon^* \equiv \varepsilon\,\mathrm{sgn}(x) = \begin{cases} -\varepsilon \text{ if } x<0 \\ 0 \text{ if } x=0 \\ +\varepsilon \text{ if } x>0 \end{cases}$$

The nonlinear vibration response becomes

$$y(x) = Zx + Z\varepsilon\,\mathrm{sgn}(x)x^2 + O(x^3) \tag{7.60}$$

The nonlinear vibration response $y(x)$ is displayed in Fig. 7.14 and compared to the linear response. In linear system, the response of Zx is proportional to the input function x. On the contrary, the nonlinear response is different to the linear term of Zx. It increases or decreases more quickly with the input amplitude $x(t)$ because the nonlinear term is proportional to x^2. As a result, the nonlinear system can only be linearized at the small amplitudes since the nonlinear term is quite small.

In case of the excessively large unbalance, the system behavior is strongly nonlinear. Thus, the response amplitude $y(x)$ increases faster than the amplitude of the linear system since the nonlinear term increases with squared input amplitude that is no longer negligible, as shown in Fig. 7.14.

7.6.2 Modulated Sideband Frequencies

It is supposed that the input $x(t)$ is a harmonic vibration with the amplitude X and single frequency ω_1.

$$x = X \cos \omega_1 t \tag{7.61}$$

Its response results from Eq. (7.59) with a second order of $x(t)$.

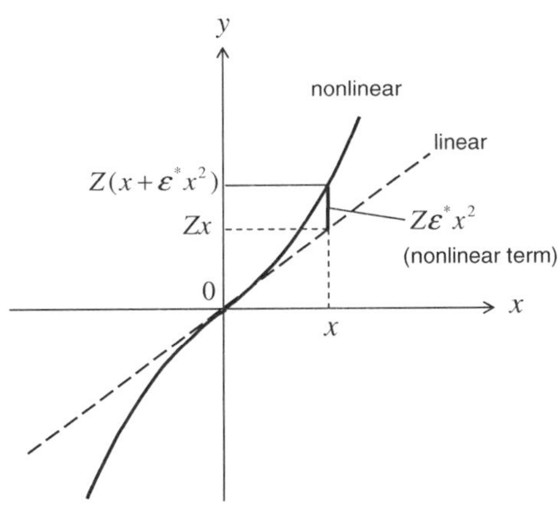

Fig. 7.14 Comparison of the linear and nonlinear responses

$$\begin{aligned} y &= ZX\cos\omega_1 t + Z\varepsilon^* X^2\cos^2\omega_1 t \\ &\equiv y_{L-term} + y_{NL-term} \end{aligned} \tag{7.62}$$

The rotor response consists of two terms; the first term $y_{L\text{-}term}$ is linear with $x(t)$ and the latter $y_{NL\text{-}term}$ is the nonlinear term of x^2. Using the trigonometric relation, the nonlinear term can be written as

$$\begin{aligned} y_{L-term} &\equiv ZX\cos\omega_1 t; \\ y_{NL-term} &\equiv Z\varepsilon^* x^2 = Z\varepsilon^* X^2\cos^2\omega_1 t \\ &= \frac{1}{2}Z\varepsilon^* X^2(1+\cos 2\omega_1 t) \end{aligned} \tag{7.63}$$

Substituting Eqs. (7.62 and 7.63), one obtains the rotor response.

$$y = ZX\cos\omega_1 t + \frac{1}{2}Z\varepsilon^* X^2 + \frac{1}{2}Z\varepsilon^* X^2\cos 2\omega_1 t \tag{7.64}$$

Compared to the linear system, the nonlinear response has two new additional terms: the rectification term of X^2 and the second-order harmonic term of $2\omega_1$.

Now, we consider the input function $x(t)$ of a periodic vibration containing two frequency components ω_1 and ω_2 and corresponding amplitudes X_1 and X_2.

$$x = X_1\cos\omega_1 t + X_2\cos\omega_2 t; \quad \omega_1 > \omega_2 \tag{7.65}$$

For the simplicity, we consider only the nonlinear term of the rotor response in the following calculation.

$$\begin{aligned} y_{NL-term} &\equiv Z\varepsilon^* x^2 = Z\varepsilon^*(X_1\cos\omega_1 t + X_2\cos\omega_2 t)^2 \\ &= Z\varepsilon^*(X_1^2\cos^2\omega_1 t + X_2^2\cos^2\omega_2 t + 2X_1X_2\cos\omega_1 t\cdot\cos\omega_2 t) \\ &\equiv y_1 + y_2 + y_3 \end{aligned} \tag{7.66}$$

Similarly, the first two terms y_1 and y_2 provide two additional second-order harmonics of the frequencies $2\omega_1$ and $2\omega_2$, respectively. Furthermore, one obtains the third term y_3 that generates sidebands of ω_1 and ω_2 after a few trigonometric calculating steps.

$$\begin{aligned} y_3 &= Z\varepsilon^* X_1X_2(2\cos\omega_1 t\cdot\cos\omega_2 t) \\ &= Z\varepsilon^* X_1X_2[\cos(\omega_1 t - \omega_2 t) + \cos(\omega_1 t + \omega_2 t)] \\ &= Z\varepsilon^* X_1X_2[\cos(\omega_1 - \omega_2)t + \cos(\omega_1 + \omega_2)t] \\ &\equiv Z\varepsilon^* X_1X_2(\cos\omega_{LSB}t + \cos\omega_{USB}t) \end{aligned} \tag{7.67}$$

The upper and lower sideband frequencies (rad/s) are defined as

$$\omega_{USB} = \omega_1 + \omega_2; \quad \omega_{LSB} = \omega_1 - \omega_2. \tag{7.68}$$

The carrier and modulation frequencies f_c and f_m are defined as

$$f_c = \frac{\omega_c}{2\pi} \equiv \frac{\omega_1}{2\pi} = \frac{1}{4\pi}(\omega_{USB} + \omega_{LSB}); \tag{7.69}$$

$$f_m = \frac{\omega_m}{2\pi} \equiv \frac{\omega_2}{2\pi} = \frac{1}{4\pi}(\omega_{USB} - \omega_{LSB}). \tag{7.70}$$

Figure 7.15 shows the sideband frequencies of the modulation. Due to nonlinearity of the system, the lower and upper sidebands are induced by the amplitude and FMs of f_c and f_m. In fact, the sideband frequency varies in a bandwidth of frequency Δf_m, e.g., in case of oil whirl. Therefore, the sideband frequencies f_{LSB} and f_{USB} scatter in the corresponding frequency bandwidth Δf_m.

In radio engineering, the carrier frequency f_c is much higher than the modulation frequency f_m ($f_c \gg f_m$). To broadcast the modulation signal in a long distance, the carrier signal modulates itself in the double sideband (*DSB*) containing the lower (*LSB*) and upper sidebands (*USB*) with the ultrahigh frequencies f_{LSB} and f_{USB} (see Figs. 7.15 and 7.16). The sideband frequencies are modulated by the carrier and modulation frequencies in the wide band of frequency.

The modulation frequency f_m is normally the human voice frequency spreading from 300 to 3 kHz; therefore, the sidebands are generated with the maximum frequency width of the *DSB* of $2f_m$ about 6 kHz. Using demodulation or detection of the received sideband signals, the modulation signal is recovered at receiving where the signal detector and low-pass filter are used. The entire process of sending and receiving signal is carried out by the MODEM method that consists of MOdulation and DEModulation of the signals, generally applied in the wide band technique.

The sideband signal generated by the carrier and modulating signals is displayed in Fig. 7.16, in which one finds the carrier and modulation frequencies again in the

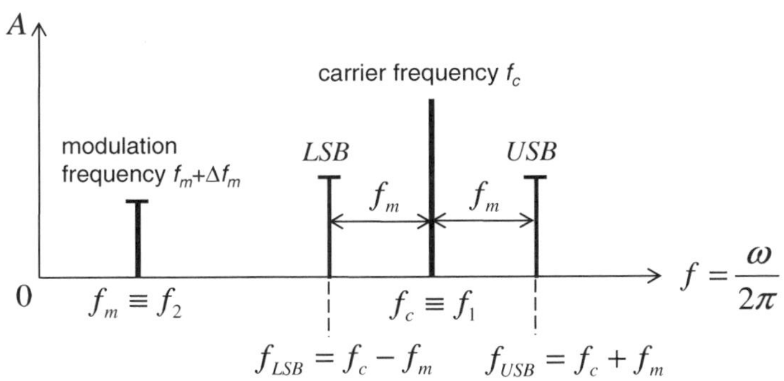

Fig. 7.15 Double sidebands of signal frequency modulations

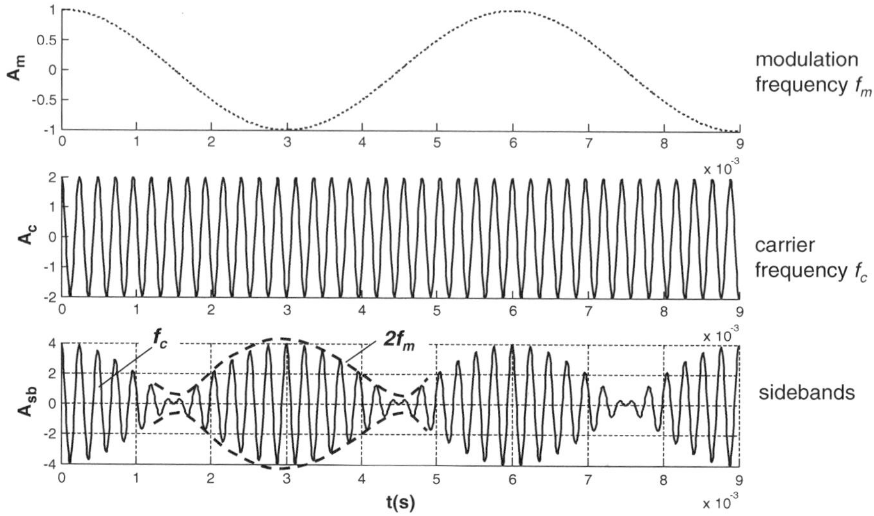

Fig. 7.16 Double sideband (*DSB*) of the amplitude modulation

sideband signal. The envelop signal of the sideband has a frequency of $2f_m$ since its period equals one-half of the period of the modulation signal (see Fig. 7.17).

The lower and upper sideband frequencies of f_{LSB} and f_{USB} are calculated from the carrier f_c and modulation f_m frequencies with $f_m \ll f_c$; hence, $f_1 \gg f_2$:

- Lower sideband frequency (*LSB*):

$$f_{LSB} = f_c - f_m \equiv f_1 - f_2 \tag{7.71}$$

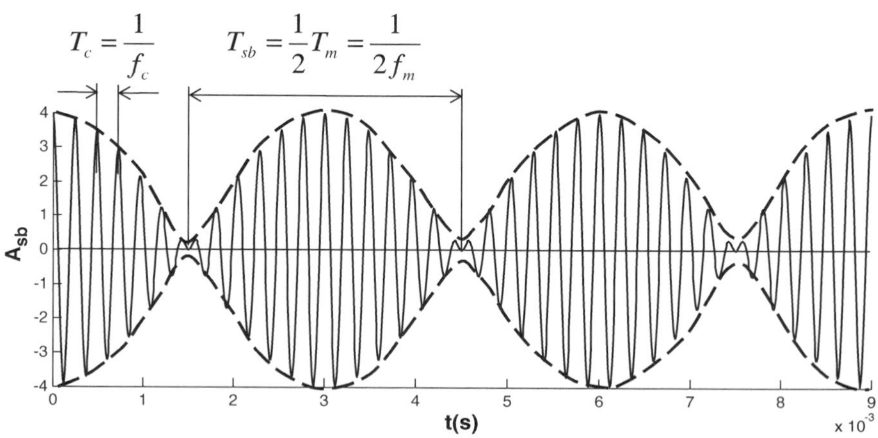

Fig. 7.17 Frequencies and periods of the sideband signal

- Upper sideband frequency (*USB*):

$$f_{USB} = f_c + f_m \equiv f_1 + f_2 \tag{7.72}$$

To sum up, the nonlinear rotordynamic system induces some new additional terms due to the amplitude and frequency modulations as follows:

- The rectification term $\frac{1}{2}Z\varepsilon^* X^2$ in Eq. (7.64);
- Second-order harmonic frequencies of $2\omega_1$ and $2\omega_2$ with the amplitudes X_1^2 and X_2^2 in Eqs. (7.64) and (7.66);
- Sideband frequencies modulated by the excitation frequencies ω_1 and ω_2 have the amplitudes that are proportional to the term $Z\varepsilon^* X_1 X_2$. According to Eqs. (7.67 and 7.68), the sideband frequencies are the sum or difference of the frequency components. Hence, the first-order sideband frequencies result in $3\omega_2 - 2\omega_1$, $2\omega_2 - \omega_1$, $2\omega_1 - \omega_2$, $3\omega_1 - 2\omega_2$, $4\omega_1 - 3\omega_2$, etc., cf. Eqs. (7.74a and 7.74b) and Fig. 7.21.

In case of the unbalanced and misaligned rotor, the sidebands induced by the FMs of misalignment, unbalance, and inner oil whirl are found in Waterfall diagram of noise spectra (see Fig. 7.18).

Beat phenomenon, a special case of the modulation, is caused by the modulation of two harmonic vibrations with the frequency components ω_1 and ω_2 that are close to each other. If their amplitudes are equal, only the amplitude modulation occurs. Otherwise, both amplitude and FMs take place at the same time.

Let $x(t)$ be the input signal that is the sum of two sinus functions with the same amplitude a (= a_1 = a_2).

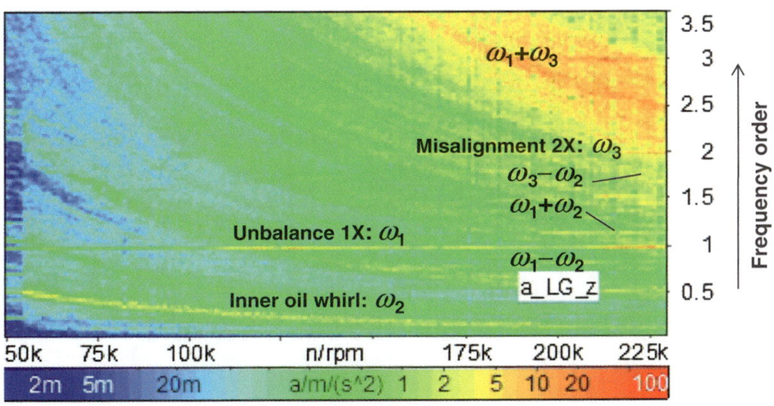

Fig. 7.18 Sideband frequencies in Waterfall diagram

$$\begin{aligned} x &= a_1 \sin \omega_1 t + a_2 \sin \omega_2 t \\ &= a(\sin \omega_1 t + \sin \omega_2 t) \\ &= 2a \sin\left[\frac{1}{2}(\omega_1 + \omega_2)\right] t \cdot \cos\left[\frac{1}{2}(\omega_1 - \omega_2)\right] t \\ &\equiv 2a \sin \omega_c t. \cos \omega_m t \end{aligned} \tag{7.73}$$

The carrier and modulation frequencies f_c and f_m are defined as

$$f_c = \frac{\omega_c}{2\pi} = \frac{1}{4\pi}(\omega_1 + \omega_2);$$
$$f_m = \frac{\omega_m}{2\pi} = \frac{1}{4\pi}(\omega_1 - \omega_2).$$

As an example, we choose two frequency components f_1 = 100 Hz and f_2 = 85 Hz for the beat modulation shown in the frequency domain (see Fig. 7.19). The carrier frequency f_c is the average of frequencies f_1 and f_2 and equals 92.5 Hz, and the modulation frequency f_m equals 7.5 Hz. The beat frequency $2f_m$ (= 15 Hz) is defined by the frequency difference of f_1 and f_2. Figure 7.20 shows that the closer the frequency components are, the smaller the beat frequency is, and the larger the beat signal period is. The beat signal period is inversely related to the beat frequency $2f_m$; hence, the beat period is about 0.067 s at the beat frequency 15 Hz. The carrier frequency has a period of 0.011 s at 92.5 Hz in the time domain.

The generated first-order frequencies of the *LSB* and *USB* sidebands result from the sum or difference of the frequency components of f_1 and f_2, which are located on the LHS and RHS of the carrier frequency f_c, as plotted in Fig. 7.21.

$$f_{LSB} = \begin{cases} f_1 - 2f_m = f_c - f_m = f_2 \\ f_1 - 4f_m = f_c - 3f_m = 2f_2 - f_1 \\ f_1 - 6f_m = f_c - 5f_m = 3f_2 - 2f_1 \\ \ldots \end{cases} \tag{7.74a}$$

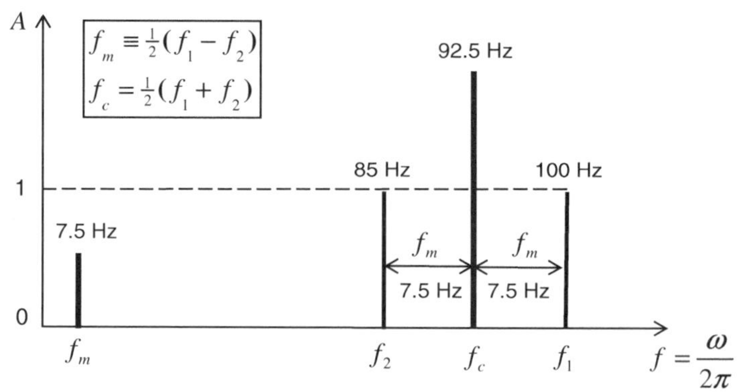

Fig. 7.19 Beat frequency $2f_m$ of the frequencies f_1 and f_2

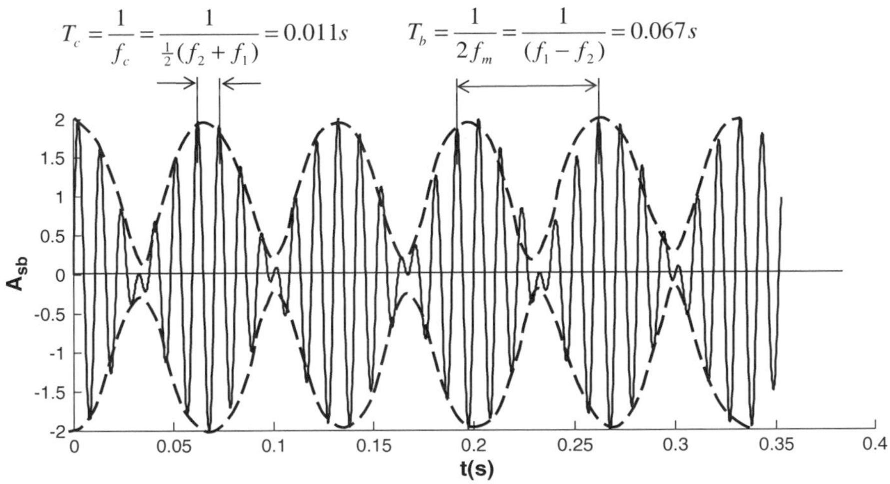

Fig. 7.20 Beat signal response in the time domain

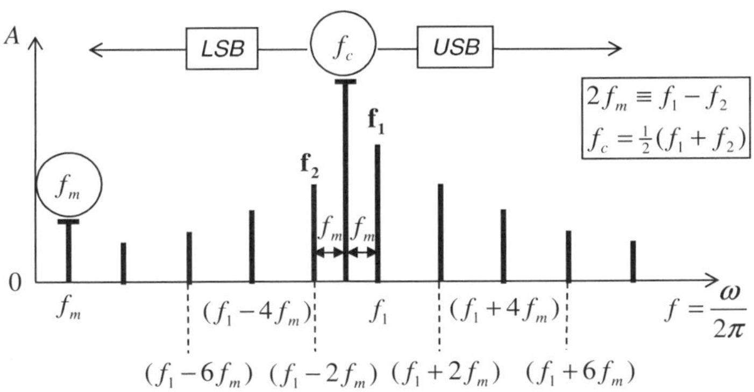

Fig. 7.21 *LSB* and *USB* sideband frequencies

and

$$f_{USB} = \begin{cases} f_1 + 0f_m = f_c + f_m = f_1 \\ f_1 + 2f_m = f_c + 3f_m = 2f_1 - f_2 \\ f_1 + 4f_m = f_c + 5f_m = 3f_1 - 2f_2 \\ f_1 + 6f_m = f_c + 7f_m = 4f_1 - 3f_2 \\ \cdots \end{cases} \tag{7.74b}$$

However, the nonlinear terms of the rotor response are not only proportional to squared input amplitude x^2 but also the high-order input amplitude, such as x^3 or

Table 7.1 Generated sideband frequencies by frequency modulations

Lower sideband (*LSB*)	f_c	Upper sideband (*USB*)	*f* Order
$2f_2 - f_1 \leftarrow f_2$	$(f_1 + f_2)/2$	$f_1 \rightarrow 2f_1 - f_2$	1X
$3f_2 - f_1$ $2f_2$	$(f_1 + f_2)$	$2f_1$ $3f_1 - f_2$	2X
$4f_2 - f_1$ $3f_2$ $2f_2 + f_1$	$3(f_1 + f_2)/2$	$2f_1 + f_2$ $3f_1$ $4f_1 - f_2$	3X

higher orders compared to Eq. (7.59). According to Ehrich [5], the truncation of the modulation frequencies f_1 and f_2 could induce two new second harmonic frequencies $(f_1 - f_2)$ and $(f_1 + f_2)$ that in turn generate the third harmonics $(2f_2 + f_1)$ and $(2f_1 + f_2)$ in a strongly nonlinear rotor system, as shown in Table 7.1.

To generate the *LSB* frequencies, one superimposes the excitation frequency f_2 on $(2f_2 - f_1)$ to get the second-order frequency $(3f_2 - f_1)$ following the thin arrow, and superimposing f_2 again on the second-order frequency to generate the third-order frequency $(4f_2 - f_1)$. Another way, following the dark arrow, one obtains the new high-order harmonic frequencies $2f_2$ and $3f_2$ adding f_2 to itself and $2f_2$, respectively. In order to receive the third-order frequency $(2f_2 + f_1)$, one carries out the same way adding f_2 to $(f_1 + f_2)$ in the direction of the dashed arrow.

Similar to the *LSB* frequencies, the *USB* frequencies are generated in the first, second, and third orders (see Table 7.1).

7.7 Induced Airborne Noises in Automotive Turbochargers

Noises induced in the turbocharger transmit through the bearing oil films, bearing housing, compressor and turbine housings, exhaust gas manifold, engine block, and exhaust gas pipe into the vehicle cabin. The induced noises excite vibrations of the exhaust gas pipe and chassis and generate some airborne noises that are considered as uncomfortable for drivers and passengers and as malfunction of the turbocharger as well. They are undesirable and therefore must be reduced or avoided in the passenger vehicles.

7.7.1 Classification of Noises

The induced noises in automotive turbochargers are caused by the airflow in the compressor wheel, rotation of the compressor and turbine wheels, rotor unbalance, and oil whirling in the radial bearings. They excite the vibrations of the intake air filter, charge-air intercooler, and exhaust gas system; therefore, they generate the airborne noises in the environment [15].

- *Pulsation noise* is caused by the different volume chambers of the compressor wheel. Its frequency spreads from 1200 to 4500 Hz that occurs at the engine speed range from 1500 to 3500 rpm in second gear with high loads and depends on the number of unequal chambers of the compressor wheel.
- *Rotation noise (rotating-blades-related noise)* is generated by rotating blades of the compressor or turbine wheels at the engine speed range from 1400 to 2500 rpm at second and third gears. These noises have very high frequencies between about 8000 and 18,000 Hz or higher that result from the number of blades and rotor speed. Although adult human ears could not recognize noises above 16 kHz, animals in the car could hear such high-frequency noises.
- *Growling noise (compressor-stall-related noise*) is induced by partial recirculation of the charge air in the compressor wheel due to flow separation at the blades. Its frequency spreads in a wide band between 1200 and 3500 Hz containing partly metallic noise. It occurs at the engine speed range between 1400 and 2500 rpm in second and third gears.
- *Whining noise (compressor-surge-related noise)* is induced at the surge operating condition in the compressor wheel where the charge air totally recirculates to the compressor inlet. The whining noise occurs when the gas pedal is suddenly released. Therefore, the required mass flow rate of the charge air is strongly reduced at a still high rotor speed of the turbocharger leading to the surge operating condition in the compressor. Its frequency spreads in a wide band from nearly 800 to 2700 Hz containing partly metallic noise that occurs at the engine speed range from 1400 to 2500 rpm in second and third gears.
- *Unbalance whistle* is caused by unbalanced rotor and unbalance change of the rotor after the long operating time interval. Its harmonic frequency (1X) is between 1200 and 4500 Hz. It happens at the engine speed range between 1500 and 3500 rpm in second gear with the relatively low loads.
- *Constant tone* is induced by the inner oil whirl due to oil whirling in the journal bearing. Its frequency is between about 600 and 1100 Hz in the human audible range. The frequency order of the inner oil whirl in the rotating floating ring bearings reduces from about 0.35–0.25 as the rotor speed increases according to Eq. (7.29b). Therefore, the induced frequency of the inner oil whirl varies in the small range between 600 and 1100 Hz. Thus, its frequency can be considered quasi-constant in the operating speed range of automotive turbochargers. The constant tone often occurs at the engine speed range between 1500 and 3500 rpm in second to fifth gear with the middle to high loads.

Turbocharger engineers are very concerned about both last airborne noises (unbalance whistle and constant tone) that must be reduced or avoided in the automotive turbocharger. Both noises are considered as disturbing factor for drivers and passengers, and malfunction of the turbocharger.

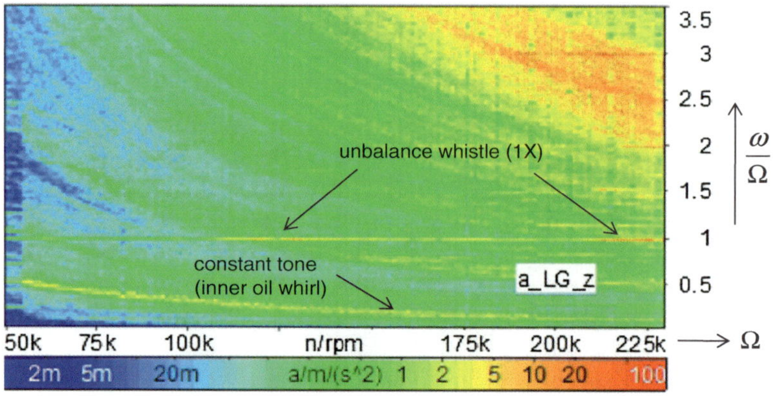

Fig. 7.22 Unbalance whistle and constant tone in Waterfall diagram

7.7.2 Unbalance Whistle and Constant Tone

The malfunctions of the unbalance whistles and constant tone are found in Waterfall diagram of noise spectra, as shown in Fig. 7.22.

(a) Unbalance whistles (1X)

It is caused by the residual unbalance of the rotor after balancing. The possible reasons for the unbalance whistle are either the rotor is not well balanced or the unbalance change takes place after a long operating time [18, 19].

In order to reduce the unbalance whistle at the harmonic frequency 1X, the rotor is balanced by the low-speed (shop balancing) and high-speed balancing (trim balancing) at the two balancing planes. They will be discussed in more detail in the next chapter. Moreover, one should check whether the compressor wheel is loose in the shaft because the rotor is over speeding or the screw-nut is not enough tightened or loose at the compressor wheel. Furthermore, the rotor unbalance significantly increases when the compressor or turbine wheels were damaged or partly broken by any reason during operation, such as impact of hard particles on them, blade contact on the housing, or LCF.

(b) Constant tone (subsynchronous noise)

To study the root cause of the constant tone, we return to Fig. 7.9 and find that the oil inflow rate at the convergent wedge is larger than the outflow rate at any time t. As a result, the journal has to move away from the bearing with a velocity v_i perpendicular to the journal eccentricity e in order to fulfill the mass flow rate balance of the incompressible oil.

Furthermore, oil could only flow in the axial direction along the inner bearing width to outside; however, the axial flow rate $\dot{Q}_{ax}$ is quite small. Therefore, the journal whirling with the whirl velocity ω_i induces the constant tone. As long as the damping force in the bearing is larger than the destabilizing force caused by the cross-coupled stiffness coefficient of the inner oil film, the journal whirling amplitude is relatively small in a quasi-stable operating condition; hence, the constant tone is inaudible in the vehicle cabin (see Sect. 7.5.2). In this case, the whirling energy of the rotor, that is proportional to the rotor speed, is dissipated by the damping energy of oil in the bearing. In fact, the damping force of the bearing decreases with the rotor speed because the oil temperature in the inner bearing rises. Therefore, its viscosity strongly reduces at high oil temperatures. The journal begins whirling in the bearing when the damping force is smaller than the destabilizing force in the inner bearing in which the non-dissipated whirling energy increasing with the rotor speed excites the rotor to the oil whirl instability.

On the contrary, the outer oil whirl does not induce any subsynchronous noise because its frequency is extremely low so that the intensity of noise that is proportional to squared frequency is negligible. As a result, the outer oil whirl is inaudible although its deflection amplitude is larger due to the conical vibration mode.

There are two possible measures to reduce or avoid the constant tone minimizing the inner oil whirl.

- **Passive measures**

The outer bearing clearance should be large enough to suppress the constant tone amplitude as it transmits through the outer oil film of the bearing. Additionally, the inner bearing clearance should be as small as possible since the smaller the inner bearing clearance, the less the inner oil whirl becomes. However, an excessively large bearing clearance causes rotor instability since its bearing stiffness coefficient reduces. A very small inner bearing clearance could cause wear between the journal and bearing ring operating with contaminated oil. Therefore, the combination of a large outer bearing clearance and a small inner bearing clearance must be carefully chosen so that less inner oil whirl is induced in the inner clearance without wear and more damping is in the outer clearance.

Another possible measure is the large outer diameter of the bearing; hence, the damping effect of the outer oil film increases to damp the constant tone. However, the ring speed ratio RSR of the bearing reduces leading to increasing bearing friction on the rotor, as discussed in Chap. 6.

- **Active measures**

The main reason of the oil whirl is the surplus of oil flow rate in the converged wedge of the inner bearing clearance. In order to minimize the constant tone, the surplus of oil flow rate must be lessened to keep the inner oil whirl as small as possible; therefore, the rotor remains in the stable condition.

According to Eq. (7.34), the surplus of oil flow rate in the inner bearing is

$$\begin{aligned} \Delta\dot{Q} &= \dot{Q}_{in} - \dot{Q}_{out} \\ &\approx 2RL_i e\lambda_i \Omega(1 + RSR) \end{aligned} \tag{7.75}$$

Applying the continuity equation to the convergent wedge, one obtains the mass balance of oil flow rates.

$$\sum_i \dot{Q}_i = \Delta\dot{Q} - \dot{Q}_w - \dot{Q}_{ax} = 0 \tag{7.76}$$

The whirling-related oil flow rate is written as

$$\dot{Q}_w = 2RL_i e\omega_i \tag{7.77}$$

Substituting Eqs. (7.75, 7.76, and 7.77), one obtains the whirling-related oil flow rate in terms of the surplus of oil flow rate and axial flow rate.

$$\begin{aligned}\dot{Q}_w &= 2RL_i e\omega_i \\ &= 2RL_i e\lambda_i\Omega(1 + RSR) - \dot{Q}_{ax}\end{aligned} \tag{7.78a}$$

Thus, the frequency of the inner oil whirl results as

$$\omega_i = \lambda_i\Omega(1 + RSR) - \frac{\dot{Q}_{ax}}{2RL_i e} \tag{7.78b}$$

The constant tone can be avoided if the whirl velocity ω_i or whirl eccentricity e in Eq. (7.78a) would equal zero.

$$\dot{Q}_w = 2RL_i e\lambda_i\Omega(1 + RSR) - \dot{Q}_{ax} = 0 \tag{7.78c}$$

Equation (7.78c) shows that the whirling-related oil flow rate can be minimized either reducing the surplus of oil flow rate (the first term on the RHS) or increasing the axial oil flow rate (the second term on the RHS) to compensate the surplus oil flow rate. As a result, the whirling-related oil flow rate is smaller leading to reducing the constant tone.

For reducing the surplus oil flow rate, some measures are possible according to Eq. (7.75). They are namely, reducing the inner bearing clearance c to increase the bearing stiffness so that the journal eccentricity $e = c \cdot \varepsilon$ becomes much smaller; shortening the inner bearing width L_i; reducing the journal radius R; decreasing the ring speed ratio RSR. However, all these measures must be thoroughly investigated since they could cause the rotor instability, bearing wear due to reducing the bearing load capacity, increase of the bearing friction due to reducing the bearing ring speed, and oil coking in the small inner bearing clearance.

For increasing the axial oil flow rate, axial oil grooves are installed in the inner bearing wall to compensate the surplus oil flow rate (see Fig. 7.23). Therefore, the journal whirling amplitude is reduced; the constant tone is minimized or avoided. However, wear could occur in the bearing because lubricating oil does not maintain sufficiently enough in the bearing clearance, especially at the start–stop driving condition or highly unbalanced rotor.

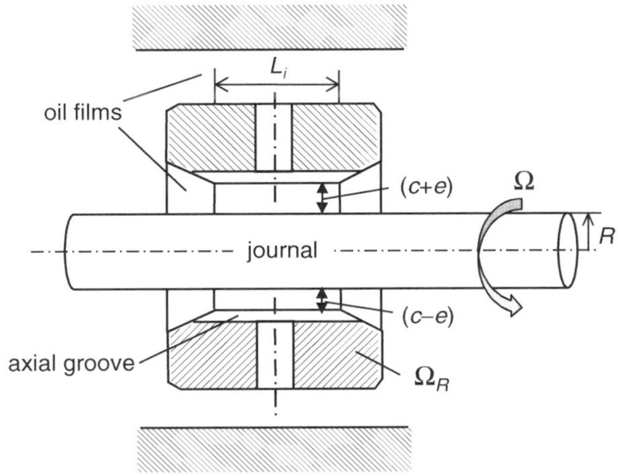

Fig. 7.23 Rotating floating ring bearing with axial grooves

7.8 Aliasing in DFT and Nyquist Frequency

FFT developed by Cooley and Tukey in 1965 is the computing algorithm for discrete Fourier transforms (DFT) that are based on the Fourier transform (FT). The DFT have been used to analyze digitalized time signals of the rotor responses not only in the rotordynamic computation but also in the measurement.

The computing operations of the discrete Fourier transform (DFT) using FFT are reduced from N^2 to $N \cdot \log_2 N$ where N is the discrete values of the continuous time signal. Therefore, the DFT using the FFT algorithm computes much faster and increases the computing accuracy due to round-off errors compared to the DFT using the FT. As an example, at a given number of sampling points $N = 2^{10}$ (= 1024 time discrete values), the required computing operations using the FFT is 1.024×10^4 compared to 1.048×10^6 using the FT. Thus, the number of computing operations using the FFT is reduced by the factor of $N/\log_2 N$, approximately 100 times; and nearly 2200 times for $N = 2^{15}$ time discrete values.

The DFT based on the FFT algorithm calculates the time discrete values partitioning the original full sequence into a number of shorter sequences. Further details about DFT can be found in [13].

7.8.1 Discrete Fourier Transform (DFT)

The periodic time signal $x(t)$ is discretized by the DFT and can be written in the discrete time series (see Fig. 7.24).

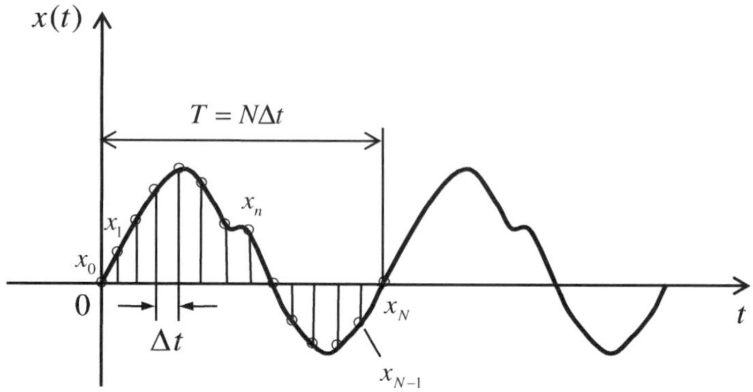

Fig. 7.24 Discrete periodic time signal

$$x(t) = a_0 + 2\sum_{k=1}^{\infty}\left(a_k \cos\frac{2\pi kt}{T} + b_k \sin\frac{2\pi kt}{T}\right) \tag{7.79}$$

where the Fourier coefficients a_k and b_k are defined as

$$\begin{aligned} a_k &= \frac{1}{T}\int_0^T x(t)\cos\frac{2\pi kt}{T}\mathrm{d}t; \quad k \geq 0; \\ b_k &= \frac{1}{T}\int_0^T x(t)\sin\frac{2\pi kt}{T}\mathrm{d}t; \quad k \geq 1. \end{aligned} \tag{7.80}$$

Instead of two Fourier coefficients, one complex Fourier coefficient X_k combined a_k with b_k is used in the DFT [13].

$$X_k \equiv a_k - jb_k \tag{7.81}$$

Using Euler's theorem, the new Fourier coefficient X_k becomes

$$X_k = \frac{1}{T}\int_0^T x(t)e^{-j\left(\frac{2\pi kt}{T}\right)}\mathrm{d}t; \quad k \geq 0 \tag{7.82}$$

The Fourier coefficient X_k is written in the discrete time series of $\{x_n\}$ for n = 0, 1,..., N − 1.

$$X_k \approx \frac{1}{T}\sum_{n=0}^{N-1} x_n e^{-j\left(\frac{2\pi n \Delta t k}{T}\right)} \Delta t \tag{7.83}$$

where Δt is the discrete time interval; N is the number of discrete sampling points in N discrete intervals.

Substituting $T = N\Delta t$ into Eq. (7.83), the Fourier coefficients in the DFT are formulated in the discrete time series $\{x_n\}$ for $n = 0, 1,\ldots, N - 1$ (see Fig. 7.24).

$$X_k = \frac{1}{N}\sum_{n=0}^{N-1} x_n e^{-j\left(\frac{2\pi n k}{N}\right)}; \quad k = 0, 1, \ldots, N - 1 \tag{7.84}$$

Inversing the DFT (IDFT: inverse discrete Fourier transform), one obtains the values of x_n.

$$x_n = \sum_{k=0}^{N-1} X_k e^{j\left(\frac{2\pi k n}{N}\right)}; \quad n = 0, 1, \ldots, N - 1 \tag{7.85}$$

The Fourier coefficients X_k result in the DFT as

$$X_k = \frac{1}{N}\sum_{n=0}^{N-1} x_n e^{-jn\omega_k \Delta t}; \quad k = 0, 1, \ldots, N - 1 \tag{7.86}$$

in which the periodic frequency is defined as

$$\omega_k = \frac{2\pi k}{T} = \frac{2\pi k}{N\Delta t} \tag{7.87}$$

According to Eq. (7.84), the amplitude $|X_k|$ of the Fourier coefficient X_k is an even function; hence, it is symmetric about $\omega_k = 0$ (see Fig. 7.25). It is written as

$$|X_k| = |X_{-k}| \tag{7.88}$$

The amplitudes $|X_k|$ of the time discrete components x_n at the frequencies ω_k for $k = 0, 1,\ldots, N$ are computed according to Eq. (7.86). At the higher frequencies $\omega_k > \omega_{\text{lim}}$ ($k > N/2$), the amplitudes $|X_k|$ of the Fourier coefficients X_k are symmetrical about the limit frequency ω_{lim} (see Fig. 7.25). Due to aliasing of the amplitude, the Fourier coefficients X_k computed by the DFT are correct for angular frequencies up to ω_{lim} [13].

$$\omega_k \le \omega_{\text{lim}} \equiv \frac{\pi}{\Delta t}; \quad k = 0, 1, \ldots, N/2 \tag{7.89}$$

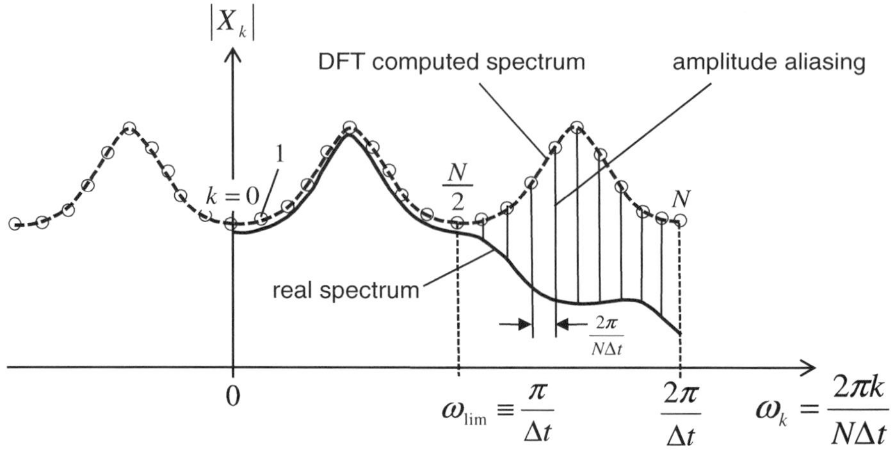

Fig. 7.25 Aliased amplitudes between the DFT-computed and real spectra

7.8.2 Aliasing in DFT

If any frequency of the real signal exceeds the limit frequency ω_{lim} in Eq. (7.89), the Fourier coefficients computed by the DFT have no longer been correct leading to aliasing (distortion) of the amplitudes between the DFT computed and real spectra in the frequencies above ω_{lim}, as shown in Fig. 7.25.

7.8.3 Nyquist Frequency

For anti-aliasing of the DFT-computed amplitudes, the maximum angular frequency $\omega_{\max}$ (rad/s) of the real time signal must be below the limit frequency.

$$\omega_{\max} < \omega_{\text{lim}} \equiv \frac{\pi}{\Delta t} \tag{7.90}$$

Thus, the maximum frequency $f_{\max}$ (1/s) of the vibration response should be limited by

$$\begin{aligned} f_{\max} &= \frac{\omega_{\max}}{2\pi} \\ &< \frac{\omega_{\text{lim}}}{2\pi} = \frac{1}{2\Delta t} \end{aligned} \tag{7.91}$$

The frequency $1/(2\Delta t)$ on the RHS term of Eq. (7.91) is called the Nyquist frequency where Δt is the time interval between two discrete time values of the vibration response (see Fig. 7.24).

Hence, the Nyquist criterion is satisfied for anti-aliasing when the maximum sampling frequency is below the Nyquist frequency.

$$f_{Nyq} \equiv \frac{1}{2\Delta t} > f_{\max} \tag{7.92}$$

Using Eq. (7.92), the sampling frequency results from the tasting time as

$$f_S = \frac{1}{\Delta t} = 2f_{Nyq} > 2f_{\max} \tag{7.93}$$

According to Eq. (7.93), the sampling frequency (twice the Nyquist frequency) must be chosen at least higher than two times of the maximum signal frequency.

To satisfy the Nyquist criterion for anti-aliasing, the discrete time interval Δt at sampling must be smaller than $\Delta t_{\max}$.

$$\begin{aligned} \Delta t &= \frac{1}{2f_{Nyq}} \\ &< \frac{1}{2f_{\max}} = \Delta t_{\max} \end{aligned} \tag{7.94}$$

The sampling time interval Δt is chosen for the integrating time step in computation of nonlinear rotordynamics. In other words, the time interval must be small enough to cover the entire frequency range of the vibration response. Hence, the aliasing of the DFT computed amplitudes is prevented in the computation. Moreover, at measurements of the vibration responses, the sampling time interval is chosen according to the Nyquist criterion, as given in Eqs. (7.92, 7.93, and 7.94) for the anti-aliasing of the measured amplitudes.

The diagram in Fig. 7.26 shows how to select the discrete time interval. At the maximum frequency $f_{\max} = 0.1$ kHz of the sampling time signal, the intersection between the Nyquist frequency $f_{Nyquist}$ and maximum frequency line $f_{\max}$ gives the maximum time interval $\Delta t_{\max}$. To fulfill the Nyquist criterion, the discrete time interval Δt must be smaller than $\Delta t_{\max}$ so that $f_{Nyquist} > f_{\max}$.

If the real time signal contains a very high frequency, an extremely small integrating or sampling time interval is required for anti-aliasing leading to an extreme computing time (CPU). In this case, the maximum frequency line $f_{\max}$ moves upward in Fig. 7.26; the sampling time interval becomes smaller, leading to higher CPU. Therefore, such high frequencies are filtered from the real signal using the undersampling or band-pass sampling technique for an appropriate time interval. As a result, the maximum frequency of the real time signal after undersampling is lower than the Nyquist frequency at the appropriate time interval so that the Nyquist criterion is satisfied.

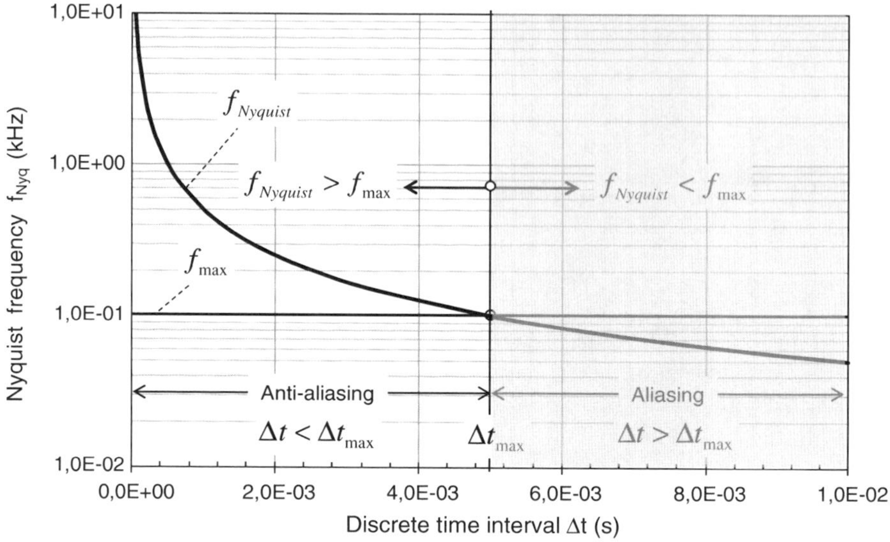

Fig. 7.26 Nyquist frequency versus discrete time interval

7.9 Computations of Nonlinear Rotordynamics

Using computation of nonlinear rotordynamics, the nonlinear rotor behavior of an automotive turbocharger with two rotating floating ring bearings is studied in the following section.

The following computational features are taken into account in the rotordynamic computation that is based on the further developed subroutine for nonlinear rotordynamics of the automotive turbochargers [11]:

- Using the finite element method for the computing model;
- Two-phase flows of the oil films in the bearing clearances;
- Rotating floating ring bearings with changing ring speed ratios;
- Nonlinear bearing forces of the rotating floating ring bearings;
- Run-up simulation to very high rotor speeds;
- Transient behavior of the rotor response;
- Frequency jumps in Waterfall diagram;
- Simulation of synchronous (unbalance) and asynchronous vibrations (inner and outer oil whirls);
- Gyroscopic effect of the flexible rotor;
- Unbalance forces and moments acting upon the rotor;
- Damping effects of the seal rings;
- Computing the ring speed ratios of the bearing rings;
- Computing the induced temperatures in the inner and outer oil films;
- Heat convection in the oil films in the axial direction;

- Heat conduction between the oil films in the radial direction;
- Various types of lubricating oil;
- Various inlet pressures and temperatures of lubricating oil.

7.9.1 Two-Phase Reynolds Lubrication Equation

Figure 7.27 shows a two-phase oil model in the bearing clearance in which the oil mixture volume V_{mix} consists of the volumes of gas bubbles V_B and pure oil liquid V_{liq}. As a result, the oil mixture volume is the sum of gas and oil liquid volumes.

$$V_{mix} = V_{liq} + V_B.$$

The dynamic viscosity of the oil mixture results from the oil liquid viscosity and gas bubble fraction, cf. [2, 14].

$$\eta_{mix} = \eta_{liq}\left(\frac{1}{1+r}\right) = \eta_{liq}\theta \tag{7.95}$$

where $r = \frac{V_B}{V_{liq}}$ is the bubble fraction of the oil liquid.

The filling grade θ in the bearing is defined as the ratio of the oil liquid to oil mixture volumes; it can be written in the bubble fraction r.

$$\begin{aligned} \theta &\equiv \frac{V_{liq}}{V_{mix}} = \frac{V_{liq}}{V_{liq}+V_B} = \frac{1}{1+\frac{V_B}{V_{liq}}} \\ \Rightarrow \theta &= \frac{1}{1+r} \end{aligned} \tag{7.96}$$

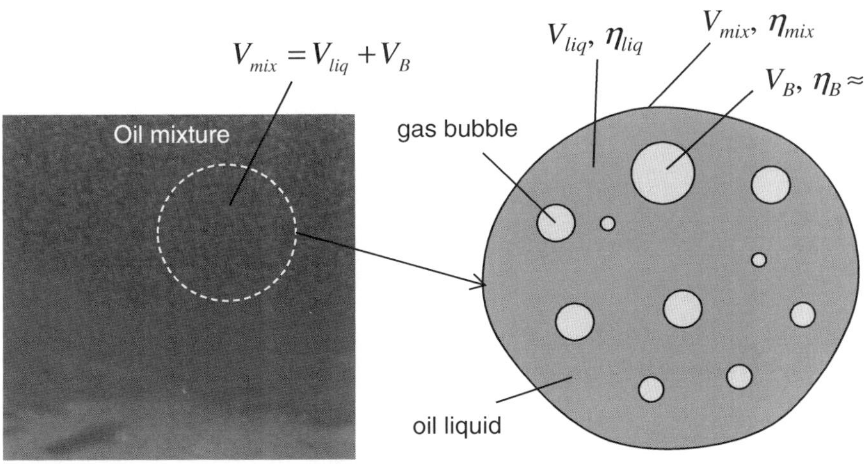

Fig. 7.27 Computing model of a two-phase oil mixture

Similarly, the oil mixture density is calculated from the oil liquid density and filling grade in bearing at $m_B \ll m_{liq}$.

$$\begin{aligned}\rho_{mix} &= \frac{m_{mix}}{V_{mix}} \approx \frac{m_{liq}}{V_{liq} + V_B} \\ &= \frac{m_{liq}}{V_{liq}} \frac{1}{1 + \frac{V_B}{V_{liq}}} = \rho_{liq}\left(\frac{1}{1+r}\right) \\ &= \rho_{liq}\theta\end{aligned} \tag{7.97}$$

Equations (7.95 and 7.97) indicate that the dynamic viscosity and density of the oil mixture depend only on the gas bubble fraction. Therefore, they are independent of the characteristics of the dissolved gas in the oil liquid.

There are two cases of the filling grade in bearing. First, $\theta = 1$ for the pure oil liquid without air releasing or cavitation. Second, $\theta = 0$ for the oil condition at 100 % gas phase caused by a full cavitation in oil; i.e., the oil liquid is totally evaporated ($V_{liq} = 0 \Rightarrow r \to \infty$). In fact, the filling grade in bearing lies between 0 and 1 ($0 < \theta \leq 1$) in the most applications.

Using the Henry-Dalton and Boyle–Mariotte's laws, the gas bubble fraction at T and p is calculated as

$$\begin{aligned}r &\equiv \frac{V_B}{V_{liq}} \\ &= r_0\left(\frac{Tp_0}{T_0 p}\right) - a_v\left(1 - \frac{Tp_0}{T_0 p}\right) \\ &= (r_0 + a_v)\left(\frac{Tp_0}{T_0 p}\right) - a_v\end{aligned} \tag{7.98}$$

where chosen between 0.08 and 0.09 for the automotive mineral oils to a pressure of 30 MPa (ISO-VG 32 to ISO-VG 220). The Bunsen solubility coefficient is defined as the maximum dissolved gas volume per unit oil liquid volume in the equilibrium condition at the gas partial pressure of 101.3 kPa;

- a_v is the Bunsen's coefficient;
- r_0 is the initial gas bubble fraction in the oil liquid at the bearing inlet;
- T and p are the oil temperature and pressure in the oil film in the operating condition;
- T_0 and p_0 are the ambient or oil inlet temperature and pressure.

Substituting Eq. (7.98) into Eq. (7.96), one obtains the filling grade in bearing

$$0 < \theta \equiv \frac{V_{liq}}{V_{mix}} = \frac{1}{(1 - a_v) + (r_0 + a_v)\left(\frac{Tp_0}{T_0 p}\right)} \leq 1 \tag{7.99}$$

The two-phase Reynolds lubrication equation is written as [2]

$$\frac{\partial}{\partial x}\left(h^3\frac{\partial p}{\partial x}\right)+\frac{\partial}{\partial z}\left(h^3\frac{\partial p}{\partial z}\right)=\frac{6\eta_{liq}}{\rho_{liq}}\left[(U_r+U_b)\frac{\partial(\rho_{liq}\theta h)}{\partial x}+2\frac{\partial(\rho_{liq}\theta h)}{\partial t}\right] \quad (7.100)$$

where;

$p(x, z)$	is the current oil film pressure in the bearing clearance;
$\eta_{liq}(T)$	is the oil liquid viscosity resulting from the Cameron and Vogel's equation;
U_r and U_b	are the circumferential velocities of the rotor and bearing ring;
$h(x, t)$	is the current oil film thickness in the bearing clearance;
θ	is the filling grade in bearing.

7.9.2 Results of Nonlinear Rotordynamics

The computed rotor of an automotive turbocharger including the compressor and turbine wheels, rotor shaft, thrust rings, seal rings, and rotating floating ring bearings is shown in Fig. 7.28. Using the finite element method, the rotor is discretized in a finite number of concentrated mass points, disks, cylinder elements, and interfaces of the bearings and seals to the rotor. As a result, the computed finite element model contains many sections that are connected with each other at their stations (see Fig. 7.29).

Applying the principle of D'Alembert to the finite element rotor, the vibration equations of the rotor with N DOF are written in the discrete matrix equation.

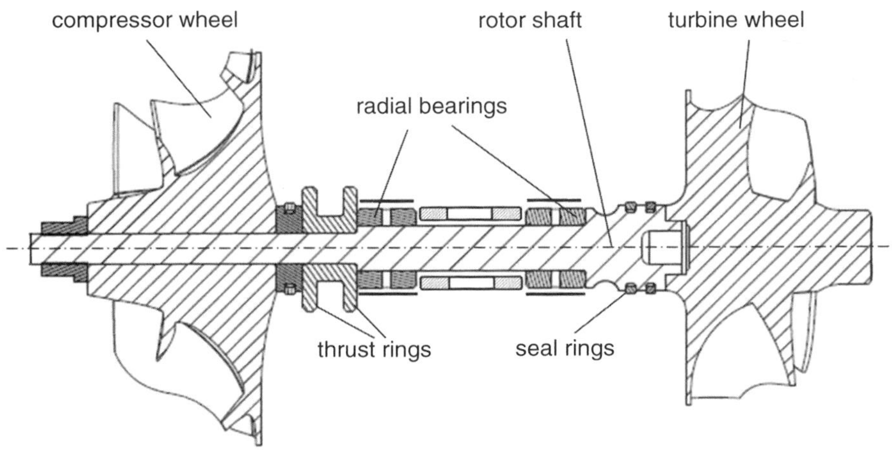

Fig. 7.28 Computed rotor of an automotive turbocharger

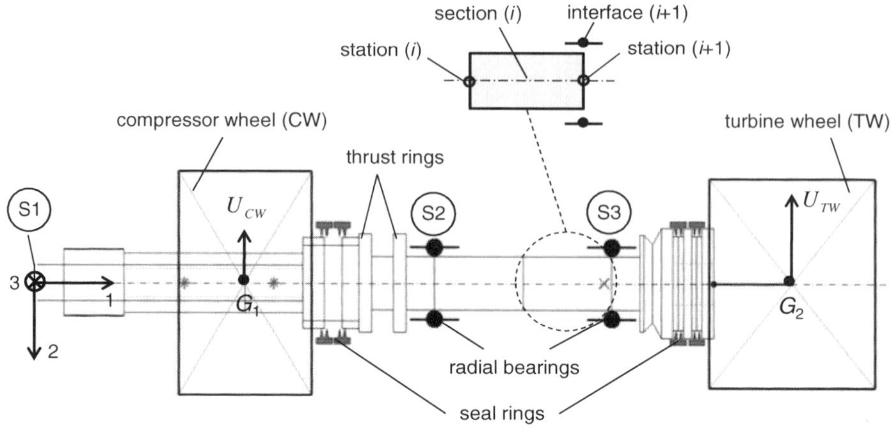

Fig. 7.29 Computed finite element model of the rotor

$$\mathbf{M}\ddot{\mathbf{x}} + \mathbf{C}_{SG}\dot{\mathbf{x}} + \mathbf{K}_S\mathbf{x} = \mathbf{f}(t)$$

where

$\mathbf{M} = (N \times N)$	mass matrix containing the masses and inertia moments of the rotor;
$\mathbf{C}_{SG} = (N \times N)$	damping coefficient and gyroscopic matrix of the rotor;
$\mathbf{K}_S = (N \times N)$	stiffness coefficient matrix containing the diagonal and cross-coupled stiffness coefficients of the shaft;
$\mathbf{x} = (N \times 1)$	vibration response vector including all DOF of the rotor with two translational and two rotational displacements at each station of the finite element;
$\mathbf{f}(t) = (N \times 1)$	vector of the unbalance and nonlinear bearing forces.

Solving the two-phase Reynolds lubrication equation in the bearing, as shown in Eq. (7.100), the bearing characteristics are computed using the impedance method. They are used to calculate the bearing forces on the RHS of the vibration equations. The nonlinear bearing forces in the entire bearing clearance result from the corresponding bearing stiffness and damping coefficients linearly interpolating the computed values of the bearing characteristics at ε, γ, and Sommerfeld number *So* in the impedance data of the bearing.

The nonlinear bearing forces contain two force components: the rotation force induced by the rotation with Ω depends on α and $So(\eta, \varepsilon, \gamma)$; and the damping force caused by the journal velocities $\dot{\varepsilon}, \dot{\gamma}$ depends on $\eta, \varepsilon, \gamma, \dot{\varepsilon}, \dot{\gamma}$.

The nonlinear bearing forces acting upon the rotor are computed at each iterating step based on the Sommerfeld number *So* of the impedance method (see Sects. 6.5.2 and 6.5.3). The coupled nonlinear vibration equations of the rotor are given in Eq. (7.4), in which we set the nonlinear bearing forces on the RHS of the equation.

Using the bimodal method, we decouple the coupled nonlinear Eqs. (7.4) in the decoupled nonlinear Eqs. (7.20). Then, the vibration responses are solved by the high-order Runge–Kutta scheme at each iterating step with the time interval given in Eq. (7.92). Using the DFT, the frequency spectrum of the rotor responses is displayed and analyzed in the spectrogram (Waterfall diagram). The orbits of the rotor at the station S1 of the compressor inlet, and at the stations S2 and S3 of the journal inside the bearings are computed in the run-up simulation, as shown in Fig. 7.28b.

Two rotating floating ring bearings and two sealing rings at each side support the computed rotor with a mass of about 150 g including the shaft with a diameter of 7 mm, compressor wheel, turbine shaft, and thrust rings. The relative inner diametral clearance of the bearing is initially about 3×10^{-3} due to the temperature difference between the inner and outer oil films, and the centrifugal force of the bearing ring, and the bearing ring expansion in radial direction. Hence, the relative inner diametral clearance increases to a maximum of about 4×10^{-3} during operation. The ratio of the outer and inner bearing clearances is varied by a factor between 3 and 4. The maximum rotor speed is limited to nearly 225,000 rpm (Ω = 3750 Hz). The compressor wheel and turbine shaft have residual unbalances between 0.1 and 0.2 g mm, respectively.

The lubricating oil SAE 5W30 is used in the run-up computation; the oil dynamic viscosity changes with the effective oil temperatures in the oil films at every iterating step using the Cameron and Vogel Eq. (6.44a). The oil condition at the inlet is given at the relative pressure of 0.3 MPa (3.0 bar relative) and the temperatures of 90 °C at S2 and 100 °C at S3.

The computing time for the run-up simulation of about 20-h-CPU is necessary on a PC with a dual-core AMD Opteron of 3 GHz with 8 GB RAM. This computing time is quite acceptable in the industry instead of a-few-week-CPU in case of instantaneously solving the Reynolds equation at each iteration step.

Figure 7.30 shows the rotor response at the station S1 in the directions 2 and 3 versus rotor speeds in the run-up simulation up to the rotor speed of 225,000 rpm. The amplitude of the rotor response results from superimposing the harmonic unbalance excitation on the subsynchronous frequency components of the inner and outer oil whirls.

At the beginning, the rotor amplitude is mainly induced by the outer oil whirl up to the relative rotor speed of nearly 30 %; its vibration mode is conical and whirls in the outer bearing clearance (see Figs. 7.32 and 7.33). From the relative rotor speed of 30 % up to 60 %, the rotor response is superimposed by the unbalance, inner, and outer oil whirls. The conical mode of the rotor response tends toward the cylindrical mode. From the relative rotor speed of 60 % to the maximum rotor speed N_{max} = 225,000 rpm, the rotor response is excited by the inner oil whirl and unbalance. Its response has the cylindrical mode, as shown in Figs. 7.33 and 7.34.

Due to the irrational frequency ratio of the inner and outer oil whirls, the rotor response is quasiperiodic vibrations, as discussed in Chap. 3. However, the rotor amplitude is stabilized on the limit cycle at any rotor speed ; e.g., at the relative

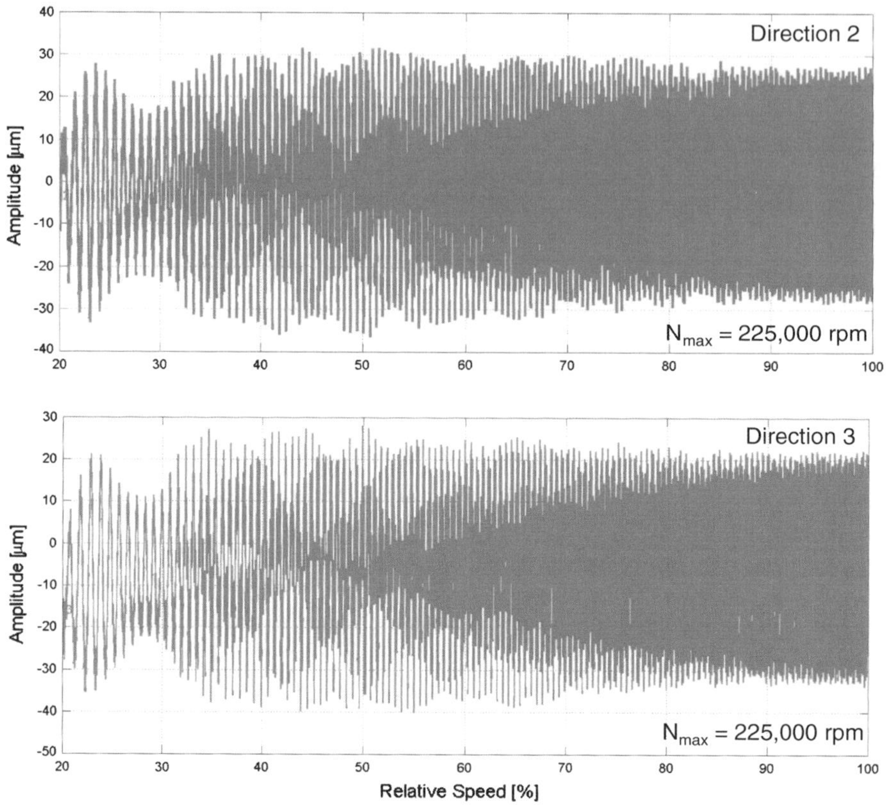

Fig. 7.30 Computed vibration response of the rotor at the station S1

rotor speed of 60 % (135,000 rpm), the limit cycle at the station S1 has a peak-to-peak amplitude of nearly 65 μm (see Fig. 7.31).

Figure 7.31 indicates that the rotor response is a forward whirl since its orbit contains only the inner loops (see Chap. 3); i.e., the whirl velocity ω has the same direction of the rotor speed Ω. Within the limit cycle the rotor orbit is unstable due to the self-excited vibrations with subsynchronous frequencies ($\omega < \Omega$) of the oil whirls. Therefore, the rotor orbit at S1 moves toward the stable limit cycle; during this time, the eccentricity of the journal increases in the bearing clearances leading to reducing the oil film thickness, in turn increasing the stiffness and damping coefficients of the bearing. As a result, the rotor orbit stabilizes on the limit cycle of the Hopf bifurcation (see Chap. 4).

From the rotor speed of 170,000 rpm (75 % relative rotor speed), the limit cycle of the rotor response is reduced from the maximum cycle limit of 70 μm to about 55 μm due to the increase of the bearing stiffness and damping coefficients, as shown in Fig. 7.31. As long as the limit cycle of the rotor response is smaller than both bearing clearances in the entire operating speed range, and each minimum oil

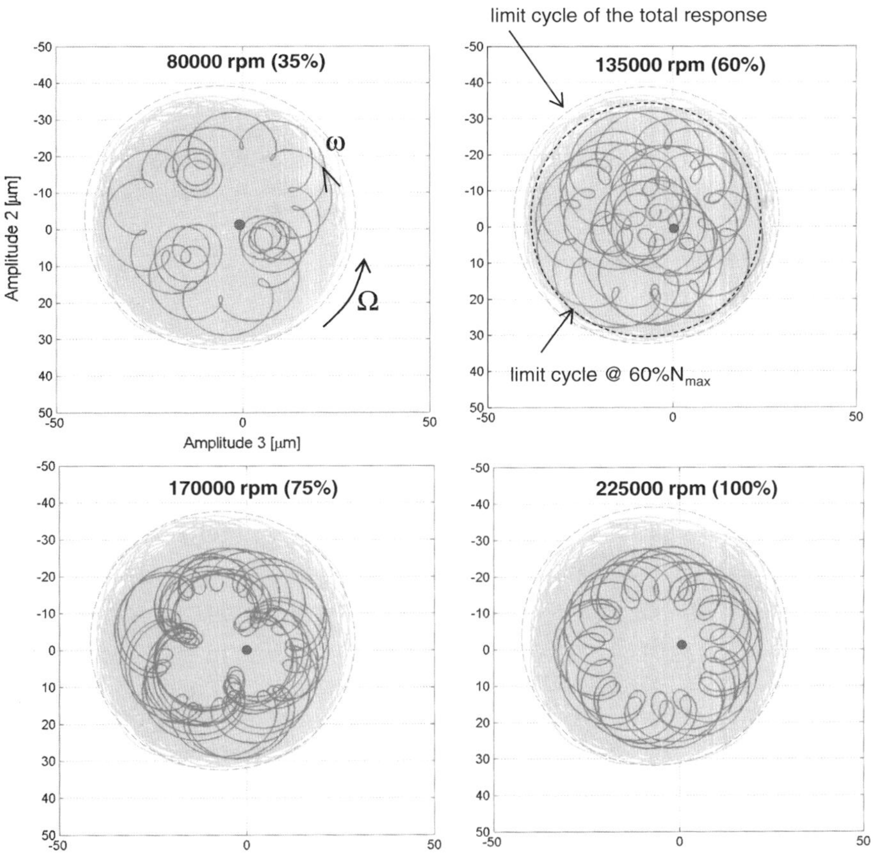

Fig. 7.31 Computed response orbits of the rotor at the station S1

film thickness is larger than the limit oil film thickness, as given in Fig. 6.11, no wear occurs in the bearing.

The vibration mode of the flexible rotor in the run-up simulation is analyzed using the vibration phases at the stations S2 and S3. At low rotor speeds up to nearly 40 % N_{max}, the rotor orbits at S2 and S3 are mostly eccentric to each other (see Fig. 7.32), i.e., the mode shape of the rotor response is conical since the rotor is still rigid at low rotor speeds. At high rotor speeds, the bearing stiffness and damping coefficients increase, the conical mode of the rotor vibration tends toward cylindrical mode. Figure 7.32 shows that orbits at S2 and S3 are nearly concentric from the relative speed of 60 % up to the maximum rotor speed. The limit cycles of the rotor responses at S2 and S3 have a peak-to-peak amplitude of about 25 μm at 60 % and reduce to 15 μm at 100 % relative rotor speed (N_{max} = 225,000 rpm).

Figure 7.33 displays the time responses of the rotor at the stations S2 and S3 with the relative rotor speeds of 20 %, 60 %, and 100 %. At small rotor speeds up to 60 % relative speed, the vibration responses at S2 and S3 in the time domain are out

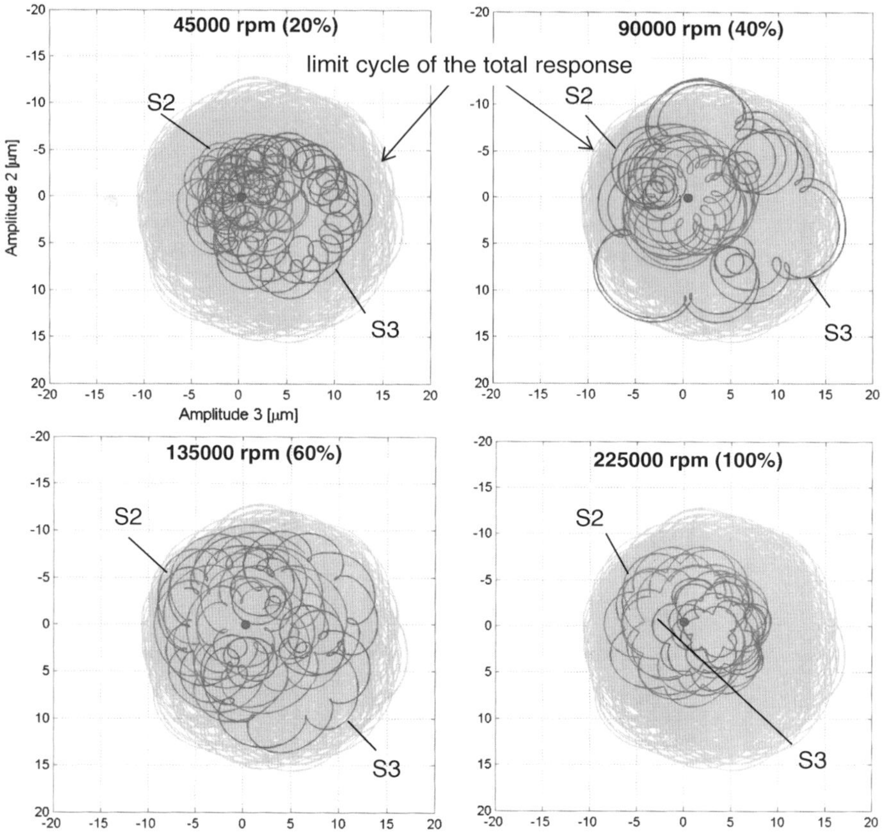

Fig. 7.32 Computed response orbits of the rotor at the stations S2 and S3

of phase (with a phase lag) leading to the eccentric orbits (conical vibration mode), as shown in Fig. 7.32. At increasing the rotor speeds from 60 % up to 100 % relative speed, the time responses at S2 and S3 are in phase (without phase lag) causing the concentric orbits of S2 and S3 (cylindrical vibration mode). In this case, the vibration mode of the rotor response is cylindrical. At each rotor speed, the rotor orbit stabilizes on the limit cycle, as shown in Fig. 7.31.

Figure 7.34 displays the vibration modes of the rotor at various rotor speeds. At low speeds, the rotor vibration has a conical mode. From the rotor speed of about 81,000 rpm (36 % relative rotor speed) to 135,000 rpm (60 % relative rotor speed), the rotor vibration changes from conical to cylindrical mode.

Applying the DFT, one obtains the frequency spectra of the rotor vibration response in the run-up simulation up to 225,000 rpm. Figure 7.35 shows Waterfall diagram of the computed rotor response.

Oil whirling in the outer bearing clearance causes the outer oil whirl. It affects the conical vibration of the rotor response at the relative rotor speed of 20 % with a

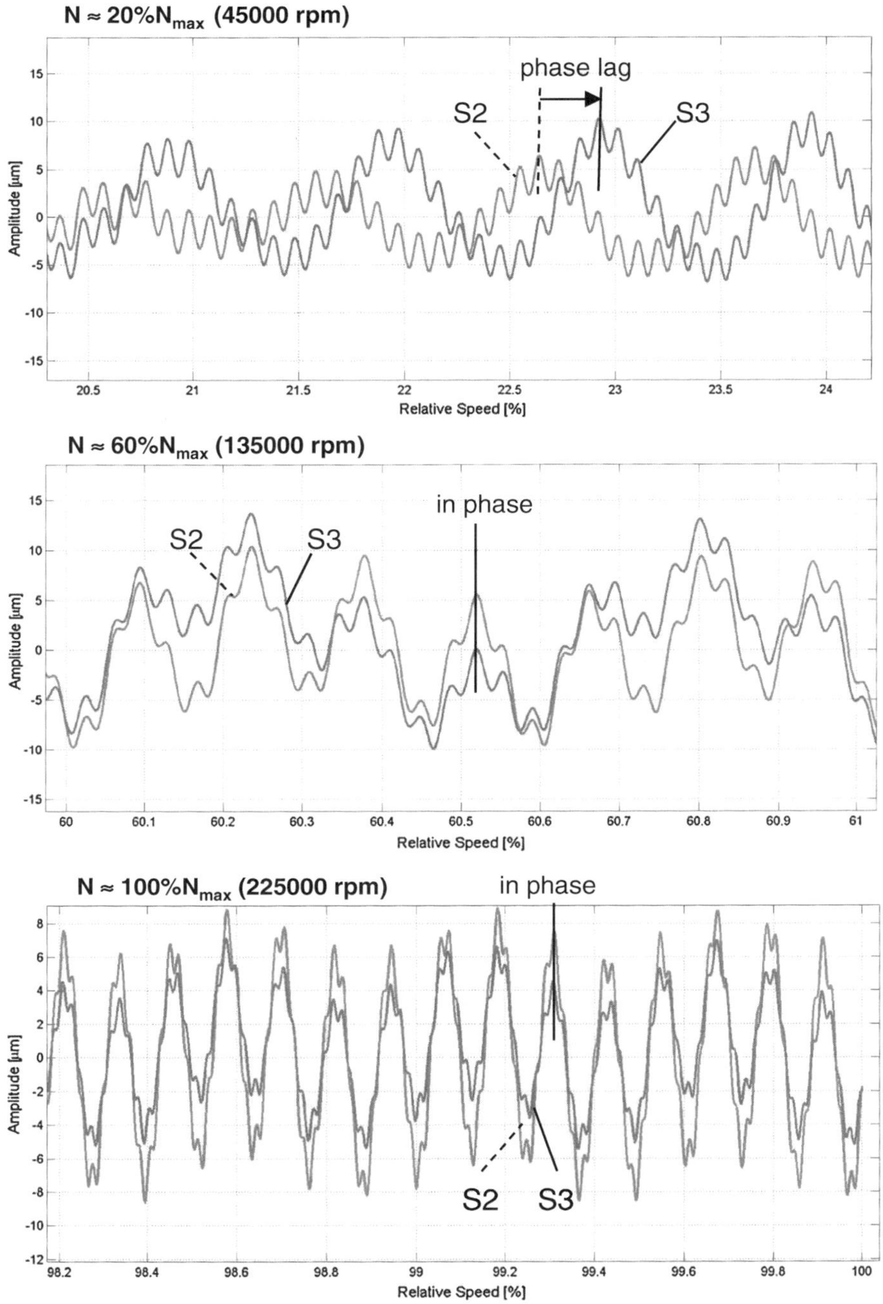

Fig. 7.33 Computed rotor responses at the stations S2 and S3 at various rotor speeds

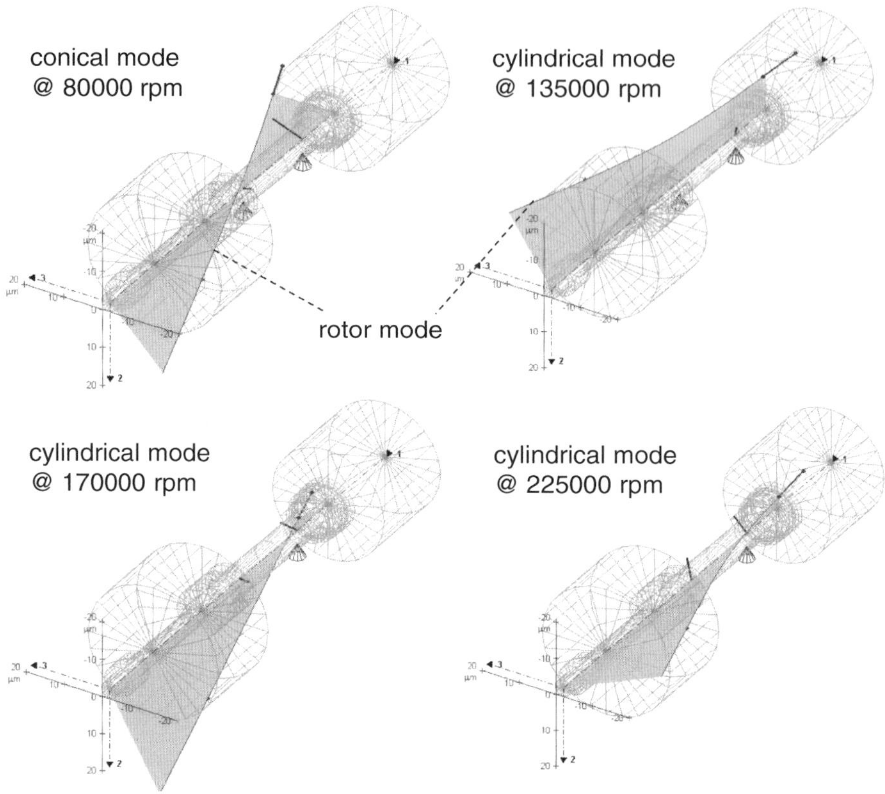

Fig. 7.34 Computed vibration modes of the rotor at various speeds

frequency order of nearly 0.15X (see Fig. 7.36). By increasing the rotor speeds, the oil temperature of the inner oil film is higher than the temperature of the outer oil film due to the bearing friction. Therefore, the ring speed ratio of the bearing reduces leading to decreasing the frequency order of the outer oil whirl to approximately 0.1X at the maximum rotor speed of 225,000 rpm.

Similarly, the inner oil whirl takes place a little later with the whirl frequency order of about 0.4X at the relative rotor speed of 30 %. Because of the reduced bearing ring speed during operation at high rotor speeds, the frequency order of the inner oil whirl drops from 0.4X to nearly 0.2X at the maximum rotor speed. Its vibration response is a cylindrical mode; the journal amplitude moves the bearing ring toward and is finally limited on the limit cycle inside both bearing clearances. The rotor response results from superimposing the harmonic unbalance excitation on the subsynchronous vibration components that are induced by the inner and outer oil whirls. The rotor orbits at the stations S1, S2, and S3 at various rotor speeds are plotted in the phase plane 2–3 (see Figs. 7.31 and 7.32).

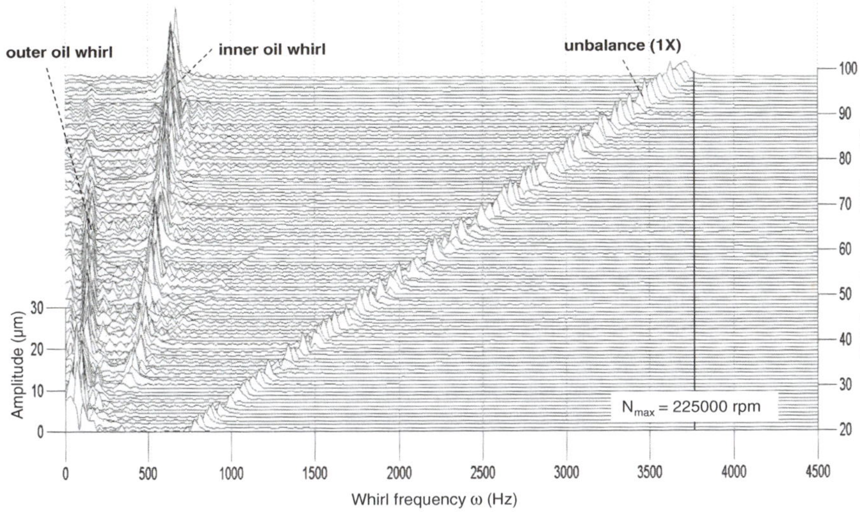

Fig. 7.35 Computed Waterfall diagram at the station S1

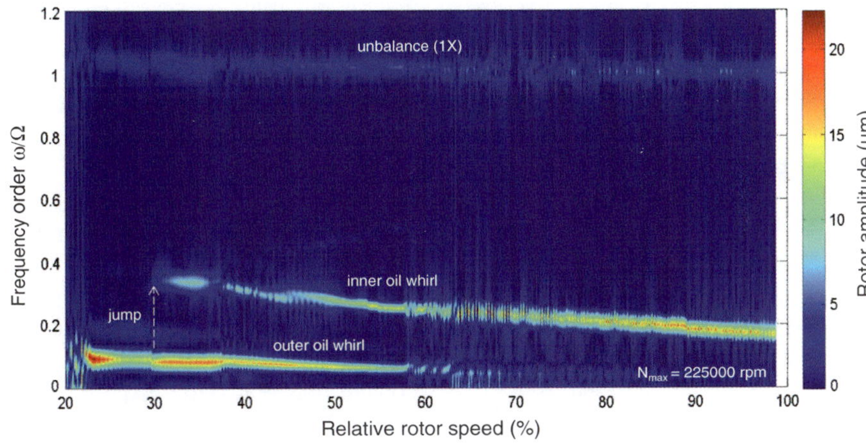

Fig. 7.36 Computed modified Waterfall diagram at the station S1

The unbalance amplitude (1X) is relatively small compared to the asynchronous amplitudes of the oil whirls in the entire rotor speed range. Figure 7.35 shows that the amplitudes of the oil whirls are much larger than the quasi-resonance amplitude of the rotor. In fact, the typical resonance of linear rotordynamics does not exist in nonlinear rotordynamics, instead only the limit cycle of the rotor response occurs at each rotor speed. The reason is that the effective stiffness coefficient of the rotor depends not only on the stiffness coefficient of the rotor shaft itself but also on the bearing stiffness coefficient. Both stiffness coefficients generate the equivalent stiffness coefficient of the rotor called *the complex rotor dynamic stiffness coefficient*. The bearing stiffness is dependent on the effective oil pressure, temperature, rotor speed, bearing ring speed, and journal eccentricity as well. At large rotor deflections, the journal eccentricity increases in the bearing clearance; the oil film thickness reduces in the bearing clearance. Therefore, the stiffness and damping coefficients of the bearing increase with the rotor deflection until the journal orbit stabilizes on the limit cycle at each rotor speed in the supercritical Hopf bifurcation. As a result, no resonance occurs in Waterfall diagram due to continuously increasing the rotor stiffness coefficient near the resonance (see Fig. 7.35).

In practice, Waterfall diagram is plotted in a two-dimensional diagram (modified Waterfall diagram) in which the amplitudes of the vibrations are displayed in the color scale on the frequency order spectra, as shown in Fig. 7.36. Instead of the whirl frequency ω, its frequency order ω/Ω is used in the modified Waterfall diagram with the advantage to easily analyze the frequency order spectra of the synchronous and asynchronous vibrations versus the rotor speed in the abscissa. The frequency jump of the inner oil whirl occurs where the conical vibration mode changes into the cylindrical mode.

Figure 7.37 shows the comparison of the rotor behaviors between the computation and measurement in Waterfall diagrams. Compared to the measurement, the computed inner oil whirl takes place a little later at the relative rotor speed of about 30 % (67,500 rpm) instead of at nearly 27 % (62,000 rpm) at the measurement. The frequency order of the computed inner oil whirl (inner OW) begins at a lower frequency order of 0.4X compared to about 0.55X in the measurement; and ends at the frequency order of 0.2X, nearly the same results in the measurement.

The reason for the low frequency order of the inner oil whirl is that the bearing ring speed in the measurement is higher than the ring speed in the computation. The unknown heat flow transferred from the bearing housing to the outer oil film of the bearing has not been taken into account in the computation. As a result, compared to the computation, the outer oil film temperature is higher in the measurement. Therefore, the ring speed ratio of the bearing in the measurement is larger leading to the higher frequency order of the inner oil whirl, see Eq. (7.26).

The frequency jump of the inner oil whirl occurs at a relative rotor speed of 30 % (about 65,000 rpm). Similarly, the computed frequency order of the outer oil whirl (outer OW) is a little lower than the frequency order as in the measurement due to the reduced bearing ring speed, especially at low rotor speeds. However, it is nearly the same frequency order of the measurement at high rotor speeds since the ring speed ratio of the bearing does not change much at the high-speed range, as shown in Fig. 6.28.

Some unknown boundary conditions that could not be taken into account in the rotordynamic computation are:

- Producing form tolerances;
- Exact inner and outer bearing clearances due to surface non-roundness of the journal, bearing, bearing foundations;

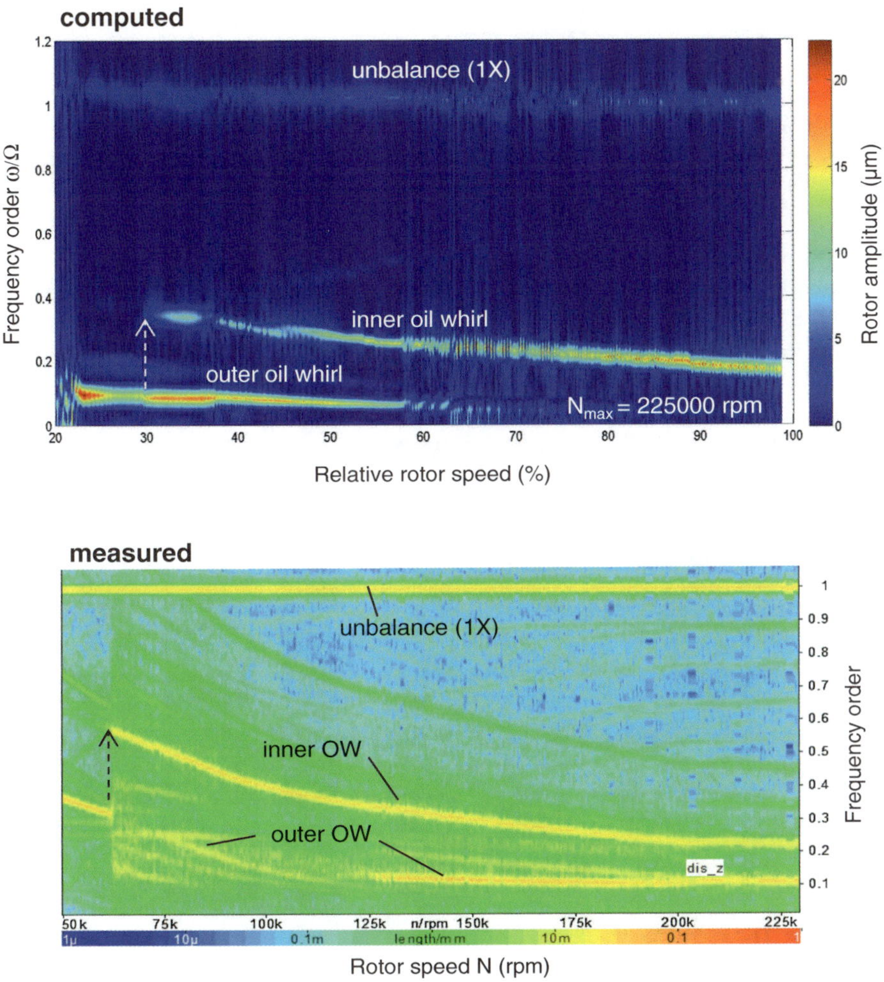

Fig. 7.37 Comparison of Waterfall diagrams at the station S1 [10]

- Non-roundness of the journal and bearing surfaces;
- Non-coaxiality of the bearing and bearing foundations;
- Change of the bearing clearances during operation;
- Tribological surface roughnesses R_z, R_a, R_{pk}, R_k, and R_{vk} of the bearing and journal;
- Real oil temperatures at the bearing inlet at the compressor and turbine sides;
- Fully or poorly lubricating oil conditions;
- Bearing inlet and outlet geometries in the bearing housing;
- Contaminated oil with particles;
- Foamy oil due to air releasing and cavitation;

- Angular position between the unbalance vectors of the compressor and turbine wheels.

However, the rotordynamic computation gives us more deeper understanding on how the rotor reacts at the given boundary conditions so that the rotor response of the turbocharger can be improved; the development time is shortened. In practice, one uses both computations and measurements to predict, analyze, and validate the amplitudes and frequency spectra of the rotor response. In fact, both methods support development of turbochargers.

Furthermore, measurements must be used to validate the lifetime of turbochargers (LCF $< 10^5$ cycles, HCF $> 10^5$ cycles) and wears in the bearings because such effects have been not yet calculated by the rotordynamic computation under the real operating conditions.

Although Campbell diagram is not valid in nonlinear rotordynamics, as discussed in Chap. 5, it is however useful to find the bending critical frequency of the rotor at the trim balancing (see Chap. 8).

The Campbell diagram in Fig. 7.38 shows the eigenfrequencies (natural frequencies) of the forward and backward whirls of the free rotor vibration. In case of linear rotordynamics, the first bending mode of the forward whirl #10 occurs at the critical speed Ω_{cr1} of approximately 147,000 rpm. The second bending resonance of the forward whirl #12 occurs outside the rotor speed range of 225,000 rpm. The first critical speed should be high enough to avoid the rotor instability during the operation of turbocharger. Note that the rotor should be designed so that the first bending critical frequency is as high as possible in order to keep the rotor from oil whip induced by the self-excited vibration in the bearing. In case of a symmetric rotor, the backward whirls #9 and #11 do not excite the rotor to the resonance. In case of unsymmetrical rotors and excitation forces with alternately changing

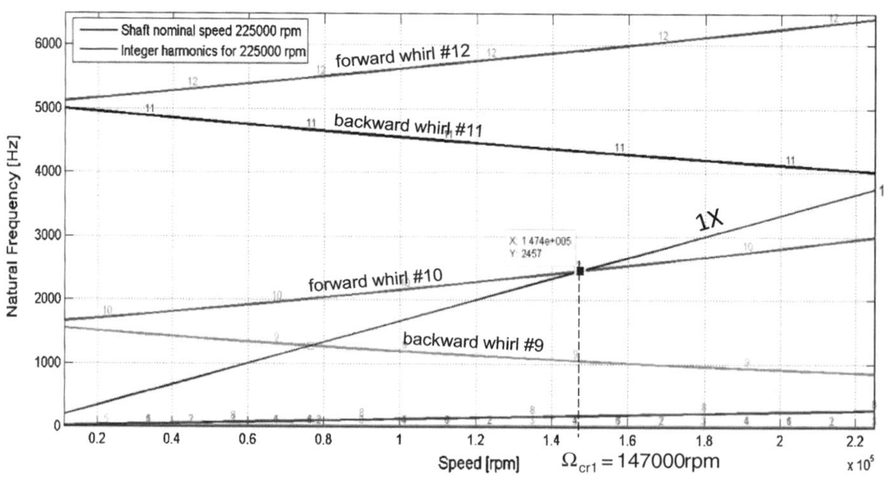

Fig. 7.38 Campbell diagram of the free rotor vibration

directions, the backward whirls could cause resonances at the intersections between the synchronous excitation line (1X) and the backward eigenfrequencies.

In case of the rotating floating ring bearing, the journal moves relatively to the bearing ring at any rotor speed; i.e., the journal pushes the bearing ring in the radial direction on the orbit. In practice, it is very difficult to measure the relative displacements between the journal and bearing ring during the operation. Hence, the rotordynamic computation is a useful tool to compute the relative displacements and the current oil film thickness in the rotating floating ring bearings at any rotor speed.

Figure 7.39 displays the relative displacements between the journal and bearing ring at various rotor speeds. At the beginning, the journal and bearing ring are nearly concentric to each other since the rotor mass is very small (nearly 150 g). If the relative displacement is positive, the journal moves closer to the bearing ring because the journal displacement is larger than the ring displacement. In other case,

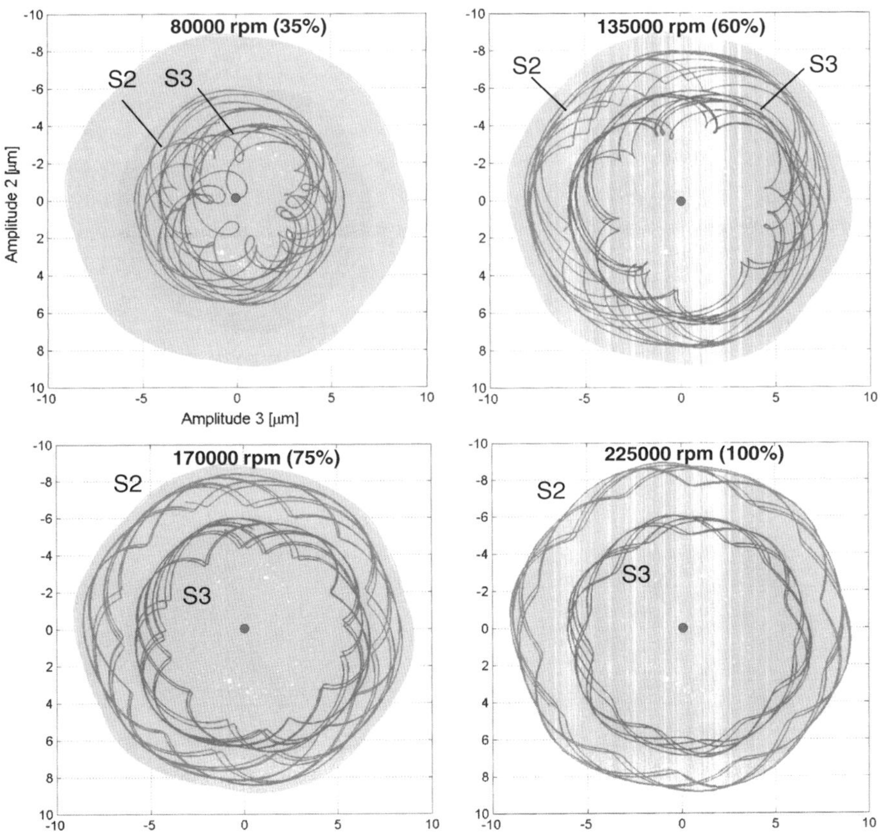

Fig. 7.39 Computed relative displacements between the rotor and bearing ring at the stations S2 and S3

the journal moves away from the bearing ring. The current oil film thickness results from the bearing clearance and relative displacement. When the oil film thickness is larger than the limit oil film thickness in Fig. 6.11, the lubrication regime is fully hydrodynamic, and therefore is outside the mixed and boundary lubrications. As a result, no wear occurs in the bearing.

The computed results show that the maximum relative displacement is about 9 μm on the limit cycle for the entire rotor speed range. It indicates that the minimum oil film thickness in the inner bearing clearance is nearly 5 μm with the inner radial clearance of 14 μm. The minimum oil film thickness of 5 μm in the bearing clearance is larger than the limit oil film thickness; hence, no wear occurs in the bearing.

Therefore, the lubrication regime in the bearing is fully hydrodynamic for all operating conditions. The difference of temperature between two oil films at high rotor speeds and the bearing ring expansion in radial direction lead to extending the inner bearing clearance and reducing the outer bearing clearance. As a result, the minimum inner oil film thickness is larger than 5 μm in practice leading to the larger contact circle.

Figure 7.40 displays the bearing force capabilities acting upon the rotor at the stations S2 and S3 running with various rotor speeds. The bearing forces keep the rotor in balance with the gyroscopic moments of the wheels, the unbalance forces of the rotor shaft, and unbalance moments of the compressor and turbine wheels.

The induced nonlinear bearing forces result from two force components of the rotation and damping forces, as discussed in Sect. 6.5.2. First, the rotation force is generated by the rotor speed Ω; it depends on the oil viscosity η, journal relative eccentricity ε, and the angular position γ. Second, the journal velocities $\dot{\varepsilon}, \dot{\gamma}$ cause the damping force that depends on $\eta, \varepsilon, \gamma, \dot{\varepsilon}, \dot{\gamma}$. At the low rotor speeds up to about 30 % relative speed, the bearing force at the turbine side (S3) is quite larger than the bearing force at the compressor side (S2). At increasing rotor speeds, the bearing force at the turbine side S3 is smaller than the bearing force at the compressor side S2. The journal amplitude at the turbine side S3 is higher than the journal amplitude at S2 (see Fig. 7.32) leading to the higher oil temperature induced in the inner oil film due to bearing friction, in turn the significant reduction of oil viscosity. It results to the lower bearing force at the turbine side S3 compared to the bearing force at the compressor side S2. In this case, the influence of the oil viscosity on the bearing force is stronger than other effects. At the small bearing force capability at the turbine side S3, the bearing could be damaged by a large excitation force in which the oil film thickness is in the mixed or boundary lubrication layer.

The maximum bearing force capabilities at each rotor speed are shown in Fig. 7.40. At the maximum rotor speed of 225,000 rpm, the peak amplitude of the bearing force at the station S2 (compressor) is about 20 N and approximately 10 N at the station S3 (turbine). To investigate further the induced bearing forces, their time signals at the stations S2 and S3 are analyzed in Waterfall diagrams using DFT, as shown in Fig. 7.41.

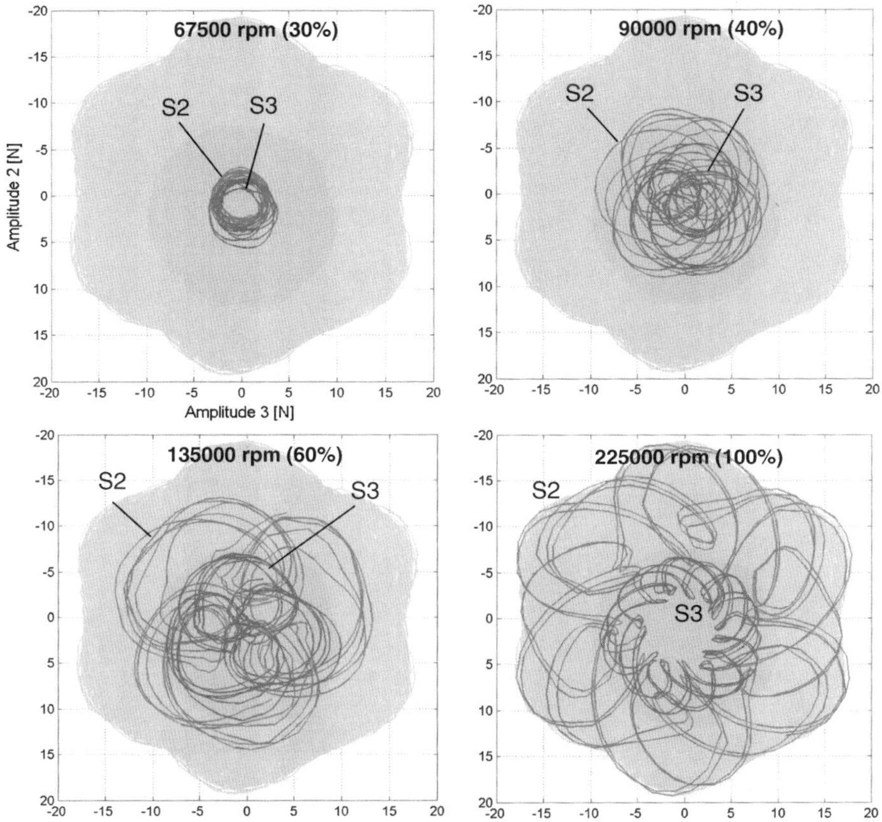

Fig. 7.40 Computed bearing force capabilities acting upon the rotor at the stations S2 (compressor) and S3 (turbine)

The bearing force consists of two force components with the different frequencies: the rotation force has the harmonic frequency ($\omega = 1X$); the damping force, the subsynchronous frequency ($\omega < 1X$). Obviously, the frequency order of the rotation force is synchronous with the rotor speed (1X) because it is proportional to the rotor speed. Similarly, the frequency order of the damping force is subsynchronous with the frequency of the inner oil whirl. The frequency order of the damping force begins with 0.4X at the relative rotor speed of 30 % and continuously decreases to the frequency order of nearly 0.2X at the maximum rotor speed. Its frequency order is the same of the frequency order of the inner oil whirl, as displayed in Fig. 7.36. The color scale denotes their force amplitudes on the frequency curves. The resulting bearing force is superimposed by the rotation and damping forces with an irrational frequency ratio; therefore, the force vibration is quasiperiodic.

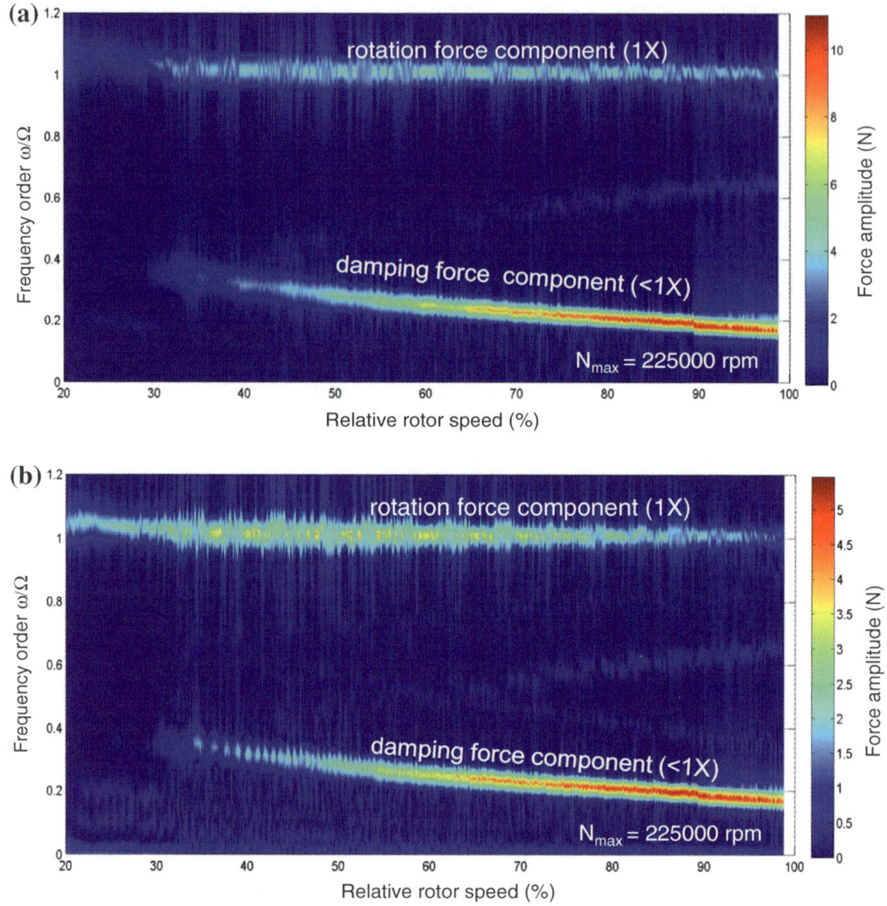

Fig. 7.41 Computed Waterfall diagrams of the bearing forces at the stations S2 and S3. **a** Bearing force at S2 (compressor side). **b** Bearing force at S3 (turbine side)

To reduce the constant tone (subsynchronous noise) induced by the inner oil whirl, one possible measure can reduce the inner bearing clearance, as discussed in Sect. 7.7.2. In case of the large diametral bearing clearance of about 27 μm, the inner oil whirl occurs with the high amplitude leading to the constant tone besides the unbalance whistle in Fig. 7.42a.

Having already discussed, at reducing the inner bearing clearance, the inner oil whirl amplitude decreases. Hence, the constant tone becomes weaker. The computed result with the diametral bearing clearance reduced by 25 % to about 20 μm shows that the inner oil whirl disappears nearly in the entire speed range. However, it only occurs at very small amplitudes from the relative rotor speed of 90 % (see Fig. 7.42b). On the contrary, the outer oil whirl does not induce any

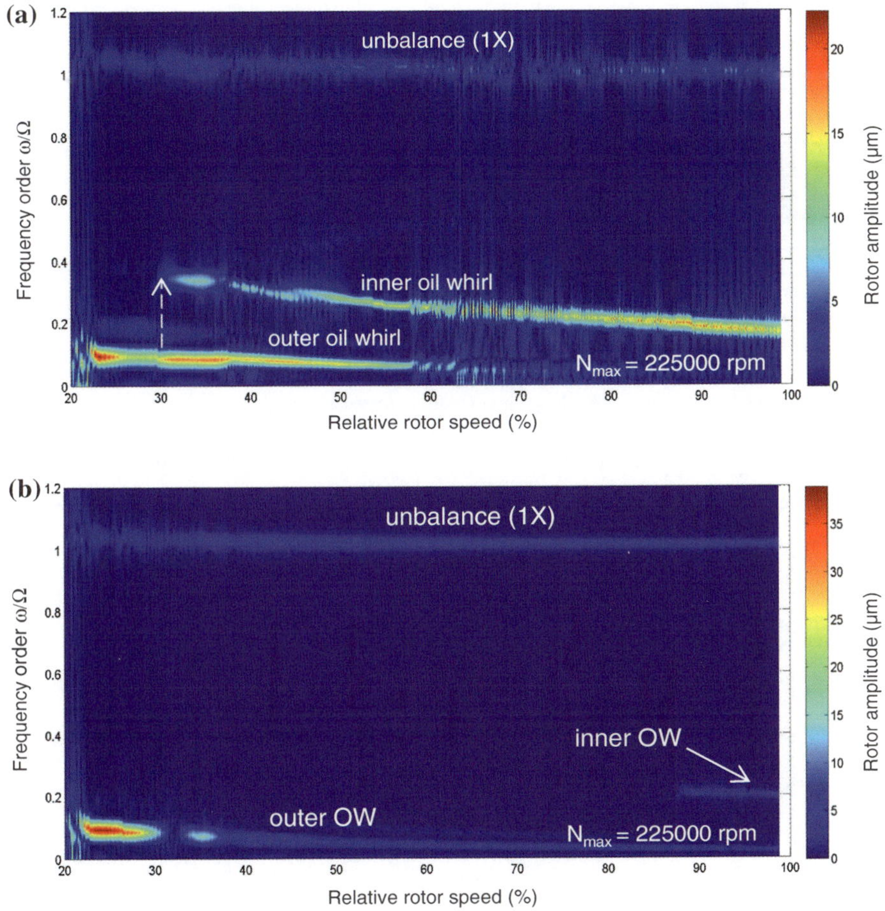

Fig. 7.42 Computed Waterfall diagrams of the rotor responses at different bearing clearances. **a** Large inner bearing clearance. **b** Small inner bearing clearance

subsynchronous noise because its frequency is extremely low so that the intensity of noise is negligible. As a result, the outer oil whirl is inaudible although its whirl amplitude is larger due to the conical vibration mode.

References

1. Adams Jr, M.L.: Rotating Machinery Vibration. CRC, Taylor and Francis, London (2001)
2. Bartel, D.: Simulation von Tribosystemen (in German). Springer-Teubner (2010)
3. Bently, D.E., Hatch, C.: Fundamentals of Rotating Machinery Diagnostics. Bently Pressurized Bearing Press, Minden (2002)
4. Childs, D.: Turbomachinery Rotordynamics. Wiley, New York (1993)

5. Ehrich, F.: Handbook of Rotordynamics. Krieger Publishing Company, Florida (2004)
6. Gasch, R., Nordmann, R., Pfuetzner, H.: Rotordynamik, 2. Auflage. Springer, New York (2006)
7. Genta, G.: Dynamics of Rotating Systems. Springer, New York (2005)
8. Hori, Y.: Hydrodynamic lubrication. Springer, New York (2006)
9. Kraemer, E.: Rotordynamics of Rotors and Foundations. Springer, New York (1993)
10. Kothe, P., Ahrens, T., Kleinschmidt, R.: Measurements of Rotordynamics of Turbochargers. Internal Report BMTS (2010)
11. MADYN 2000: Computational Program for Rotordynamics. Delta JS (2011)
12. Muszýnska, A.: Rotordynamics. CRC, Taylor and Francis, London (2005)
13. Newland, D.E.: An Introduction to Random Vibrations, Spectral & Wavelet Analysis, 3rd edn. Dover Publications Inc., New York (2005)
14. Nguyen-Schäfer, H., Sprafke, P.: Numerical study on interaction effects of the bubbles induced by air-release and cavitation in hydraulic systems. In: Tenth Bath International Fluid Power Workshop, Bath, UK. Research Studies Press (1997)
15. Nguyen-Schäfer, H.: Aero and Vibroacoustics of Automotive Turbochargers. Springer, Berlin (2013)
16. Nordmann, R.: Schwingungsberechnung von nichtkonservativen Rotoren mit Hilfe von Links- und Rechts-Eigenvektoren. VDI Bericht 269 (1976)
17. Rao, J.S.: Rotordynamics, 3rd edn. New Age International Publishers, New Delhi (2007)
18. Rieger, N.F.: Rotordynamics 2—Problems in Turbomachinery. CISM Courses and Lectures No. 297 (1988)
19. Rieger, N.F.: Balancing of Rigid and Flexible Rotors. U.S. DoD (1986)
20. Vance, J.: Rotordynamics of Turbomachinery. Wiley, New York (1988)
21. Yamamoto, T., Ishida, Y.: Linear and Nonlinear Rotordynamics. Wiley, New York (2001)

Chapter 8
Rotor Balancing in Turbochargers

8.1 Reasons for Rotor Balancing

In the following section, we focus only on rotor balancing in turbochargers and not on industrial turbomachines that can be found in [1, 4, 6, 10].

Production process of the compressor wheel and turbine shaft causes a primary unbalance where the mass center does not locate in the geometrical axis of the compressor wheel or turbine shaft. Excessively large unbalance force and moment induce large amplitudes of the rotor response leading to the bearing wear, rub contact, and seizure of the journal, bearings, and as well as the compressor, turbine wheels at their housings. Additionally, the rotor unbalance generates unbalance whistle that is synchronous with the rotor frequency (frequency order 1X). The unbalance whistle is one of the undesirable airborne noises in automotive turbochargers.

There are two possibilities of producing the turbocharger rotor: either production of the rotor *without* or *with* the rotor balancing. If we choose the first *without* rotor balancing, the production of the compressor wheel and turbine shaft must be extremely precise so that the mass center eccentricity of the rotor is of the order of a few microns ($\sim 10^{-6}$ m) from its geometric axis. It would lead to large deficient producing rates and therefore very high production cost. Note that the *production cost* is one of the most important competition key products in the industry; and "*cost, cost cut, and cost breakdown*" always accompany the products on their lifetimes (PPR: price performance ratio). Therefore, the economical way is to produce turbochargers *with* rotor balancing at which the production process is less precise. Despite the less precise production process, the rotor reaches an acceptable residual unbalance after the balancing procedure that brings the mass center of the rotor possibly as close as possible to the rotating geometric axis.

H. Nguyen-Schäfer, *Rotordynamics of Automotive Turbochargers*,
Springer Tracts in Mechanical Engineering, DOI 10.1007/978-3-319-17644-4_8

8.2 Kinds of Rotor Balancing

Two kinds of rotor balancing are usually applied in the mass production of automotive turbochargers.

- *Low-speed balancing* (*shop balancing*) is used in rigid rotors at a low balancing speed up to 3400 rpm that depends on the balancing machine types to reduce the primary unbalance caused by the process of mass production. Generally, the whole rotor is not balanced at the shop balancing, but only the compressor wheel and turbine shaft are separately balanced in the rigid state with low-speed balancing. Hence, it is also called *single part balancing*. The low-speed balancing is generally carried out with two balancing planes at the nose and back face of the wheel. The goal of this balancing is to reduce the unbalance force and moment to minimize the induced unbalance excitations and therefore to prevent the radial bearings from damage due to contact rub and seizure between the journal and bearings caused by excessive rotor unbalances.
- *High-speed balancing* (*trim balancing*) is applied to flexible rotors at a high balancing speed (generally above the first critical speed) at which the rotor deflects in the lateral direction due to large unbalance forces and moments. Besides the primary unbalance due to mass production, an additional unbalance results from mounting the compressor wheel on turbine shaft and as well as the deformed rotor at high rotor speeds during operation. These additional unbalances lead to unbalance whistle; therefore, they must be removed using trim balancing of the rotor. High-speed balancing is only carried out with two balancing planes at the screw-nut of the compressor wheel and the hub surface between the blades at the compressor wheel outlet. The turbine wheel could not be high-speed balanced since the balancing planes on the turbine wheel are not accessible. The purpose of trim balancing is only to reduce the unbalance whistle during operation at high rotor speeds, especially in passenger vehicles. Generally, trim balancing is not carried out in turbochargers of commercial vehicles and industrial applications unless the customers explicitly require; e.g., turbochargers for public transport buses and delivery vehicles of goods.

8.3 Two-Plane Low-Speed Balancing of a Rigid Rotor

The production of the compressor wheel and turbine shaft induces static, couple, and dynamic unbalances. Static unbalance occurs when the polar mass-inertia axis I_p differs from rotation axis by eccentricity ε. On the contrary, couple unbalance occurs when the polar mass-inertia axis I_p differs from rotation axis by a misalignment angle α; however, they intersect at the mass center G of the wheel on the rotation axis [2, 3, 9].

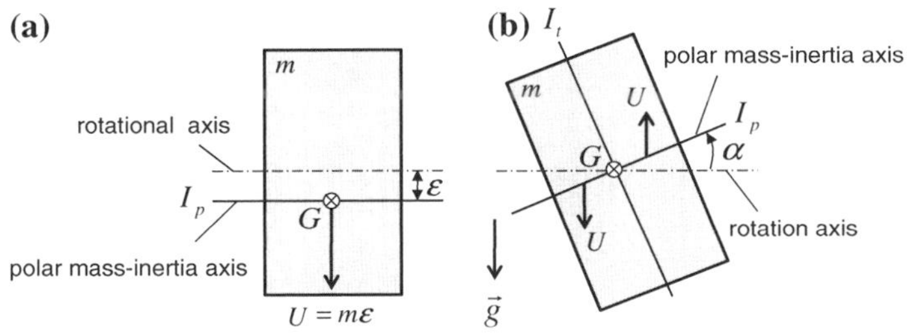

Fig. 8.1 **a** Static unbalance; **b** Couple unbalance

In static unbalance, the mass center G always lies below the rotation axis due to gravity at the equilibrium position (see Fig. 8.1a). This means the rotor component moves from any initial position to the equilibrium due to its weight. Hence, it is also called as *static unbalance* since the rotor moves itself to the equilibrium position in the static unbalance. By contrast, in couple unbalance the rotor does not move itself to the equilibrium position because the mass center G is always in the rotation axis. The couple unbalance is only recognized in the rotating condition in which an unbalance moment acts upon the rotor although the static unbalance equals zero, as shown in Fig. 8.1b.

Static unbalance is caused by production mistakes, large eccentricity at welding the turbine wheel on the rotor shaft, nonhomogeneous material, and residual unbalance after shop balancing. Moreover, it is also induced by thermal deformation, wear, and plastic deformation that are generated at over-speeding the rotor, over-heating the turbine wheel, and the impact of hard particles in the mixed charge air on the compressor wheel in case of low-pressure EGR after a long operating time. On the contrary, the couple unbalance is created by a large non-coaxiality at welding the turbine wheel on the rotor shaft, thermal and plastic deformations of the rotor due to over-speeding or loose screw-nut of the compressor wheel during operation of turbochargers. In practice, we have both static and couple unbalances at the production of the turbine shaft that includes the turbine wheel and rotor shaft leading to the dynamic unbalance of the rotor.

Dynamic unbalance is similar to couple unbalance, but the mass center of the rotor does not lie on the rotation axis (see Fig. 8.2). Therefore, dynamic unbalance can be decomposed in static and couple unbalances. In fact, dynamic unbalance often occurs in the mass production of the turbine shaft at welding the turbine wheel on the rotor shaft. The eccentricity ε of the static unbalance and the non-coaxiality with misalignment angle α of couple unbalance are combined together leading to dynamic unbalance (see Fig. 8.2). The unbalance of the turbine shaft is the largest part of the entire rotor unbalance. It consists of primary unbalance of the turbine wheel and eccentricity and non-coaxiality between the shaft and turbine wheel [2, 3, 9].

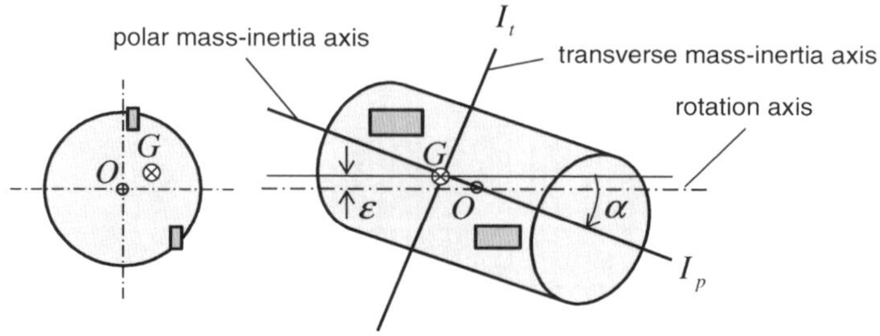

Fig. 8.2 Dynamic unbalance of rotor

Fig. 8.3 Unbalance radius ε of rotor

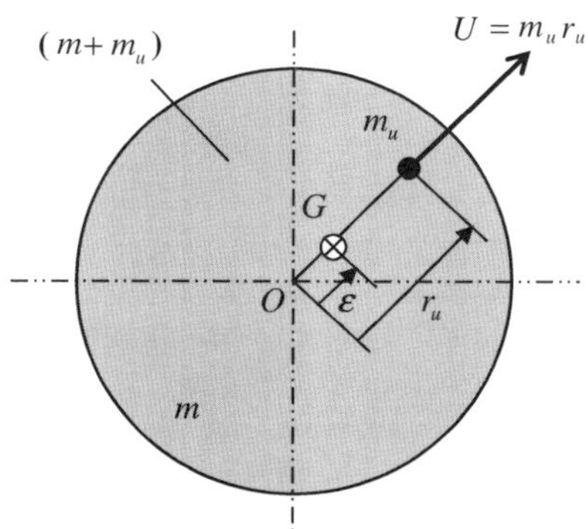

Considering the rotor with unbalance mass m_u at a radius r_u, as shown in Fig. 8.3, static unbalance of the rotor is written as

$$U = m_u r_u \tag{8.1}$$

Due to unbalance mass m_u, the resulting mass center G of the rotor locates at the unbalance radius ε. The rotor unbalance results as

$$U = (m + m_u)\varepsilon \tag{8.2}$$

Substituting Eqs. (8.1 and 8.2), one obtains the unbalance radius at $m_u \ll m$.

$$\varepsilon = \frac{m_u r_u}{(m + m_u)} \approx \frac{U}{m} \Rightarrow U \approx m\varepsilon \tag{8.3}$$

Figure 8.4 shows an unbalanced compressor wheel with the unbalance vector U_ε at the mass center G. At the shop balancing, the rotor is balanced by two balance vectors U_1 and U_2 at two balancing planes: the nose face (plane 1) and the back face (plane 2) of the compressor wheel. Generally, the balance vectors are induced by removing masses of material m_1 and m_2 at given radii of r_1 and r_2 in the opposite

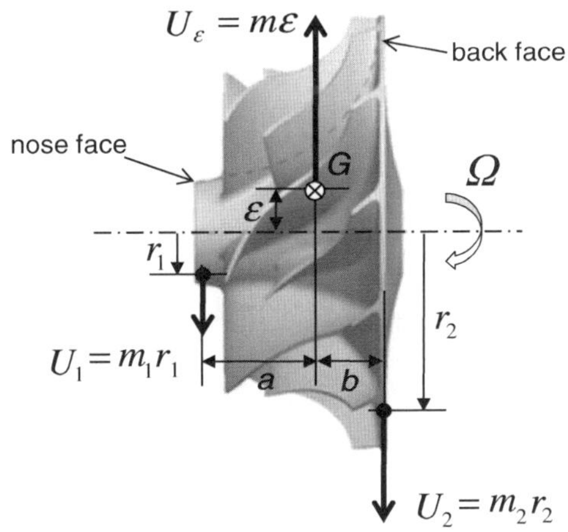

Fig. 8.4 Two-plane low-speed balancing of compressor wheel

direction of the balance vectors U_1 and U_2, respectively. The unbalance vector U_ε is called the primary unbalance of the compressor wheel due to its mass production of molding or milling.

The balancing masses of m_1 and m_2 are computed from the equations of unbalance vectors and moments.

$$\begin{aligned} \sum U_i &= U_\varepsilon - U_1 - U_2 = m\varepsilon - m_1 r_1 - m_2 r_2 = 0 \\ \sum M_i &= U_1 a - U_2 b = m_1 r_1 a - m_2 r_2 b = 0 \end{aligned} \tag{8.4}$$

Thus,

$$\begin{aligned} m_1 &= \frac{m\varepsilon b}{r_1(a+b)} = \frac{U_\varepsilon b}{r_1(a+b)}; \\ m_2 &= \frac{m\varepsilon a}{r_2(a+b)} = \frac{U_\varepsilon a}{r_2(a+b)} \end{aligned} \tag{8.5}$$

Determining the balancing radii of r_1 and r_2 at the balancing planes, the removed masses m_1 and m_2 in the opposite directions of the balance vectors U_1 and U_2 of the compressor wheel are computed according to Eq. (8.5). The balancing procedure is carried out at low balancing speed up to 3400 rpm, depending on the wheel size and balancing machine type.

To balance the compressor wheel, a well-balanced master shaft is required to mount the compressor wheel on it. The compressor wheel unit, including the master shaft and compressor wheel, is rotated on the air bearings in the balancing machine. Despite using the master shaft, it has a small static unbalance that can be eliminated by calibrating. To calibrate the master shaft at the first time of balancing a compressor wheel, we mount it on the shaft and move the compressor wheel at three marked angular positions θ of 0°, 120°, and 240°, as shown in Fig. 8.5. The

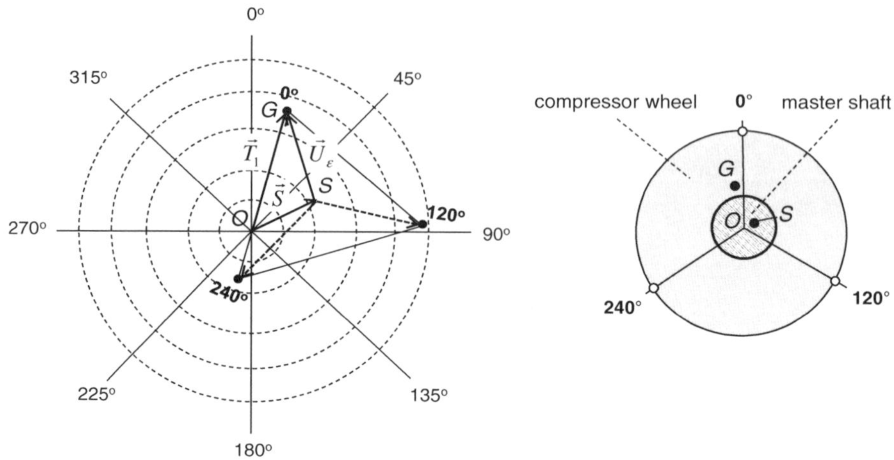

Fig. 8.5 Calibrattion of the master shaft at three positions

unbalance vector T_i of the compressor wheel unit has been successively measured three times; each time at the given angular position θ_i.

At $\theta_1 = 0°$ ($i = 1$), the measured unbalance vector can be written as

$$\vec{T}_1 = \vec{S} + \vec{U}_\varepsilon \tag{8.6}$$

where

$\vec{S}$ is the unbalance vector of the master shaft;

$\vec{U}_\varepsilon$ is the primary unbalance vector of the compressor wheel.

By the two other angular positions, the primary unbalance amplitudes are unchanged, but their vector directions change at positions 120° and 240°. Therefore, the vector $\vec{S}$ determines the unbalance vector of the master shaft where S is the center of the equilateral triangular. The primary unbalance of the master shaft is stored in the balancing machine for further balancing the compressor wheels.

The primary unbalance vector of the compressor wheel results as

$$\vec{U}_\varepsilon = \vec{T}_1 - \vec{S} \tag{8.7}$$

The static unbalance U_ε of the wheel causes the unbalance force F_u acting upon the compressor wheel.

$$F_u = m\varepsilon\Omega^2 = U_\varepsilon\Omega^2 \tag{8.8}$$

where Ω is the rotor speed.

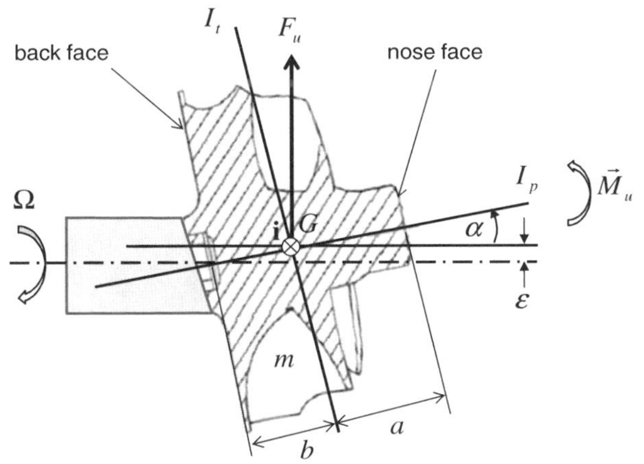

Fig. 8.6 Dynamic unbalance of the turbine shaft with $I_t > I_p$

The primary unbalance of the turbine shaft normally contains two main parts: dynamic unbalance due to welding the turbine wheel to the rotor shaft and unbalance of the turbine wheel itself due to molding process. Figure 8.6 displays the turbine shaft with dynamic unbalance caused by eccentricity ε and misalignment angle α between the shaft and turbine wheel. The dynamic unbalance induces the *moment unbalance* U_m and the *unbalance moment* M_u acting upon the turbine shaft described in Eqs. (5.49 and 5.50).

The moment unbalance vector is perpendicular to the rotation axis of the rotor and is written at a very small misalignment angle α as

$$\vec{U}_m = -(I_t - I_p)\alpha\,\mathbf{i} \tag{8.9}$$

where

U_m is the moment unbalance (g mm^2) different from the static unbalance U_ε (g mm);

I_t, I_p are the transverse and polar mass moments of inertia (kg mm^2), respectively.

Similar to static unbalance, moment unbalance U_m generates unbalance moment M_u acting on the rotor in dynamic unbalance. The unbalance moment is written using Eq. (5.49) as

$$\vec{M}_u = \vec{U}_m\Omega^2 = -(I_t - I_p)\alpha\,\Omega^2\mathbf{i} \tag{8.10}$$

where M_u is the unbalance moment (Nm), as shown in Fig. 8.6.

The unbalance force $\vec{F}_u$ and unbalance moment $\vec{M}_u$ are the external excitation sources acting upon the rotor in the vibration equations of Eqs. (5.52 and 5.53).

The turbine shaft is balanced with low-speed balancing up to about 3400 rpm at the nose and back faces. At first, its unbalance is determined by the measuring

Fig. 8.7 Unbalance measurement in a Schenck machine (Courtesy BMTS)

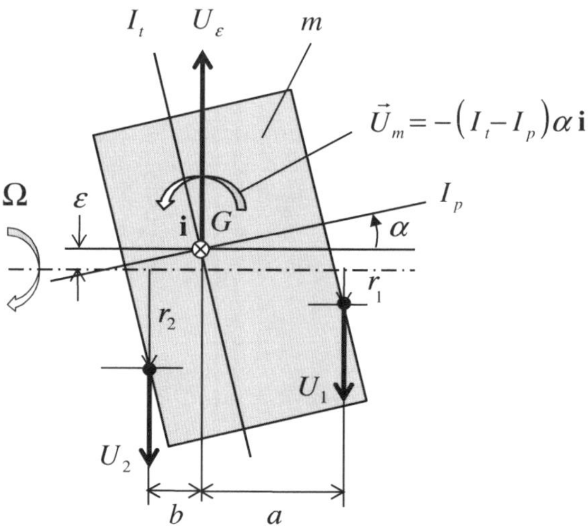

Fig. 8.8 Low-speed balancing at two planes with $I_t > I_p$

Schenck machine supported on the air bearings (see Fig. 8.7). Then, the turbine shaft is balanced in the balancing machine by removing material at the nose and back faces in the opposite directions of the balance vectors.

The balance vectors result from the equations of unbalance vectors and unbalance moments at the initial condition (see Fig. 8.8).

$$\begin{aligned} \sum U_i &= -U_1 - U_2 + U_\varepsilon = -U_1 - U_2 + m\varepsilon = 0; \\ \sum M_i &= -U_1 a + U_2 b + U_m = -U_1 a + U_2 b + (I_t - I_p)\alpha = 0 \end{aligned} \tag{8.11}$$

The balance vectors at the nose and back faces are calculated from Eq. (8.11) as

$$\begin{aligned} U_1 &= \frac{m\varepsilon b + (I_t - I_p)\alpha}{(a+b)} = m_1 r_1 \\ U_2 &= \frac{m\varepsilon a - (I_t - I_p)\alpha}{(a+b)} = m_2 r_2 \end{aligned} \tag{8.12}$$

Similar to the compressor wheel, at the balancing radii r_1 and r_2 in the balancing planes, the removed masses m_1 and m_2 in the opposite directions of U_1 and U_2 of the turbine wheel are determined by Eq. (8.12). A large misalignment angle induces the rotor response with a superharmonic frequency order of 2X in the Waterfall diagram, cf. Fig. 7.18. Note that the misalignment angle α still remains after low-speed balancing in the dynamic unbalance.

Theoretically, the unbalances would be fully eliminated if the removed masses were exactly removed at the given balancing radii. In practice, the removed mass has a tolerance of $\pm\Delta m$ and the removed mass scatters around the given balancing position in the circumferential and radial directions. Therefore, it always remains the small residual unbalances $U_{1,r}$ and $U_{2,r}$ at the balancing planes, as shown in Fig. 8.9. However, the residual unbalance at the balancing planes must be less than the permissible unbalance U_{lim} according to the ANSI Standard for Balance Quality of Rotating Rigid Body, ANSI S2.19-1975 or DIN-ISO 1940-1. The balancing quality grade of the automotive turbochargers of passenger and commercial vehicles is chosen between G40 and G100 of the DIN-ISO 1940-1, which gives the permissible residual unbalance.

The reason for the high G level is on the one hand the very high rotor speed of automotive turbochargers; on the other hand, the possible balance radius e_{lim}.

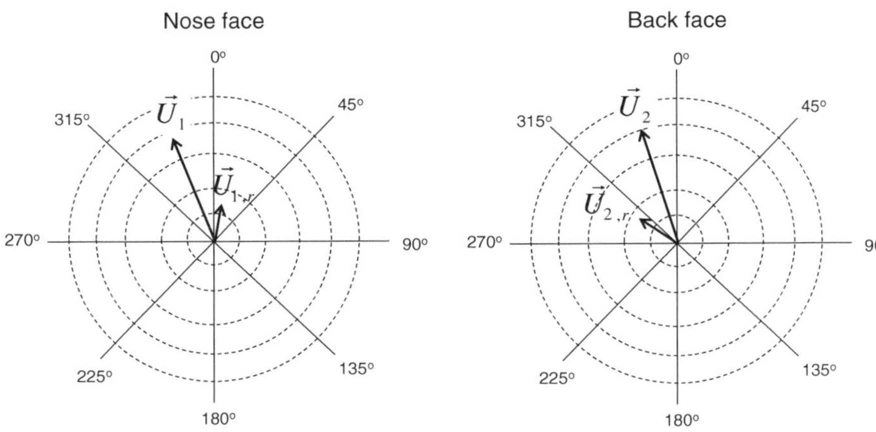

Fig. 8.9 Residual unbalance vectors $\mathbf{U}_{1,r}$ and $\mathbf{U}_{2,r}$ of the wheel

Choosing the balancing quality grade of G100 for a small turbocharger of passenger vehicles, one obtains

$$e_{\text{lim}}\Omega = 100\ \frac{\text{mm}}{\text{s}}$$

where e_{lim} is the acceptable balance radius (mm); Ω is the maximum rotor speed (rad/s).

At the maximum rotor speed of $N_{\max}$ = 280,000 rpm ($\Omega \approx$ 29,320 rad/s), the acceptable balance radius at G level of G100 is calculated as

$$e_{\text{lim}} = \frac{100\,\frac{\text{mm}}{\text{s}}}{29,320\,\frac{\text{rad}}{\text{s}}} \approx 3.4 \times 10^{-3}\ \text{mm}$$

If we use a lower G level for high rotor speeds, the balancing radius becomes much smaller and is impossible for rotor balancing. Therefore, the balancing quality grade G40 is used in the turbochargers of commercial vehicles in which their rotor speeds are not too high due to the large wheels.

With rotor mass $m \approx$ 120 g, the permissible residual unbalance at G100 for small turbochargers results in

$$\begin{aligned} U_{\text{lim}} &= m e_{\text{lim}} \\ &= (120\ \text{g}) \cdot (3.4 \times 10^{-3}\ \text{mm}) \approx 0.4\ \text{g mm} \end{aligned}$$

Thus, the residual unbalance for the compressor wheel and turbine shaft should be smaller than one-half of the permissible residual unbalance of the rotor.

$$U_{CW},\ U_{TR} \le \frac{U_{\text{lim}}}{2} \approx 0.2\ \text{g mm} \tag{8.13}$$

The results of Fig. 8.9 show that the directions of residual unbalance vectors after low-speed balancing (shop balancing) differ from the balance vectors $\vec{U}_1$ and $\vec{U}_2$ because the removed masses scatter around the given balancing points in the circumferential and radial directions. Hence, the residual unbalance vectors at the mass centers of the compressor wheel (CW) and turbine shaft (TS) result in

$$\begin{aligned} \vec{U}_{CW} &= \vec{U}_{1,rC} + \vec{U}_{2,rC} \\ \vec{U}_{TS} &= \vec{U}_{1,rT} + \vec{U}_{2,rT} \end{aligned} \tag{8.14}$$

At assembling the compressor wheel in the turbine shaft, the residual unbalance vectors $\mathbf{U}_{CW}$ and $\mathbf{U}_{TS}$ are set in fact by an arbitrary angle θ between 0° and 180° (see Fig. 8.10). If the residual unbalance vectors have the same direction (θ = 0°), the

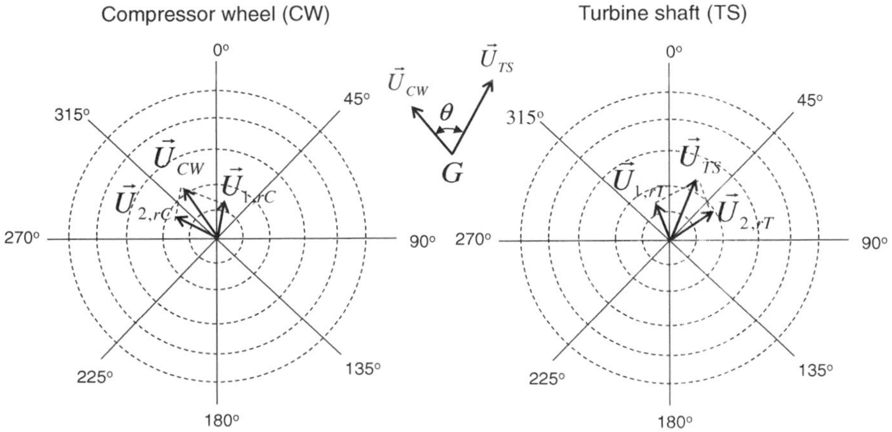

Fig. 8.10 Residual unbalance vectors in the compressor wheel and turbine shaft

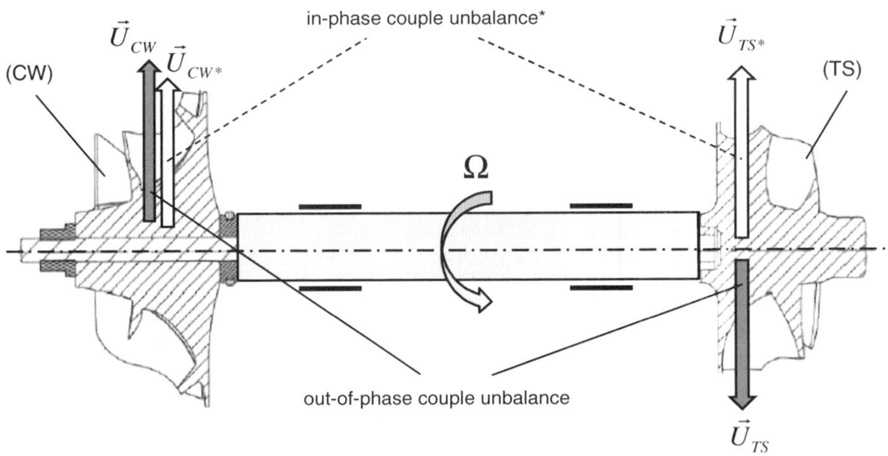

Fig. 8.11 In-phase and out-of-phase couple unbalances

rotor is called the in-phase couple unbalance with the residual unbalance vectors $\mathbf{U}_{CW*}$ and $\mathbf{U}_{TS*}$. At $\theta = 180°$, the residual unbalance vectors are opposite to each other, the rotor unbalance is called the out-of-phase couple unbalance with the residual unbalance vectors $\mathbf{U}_{CW}$ and $\mathbf{U}_{TS}$ (see Fig. 8.11).

8.4 Two-Plane High-Speed Balancing of a Flexible Rotor

Assembling the compressor wheel on the turbine shaft causes additional static and dynamic unbalances. Due to noise (unbalance whistle) by the unbalance induced at high rotor speeds, the rotor must be high-speed balanced (trim balancing). Moreover, the rotor is flexible at high-speed operating conditions that induce unbalance and therefore unbalance whistle. For this reason, the rotor should be high-speed balanced with two balancing planes at the screw-nut of the compressor wheel (balancing plane 1*) and the hub surface between the blades at the compressor wheel outlet (balancing plane 2*), as shown in Fig. 8.12. To accelerate the rotor up to 220,000 rpm, we use a nozzle ring that is supplied by a high air pressure of about 0.3 MPa absolute (3 bars abs). The nozzle ring is installed at the inlet of the turbine wheel. The pressure energy of the pressurized air is transformed into kinetic energy with the sonic speed (Mach number = 1) at the outlet of the nozzle ring. As a result, the high-speed air jets of the nozzle ring propel the turbine wheel. Due to cold air at low enthalpy, the balancing speed is, however, limited by about 220,000 rpm for a small PV turbocharger; and 170,000 rpm for a large CV turbocharger.

There are two common balancing methods with high speeds at the two planes: Modal Balancing Theory (MB) and Influence Coefficient Method Balancing (ICM).

8.4.1 Modal Balancing Theory

The modal balancing based on the bending modes of the flexible rotor at high speeds was developed between 1959 and 1967 by many pioneers such as Bishop, Parkinson, Gladwell, Kennedy-Pancu, Lindley, and Rieger in [7, 8]. Due to unbalance, the rotor deflects with the lateral amplitude y at the position z in the rotation axis, as shown in Fig. 8.12.

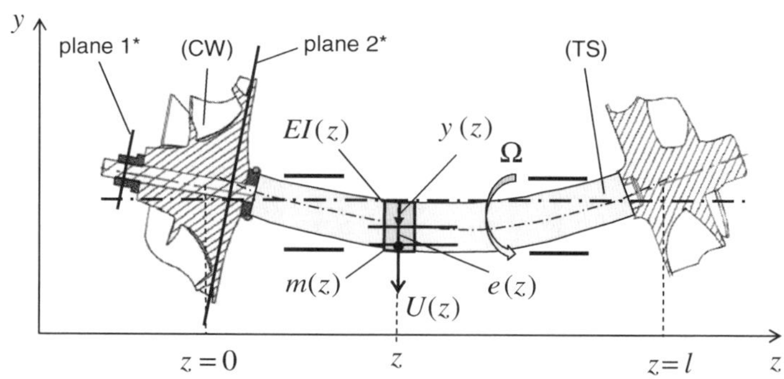

Fig. 8.12 Two-plane modal balancing of a flexible rotor

The rotor deflection shape is developed by the series of modal deflections $Y_i(z)$ according to Rieger [8].

$$y(z) = \sum_{i=1}^{n=\infty} \phi_i Y_i(z) = \phi_1 Y_1(z) + \cdots + \phi_n Y_n(z) \tag{8.15}$$

where

ϕ_i is the deflection coefficient of the vibration mode i;

Y_i is the modal deflection of the rotor of the vibration mode i.

Similarly, the mass eccentricity of the rotor is written in terms of $Y_i(z)$ as

$$e(z) = \sum_{i=1}^{n=\infty} e_i Y_i(z) = e_1 Y_1(z) + \cdots + e_n Y_n(z) \tag{8.16}$$

The induced rotor unbalance at the axial position z is written in terms of $Y_i(z)$ as

$$U(z) = \sum_{i=1}^{n=\infty} \lambda_i Y_i(z) = \lambda_1 Y_1(z) + \cdots + \lambda_n Y_n(z) \tag{8.17}$$

where λ_i is the unbalance coefficient of the vibration mode i at the position z.

The virtual kinetic work of the rotor deflection is calculated as

$$\begin{aligned} \delta W_{kin} &= \int_0^l m(z) \cdot [y(z) + e(z)] \Omega^2 Y_i(z) \delta\phi_i dz \\ &\equiv \int_0^l m(z) \cdot [y(z) + e(z)] \Omega^2 \delta y_i dz \end{aligned} \tag{8.18}$$

where

δy_i is the virtual displacement of the rotor at the axial position z;

$m(z)$ is the mass distribution of the rotor at the axial position z;

Ω is the balancing speed of the rotor

Using orthogonality of the vibration modes, one obtains

$$\int_0^l m(z) \cdot Y_i(z) \cdot Y_j(z) dz = \begin{cases} 0; & i \neq j \\ M_i; & i = j \end{cases} \tag{8.19a}$$

within

$$M_i \equiv \int_0^l m(z) \cdot Y_i^2(z)dz \tag{8.19b}$$

and

$$\int_0^l EI(z) \cdot Y_i''(z) \cdot Y_j''(z)dz = \begin{cases} 0; & i \neq j \\ K_i; & i = j \end{cases} \tag{8.20a}$$

within

$$K_i \equiv \int_0^l EI(z) \cdot Y_i''^2(z)dz \tag{8.20b}$$

Substituting Eqs. (8.15, 8.16, 8.18, and 8.19a, b), the virtual kinetic work of the vibration mode i results as

$$\delta W_{kin} = M_i(\phi_i + e_i)\,\Omega^2 \delta\phi_i \tag{8.21}$$

The virtual potential work of the rotor is computed of the vibration mode i as

$$\delta W_{pot} = -\frac{\partial V}{\partial \phi_i} \delta\phi_i \tag{8.22}$$

where the potential energy V is calculated using Eqs. (8.20a and 8.20b)

$$\begin{aligned} V &= \int_0^l \frac{1}{2} EI(z) \cdot y''^2(z)dz \\ &= \int_0^l \frac{1}{2} EI(z) \cdot \left(\sum_{i=1}^{\infty} \phi_i Y_i''(z) \right)^2 dz = \frac{1}{2} \sum_{i=1}^{\infty} K_i \phi_i^2 \end{aligned} \tag{8.23}$$

Thus,

$$\delta W_{pot} = -\frac{\partial V}{\partial \phi_i} \delta\phi_i = -K_i \phi_i \delta\phi_i \tag{8.24}$$

Using the principle of virtual work at the equilibrium ($\delta W = 0$), the total virtual work of the rotor results from Eqs. (8.21 and 8.24) as

$$\begin{aligned}\delta W &= \delta W_{kin} + \delta W_{pot} \\ &= \left[M_i(\phi_i + e_i)\,\Omega^2 - K_i\phi_i\right]\delta\phi_i = 0\end{aligned} \tag{8.25}$$

The coefficients of the deflection and mass eccentricity of the rotor result from solving Eq. (8.25) as

$$\begin{aligned}&M_i(\phi_i + e_i)\,\Omega^2 - K_i\phi_i = 0 \\ &\Rightarrow\ \phi_i = \frac{1}{\left(\frac{\Omega_{cr,i}}{\Omega}\right)^2 - 1} e_i = \frac{1}{\frac{1}{\eta_i^2} - 1} e_i = \left(\frac{\eta_i^2}{1-\eta_i^2}\right) e_i \\ &\Rightarrow\ e_i = \left(\frac{1-\eta_i^2}{\eta_i^2}\right)\phi_i\end{aligned} \tag{8.26}$$

within the speed ratio of the vibration mode i is defined as

$$\eta_i = \frac{\Omega}{\Omega_{cr,i}};\ \Omega_{cr,i} = \sqrt{\frac{K_i}{M_i}} \tag{8.27}$$

Using orthogonality of $Y_i(z)$ in Eq. (8.19a), the unbalance coefficient λ_i results as

$$\begin{aligned}\int_0^l U(z)\cdot m(z)\cdot Y_j(z)dz &= \int_0^l \left(\sum_i^{\infty} \lambda_i Y_i(z)\right) m(z)\cdot Y_j(z)dz \\ &= \int_0^l \lambda_i m(z)\cdot Y_i^2(z)dz = \lambda_i M_i\end{aligned}$$

Thus,

$$\lambda_i = \frac{1}{M_i}\int_0^l m(z)\cdot U(z)\cdot Y_i(z)dz \tag{8.28}$$

At rotor balancing, one creates the balance correction $p(z)$ at $z = c$ by the load q in the small length ξ in the opposite direction of the unbalance vector $\vec{U}(z)$.

$$\begin{aligned}p &= q\cdot\xi = \sum_{i=1}^{\infty}\alpha_i Y_i(z) \\ &= -U(z) = -\sum_{i=1}^{\infty}\lambda_i Y_i(z)\end{aligned} \tag{8.29}$$

Therefore, the balancing coefficient results as

$$\alpha_i = -\lambda_i \tag{8.30}$$

The balance correction vector $p(z)$ at the position z is written as

$$p(z) = -U(z) = -m(z) \cdot e(z) \tag{8.31}$$

At the modal balance in the vibration mode i, some characteristics of the rotor are necessary to be known in advance. They are the mass distribution of the rotor $m(z)$, modal deflection amplitude $Y_i(z)$, measured mode shape of the rotor $y(z)$, rotor eigenfrequency ω_i of the vibration mode i, and critical frequencies of the rotor as well. Then, the coefficient ϕ_i of the rotor deflection is calculated from $y(z)$ and $Y_i(z)$ in Eq. (8.15) and the mass eccentricity coefficient e_i results from ϕ_i in Eq. (8.26). As a result, the mass eccentricity of the rotor $e(z)$ is computed from $e_i(z)$ and $Y_i(z)$ according to Eq. (8.16).

Using Eq. (8.31), the balancing length ξ at the position z is determined at a given removing load $q < 0$ in the same direction as the unbalance vector $\vec{U}(z)$ so that the balance vector is opposite to the unbalance vector.

$$\xi = \frac{-U(z)}{q} = \frac{-m(z) \cdot e(z)}{q} \tag{8.32}$$

The main disadvantage of the modal balancing is the necessity of knowing many unknown or not-easily-known characteristics of the rotor, such as $m(z)$, $Y_i(z)$, $y(z)$, and ω_i. Furthermore, the additional disadvantage is that the balancing correction vector depends on the balancing speed Ω, as shown in Eqs. (8.26 and 8.27). In fact, automotive turbochargers operate in the wide rotor speed range, not just at the constant speed in the normally operating condition, such as turbomachines used in power plants and chemical industries. Hence, the modal balancing of the flexible rotors is not suitable for applications to automotive turbochargers.

On the contrary, the influence coefficient method balancing can provide an easy-to-use, efficient, and well-done balancing procedure of the turbochargers.

8.4.2 Influence Coefficient Method

The influence coefficient method (ICM) was developed by Goodman, Rieger, Lund and Tonnesen, Tessarzik, Badgley, and Anderson, etc., since 1961 [8]. The ICM balancing in automotive turbochargers is normally carried out at two balancing planes in the screw-nut of the compressor wheel (plane 1*) and the hub surface between the blades at the compressor wheel outlet (plane 2*), as indicated in Fig. 8.14.

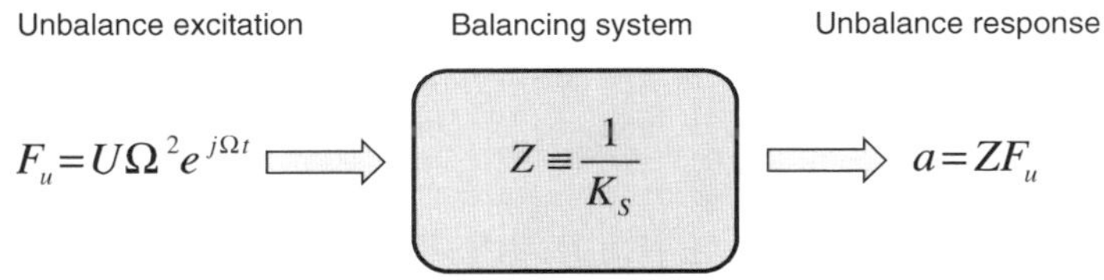

Fig. 8.13 A balancing system with transfer impedance Z

Based on the complex transfer impedance Z of the balancing system, i.e., the inversion of the complex dynamic stiffness coefficient K_S of the system, including the turbocharger rotor and less influent balancing machine, the unbalance response at the trim balancing results from the complex transfer impedance and unbalance excitation force (see Fig. 8.13).

The measured unbalance responses could be the acceleration or vibration velocity of the rotor in the measuring planes. Theoretically, the measuring planes should be the same places as the balancing planes in the compressor wheel that rotates, however, with a high rotor speed. As a result, it is time and cost intensive at trim balancing in the mass production in order to measure the rotor unbalance responses in the rotating balancing planes. Hence, in practice, they are computed from the measured rotor responses in the certain nonrotating position in the bearing housing or the compressor and turbine housings in mass production.

Figure 8.14 shows two balancing planes 1* and 2* in the compressor wheel used in ICM balancing. The reason for rotor balancing at the side of the compressor wheel (CW) rather than at the turbine wheel (TW) side is the high temperature of exhaust gas in the turbine, nearly 820 to 870 °C in diesel engines and 950 to 1050 ° C in gasoline engines.

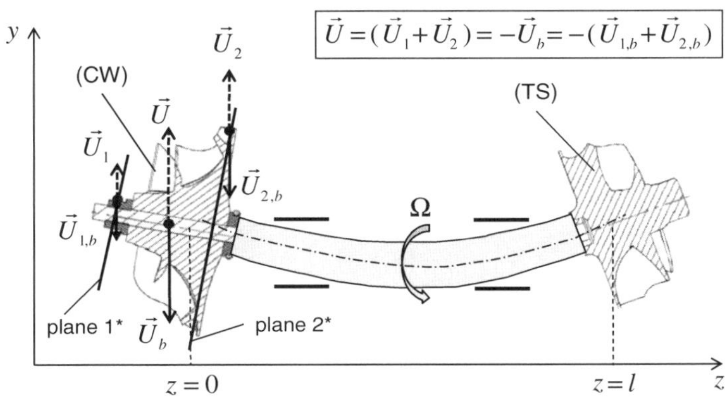

Fig. 8.14 Two-plane ICM balancing of a flexible rotor

The measured response vectors of acceleration $\mathbf{a}_1$ and $\mathbf{a}_2$ at the planes 1* and 2*, respectively, result from the unbalance vector $\mathbf{F}_u$ and the complex transfer impedance $\mathbf{Z}$ that is inversely related to the complex dynamic stiffness $\mathbf{K}_S$, as shown in Fig. 8.13 [4].

$$\mathbf{a} = \mathbf{Z}\mathbf{F}_u \Leftrightarrow \begin{bmatrix} \mathbf{a}_1 \\ \mathbf{a}_2 \end{bmatrix} = \begin{bmatrix} \alpha_{11} & \alpha_{12} \\ \alpha_{21} & \alpha_{22} \end{bmatrix} \begin{bmatrix} \mathbf{F}_1 \\ \mathbf{F}_2 \end{bmatrix} \tag{8.33a}$$

where α_{ij} are the influence coefficients with the index i for the balancing plane i, and the index j for the excitation force at the plane j.

Using Eq. (8.33a), the acceleration vectors $\mathbf{a}_1$ and $\mathbf{a}_2$ (complex numbers in the phase plane) are written in terms of the unbalance forces $\mathbf{F}_1$ and $\mathbf{F}_2$ at the balancing planes 1* and 2*.

$$\begin{aligned} \mathbf{a}_1 &= \alpha_{11}\mathbf{F}_1 + \alpha_{12}\mathbf{F}_2 \\ \mathbf{a}_2 &= \alpha_{21}\mathbf{F}_1 + \alpha_{22}\mathbf{F}_2 \end{aligned} \tag{8.33b}$$

The complex dynamic stiffness coefficient matrix of the balancing system containing the unbalanced rotor and balancing machine is nearly unchanged at the calibration with the masses $m_{1,c}$ and $m_{2,c}$.

$$\mathbf{K}_S \equiv \mathbf{Z}^{-1} = \begin{bmatrix} \alpha_{11} & \alpha_{12} \\ \alpha_{21} & \alpha_{22} \end{bmatrix}^{-1} \tag{8.34}$$

To calibrate the rotor, two given masses $m_{1,c}$ and $m_{2,c}$ are setup at the given positions (r_1,θ_1) and (r_2,θ_2) in the inlet and outlet of the compressor wheel (CW). They correspond to the calibration unbalances $\mathbf{U}_{1,c}$ and $\mathbf{U}_{2,c}$, respectively. At first, the first given mass $m_{1,c}$ is added in the position (r_1,θ_1) at the CW inlet.

The response vectors with the first calibration mass result in

$$\begin{bmatrix} \mathbf{a}_{11} \\ \mathbf{a}_{21} \end{bmatrix} = \begin{bmatrix} \alpha_{11} & \alpha_{12} \\ \alpha_{21} & \alpha_{22} \end{bmatrix} \begin{bmatrix} \mathbf{F}_1 + \mathbf{F}_{1,c} \\ \mathbf{F}_2 \end{bmatrix} \tag{8.35a}$$

From Eq. (8.35a), one gets the relation between the unbalance vectors $\mathbf{F}_1$ and $\mathbf{F}_{1,c}$.

$$\begin{aligned} \mathbf{a}_{11} &= \alpha_{11}(\mathbf{F}_1 + \mathbf{F}_{1,c}) + \alpha_{12}\mathbf{F}_2 \\ \mathbf{a}_{21} &= \alpha_{21}(\mathbf{F}_1 + \mathbf{F}_{1,c}) + \alpha_{22}\mathbf{F}_2 \end{aligned} \tag{8.35b}$$

where $\mathbf{a}_{ij}$ is the response vector at the balancing plane i with the calibration unbalance force $\mathbf{F}_{j,c}$.

After removing the first calibration mass, the second calibration mass $m_{2,c}$ is added in the position (r_2,θ_2). Similarly, the rotor response with the second calibration mass corresponding to the unbalance force $\mathbf{F}_{2,c}$ becomes

$$\begin{bmatrix} \mathbf{a}_{12} \\ \mathbf{a}_{22} \end{bmatrix} = \begin{bmatrix} \alpha_{11} & \alpha_{12} \\ \alpha_{21} & \alpha_{22} \end{bmatrix} \begin{bmatrix} \mathbf{F}_1 \\ \mathbf{F}_2 + \mathbf{F}_{2,c} \end{bmatrix} \tag{8.36a}$$

Therefore,

$$\begin{aligned} \mathbf{a}_{12} &= \alpha_{11}\mathbf{F}_1 + \alpha_{12}(\mathbf{F}_2 + \mathbf{F}_{2,c}) \\ \mathbf{a}_{22} &= \alpha_{21}\mathbf{F}_1 + \alpha_{22}(\mathbf{F}_2 + \mathbf{F}_{2,c}) \end{aligned} \tag{8.36b}$$

Subtracting Eqs. (8.35b and 8.36b) from Eq. (8.33b), one obtains

$$\begin{bmatrix} (\mathbf{a}_{11} - \mathbf{a}_1) & (\mathbf{a}_{12} - \mathbf{a}_1) \\ (\mathbf{a}_{21} - \mathbf{a}_2) & (\mathbf{a}_{22} - \mathbf{a}_2) \end{bmatrix} = \begin{bmatrix} \alpha_{11} & \alpha_{12} \\ \alpha_{21} & \alpha_{22} \end{bmatrix} \begin{bmatrix} \mathbf{F}_{1,c} & 0 \\ 0 & \mathbf{F}_{2,c} \end{bmatrix} \tag{8.37}$$

The response vectors in the balancing planes 1* and 2* are demonstrated in the polar plots (r, θ) of Fig. 8.15. The balance vectors $\mathbf{U}_{1,b}$ and $\mathbf{U}_{2,b}$ in the balancing planes 1* and 2* are equal and opposite to the unbalance vectors $\mathbf{U}_1$ and $\mathbf{U}_2$, respectively (see Fig. 8.14).

$$\begin{aligned} \mathbf{U}_b &= \begin{bmatrix} \mathbf{U}_{1,b} \\ \mathbf{U}_{2,b} \end{bmatrix} \\ &= -\mathbf{U} = -\begin{bmatrix} \mathbf{U}_1 \\ \mathbf{U}_2 \end{bmatrix} = -\frac{1}{\Omega^2}\begin{bmatrix} \mathbf{F}_1 \\ \mathbf{F}_2 \end{bmatrix} \end{aligned} \tag{8.38}$$

Equations (8.33a) and (8.34) give the relation between the excitation forces and rotor responses.

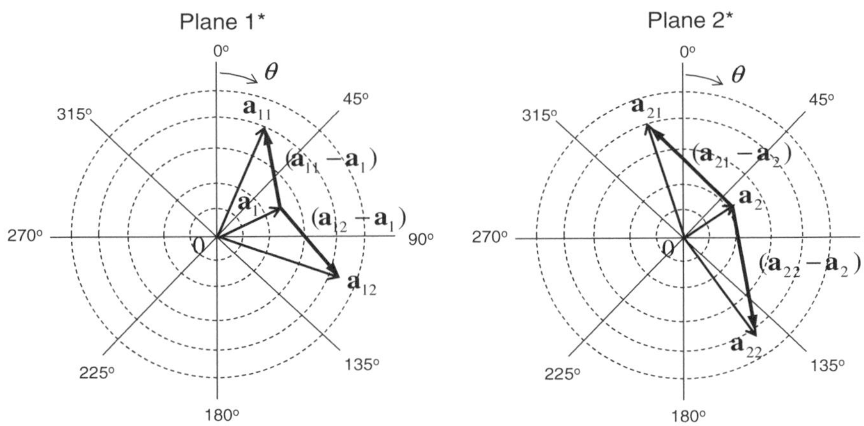

Fig. 8.15 Calibration response vectors in balancing planes 1* and 2*

$$\begin{bmatrix} \mathbf{F}_1 \\ \mathbf{F}_2 \end{bmatrix} = \begin{bmatrix} \alpha_{11} & \alpha_{12} \\ \alpha_{21} & \alpha_{22} \end{bmatrix}^{-1} \begin{bmatrix} \mathbf{a}_1 \\ \mathbf{a}_2 \end{bmatrix} = \mathbf{K}_S \begin{bmatrix} \mathbf{a}_1 \\ \mathbf{a}_2 \end{bmatrix} \tag{8.39}$$

where $\mathbf{K}_S$ is calculated from Eq. (8.37) as

$$\mathbf{K}_S = \begin{bmatrix} \alpha_{11} & \alpha_{12} \\ \alpha_{21} & \alpha_{22} \end{bmatrix}^{-1} = \begin{bmatrix} \mathbf{F}_{1,c} & 0 \\ 0 & \mathbf{F}_{2,c} \end{bmatrix} \begin{bmatrix} (\mathbf{a}_{11} - \mathbf{a}_1) & (\mathbf{a}_{12} - \mathbf{a}_1) \\ (\mathbf{a}_{21} - \mathbf{a}_2) & (\mathbf{a}_{22} - \mathbf{a}_2) \end{bmatrix}^{-1} \tag{8.40}$$

Substituting Eqs. (8.38), (8.39), and (8.40), the balance vector $\mathbf{U}_b$ results in

$$\begin{aligned} \mathbf{U}_b &= \begin{bmatrix} \mathbf{U}_{1,b} \\ \mathbf{U}_{2,b} \end{bmatrix} \\ &= -\frac{1}{\Omega^2} \begin{bmatrix} \mathbf{F}_{1,c} & 0 \\ 0 & \mathbf{F}_{2,c} \end{bmatrix} \begin{bmatrix} (\mathbf{a}_{11} - \mathbf{a}_1) & (\mathbf{a}_{12} - \mathbf{a}_1) \\ (\mathbf{a}_{21} - \mathbf{a}_2) & (\mathbf{a}_{22} - \mathbf{a}_2) \end{bmatrix}^{-1} \begin{bmatrix} \mathbf{a}_1 \\ \mathbf{a}_2 \end{bmatrix} \end{aligned} \tag{8.41}$$

Using the calibrating unbalance vectors of $\mathbf{U}_{1,c}$ and $\mathbf{U}_{2,c}$, one obtains the balance vector resulting from the rotor responses and calibrating unbalances.

$$\begin{aligned} \mathbf{U}_b &= \begin{bmatrix} \mathbf{U}_{1,b} \\ \mathbf{U}_{2,b} \end{bmatrix} \\ &= \begin{bmatrix} \mathbf{U}_{1,c} & 0 \\ 0 & \mathbf{U}_{2,c} \end{bmatrix} \begin{bmatrix} (\mathbf{a}_1 - \mathbf{a}_{11}) & (\mathbf{a}_1 - \mathbf{a}_{12}) \\ (\mathbf{a}_2 - \mathbf{a}_{21}) & (\mathbf{a}_2 - \mathbf{a}_{22}) \end{bmatrix}^{-1} \begin{bmatrix} \mathbf{a}_1 \\ \mathbf{a}_2 \end{bmatrix} \end{aligned} \tag{8.42}$$

The components of the balance vector $\mathbf{U}_b$ are calculated from the complex numbers of measured accelerations in the balancing phase planes 1* and 2* according to Eq. (8.42).

Note that the balance vector $\mathbf{U}_b$ in Eq. (8.42) is independent of the balancing speed. Hence, the balance vectors in the balancing planes 1* and 2* are determined in the wide range of the balancing speeds up to 220,000 rpm.

The ICM balancing procedures are repeated until the delivery acceleration response is below the *delivery acceleration levels* of a_{I*}, a_{II*}, and a_{III*} that are normally smaller than the *limit acceleration levels* of a_I, a_{II}, and a_{III}. The limit acceleration levels are directly determined in the car so that the unbalance whistle and constant tone are inaudible in all operating speed ranges of the turbocharger. They are divided into many speed ranges of the balancing speeds N(rpm), as illustrated in Fig. 8.16.

After a long operating period, the unbalance change of the rotor happens due to thermal and plastic deformations of the wheels. It mostly increases to nearly twice the delivery acceleration responses at the trim balancing according to experience. One-third goes to the turbine shaft due to heat-related plastic deformation and the remaining two-third goes to the compressor wheel. The unbalance change of the

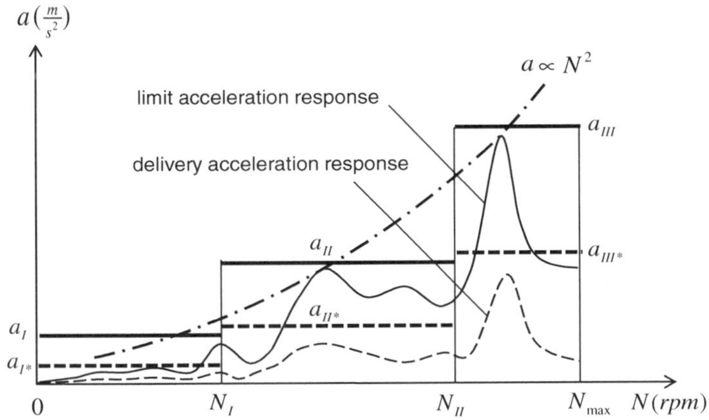

Fig. 8.16 Limit and delivery acceleration levels in ICM balancing

compressor wheel is generally created by the inappropriate new positions of eccentricity, misalignment, unbalance change caused by plastic deformation, loose screw-nut, and therefore slipping of the compressor wheel on the rotor shaft in the radial direction. As a result, the *delivery acceleration levels* at the trim balancing must be at least reduced by nearly one-half of the *limit acceleration levels* where the unbalance whistle is inaudible in the vehicle cabin.

Twice differentiating the rotor amplitude with respect to time, the acceleration amplitude measured at the trim balancing results as

$$|a| = |\ddot{y}(t)| = \Omega^2|y(t)| \propto N^2|y(t)| \tag{8.43}$$

Note that the measured acceleration amplitude is proportional not only to the rotor amplitude but also to squared rotor speed according to Eq. (8.43). Therefore, it displays not exactly the resonance behavior of the rotor of the turbocharger. The acceleration behavior is strongly affected by squared rotor speed N^2, as shown in Fig. 8.16. As discussed in the earlier chapter, the typical resonance due to unbalance does not exist in nonlinear rotordynamics, but only the limit cycle of the rotor response occurs at each rotor speed. The unbalance amplitude (1X) is relatively small compared to the asynchronous amplitudes in the entire rotor speed range (see Fig. 7.33).

According to Eq. (8.43), the acceleration amplitude at high rotor speeds (N between N_{II} and N_{max}) could be higher than the acceleration amplitude at the quasi-resonance (N_{res} between N_I and N_{II}) although its rotor amplitude is smaller than the other one. Therefore, the failure of the radial bearings at the trim balancing usually occurs at the first quasi-resonance instead of at the maximum acceleration amplitude at higher rotor speeds.

Table 8.1 Delivery and limit acceleration levels in the balancing speed ranges

Balancing speed range N (rpm)	Delivery acceleration levels a_* (m/s^2)	Limit acceleration levels a (m/s^2)
$0 < N < N_I$:	$a_{I*} < a_I$	a_I
$N_I \leq N < N_{II}$:	$a_{II*} < a_{II}$	a_{II}
$N_{II} \leq N < N_{\max}$:	$a_{III*} < a_{III}$	a_{III}

Figure 8.16 and Table 8.1 give the delivery and limit acceleration levels for the rotor acceleration responses at trim balancing [5]. The unbalance-change ratio between the delivery and limit acceleration responses must be statistically determined by means of testing many turbochargers in a long operating period. If the ratio is chosen too large, many balancing steps are required at trim balancing leading to much time needed for the balancing process. Therefore, more costs for wear of cutting tools and low production performance. Otherwise, the unbalance whistle is audible after a long operating period due to unbalance change if one selects the unbalance-change ratio too small.

8.4.3 Comparison Between Modal Balancing and ICM

Modal balancing is valid not only for rigid but also flexible rotors. However, the following characteristics that are necessary to be known in advance are the mass distribution $m(z)$, modal deflection amplitude $Y_i(z)$, vibration mode shape $y(z)$, rotor eigenfrequency ω_i, and critical frequencies Ω_{cr}. Moreover, modal balancing is dependent on the balancing speed. In practice, it is difficult to determine or know them in advance. The necessary balancing planes are at least the number of rotor vibration modes plus 1; i.e., three necessary balancing planes for the rotor with two vibration modes. However, the turbocharger has only two accessible balancing planes at the side of the compressor wheel. As a result, the modal balancing method is not practical and usable for applications to high-speed balancing of automotive turbochargers.

Using the influence coefficient method (ICM), the response accelerations are measured over the large range of high balancing speeds. The removing masses at the given radii in the balancing planes 1* and 2* are computed without condition of knowing many necessary parameters of the rotor characteristics in the modal balancing. Furthermore, ICM balancing is independent of the balancing speed; therefore, it is suitable for balancing over the large balancing speed range. However, it is needed to find a suitable position in the balancing machine or in the turbocharger to high-speed balance the flexible rotor by means of the measured response accelerations. Nevertheless, ICM balancing is a state-of-the-art method of trim balancing usually applied to automotive turbochargers rather than modal balancing.

References

1. Ehrich, F.: Handbook of Rotordynamics. Krieger Publishing Company, New York (2004)
2. Kraemer, E.: Rotordynamics of Rotors and Foundations. Springer, New York (1993)
3. Lingener, A.: Auswuchten—Theorie und Praxis, Verlag Technik, Berlin (1992)
4. Muszýnska, A.: Rotordynamics. CRC Press, Taylor and Francis, Boca Raton, pp. 555–710 (2005)
5. Nguyen-Schäfer, H.: Technical Instructions for Rotor Balancing of Automotive Turbochargers. Internal Documents, BMTS (2011)
6. Rao, J.S.: Rotordynamics, 3rd edn. New Age International Pvt. Ltd, New Delhi (2007)
7. Rieger, N.F.: Rotordynamics 2—Problems in Turbomachinery. CISM courses and lectures No. 297 (1988)
8. Rieger, N.F.: Balancing of Rigid and Flexible Rotors. U.S. DoD (1986)
9. Schneider, H.: Auswuchttechnik, 7. Springer, Auflage (2007)
10. Vance, J.: Rotordynamics of Turbomachinery. Wiley, New York (1988)

Chapter 9
Applied Tribology in the Oil-Film Bearings

9.1 Introduction

Tribology is derived from the Greek word *tribos*, which means rubbing. It deals with the tribological phenomena, such as lubrication, friction, and wears in the moving parts. Thus, tribology becomes more and more important in automotive turbochargers in terms of synthetic lubricating oils, friction reduction, adhesion and abrasion friction, and wear reduction in the oil-film bearings including journal and thrust bearings.

9.2 Characteristics of Lubricating Oils

Lubricating oils used in the industrial and automotive applications are based on mineral and synthetic oils. The mineral oils are produced by refining crude oil containing long hydrocarbon chains of paraffin and aromatic hydrocarbon rings. On the contrary, the synthetic oils are manufactured by means of polymerizing olefin ethylene gained from cracking of petroleum to produce poly-alpha-olefins (PAOs) that have similar chemical properties of the paraffinic oils (mineral oils), but they could be used in the extreme conditions of high thermal and mechanical loads. There are two kind of synthetic oils, esters (chemical compounds like alcohols or phenols), and PAOs. Synthetic oils used in the automotive engines result from combining PAOs with about 15 % of a synthetic ester [4, 7]. Some high-end synthetic engine oils are sold under the commercial names, such as Castrol Edge, Castrol Magnatec (Castrol), Mobil 1-5W30, SHC 824, and SHC 629 (ExxonMobil).

Lubricating oils are generally classified into SAE viscosity grades (Society of Automotive Engineers) for the automotive applications and ISO grades (International Organization for Standardization) for the industrial applications. The

H. Nguyen-Schäfer, *Rotordynamics of Automotive Turbochargers*,
Springer Tracts in Mechanical Engineering, DOI 10.1007/978-3-319-17644-4_9

ISO viscosity grades (VG) are based on the average kinematic viscosity in Centistokes at 40 °C oil temperature.

The kinematic viscosity is defined as the ratio of the dynamic viscosity to density.

$$\nu = \frac{\eta}{\rho} \tag{9.1}$$

where

- ν is the kinematic viscosity [m^2/s; mm^2/s; 1 cSt (Centistokes) = 1 mm^2/s];
- η is the dynamic viscosity [Ns/m^2; Pa s; 1 cP (Centipoise) = 1 mPa s];
- ρ is the oil density [kg/m^3; g/mm^3].

The SAE grade has a general form SAE xWy (e.g., SAE 5W30), where xW stands for the oil viscosity grade of SAE xW (SAE 5W); W stands for winter at −18 °C (0 °F) for low temperature applications. The index y relates to the oil viscosity grade SAE y (i.e. SAE 30) at about 100 °C (212 °F) for high temperature applications.

Table 9.1 gives the equivalent lubricating oils between the ISO VG (Viscosity Grade) and SAE specifications and their HTHS viscosities according to SAE J300. The lubricating oil of SAE 5W30 corresponds to ISO VG 22 for SAE 5W in low temperatures and ISO VG 100 for SAE 30 in high temperatures with a HTHS viscosity of 2.9 mPa s at 150 °C (302 °F) and oil share rate of 10^6 s^{-1}.

In fact, oil viscosities change almost with temperature at small or moderate oil shear rate. Generally, the higher the oil temperature is, the lower the oil viscosity becomes, and vice versa. The oil dynamic viscosity versus temperature is calculated using the Cameron and Vogel equation, as given in Eq. (6.44a). Figure 9.1 gives the dynamic viscosities depending on oil temperatures for some common lubricating oils used in the industrial and automotive engines.

The related coefficients in the Cameron and Vogel equation in Eq. (6.44a) of common oils used in the automotive engines are given in Table 9.2. The coefficients

Table 9.1 Viscosity grades and HTHS viscosities of lubricating oils of the ISO and SAE specifications (SAE J300)

ISO VG grade	SAE grade	HTHS viscosity (mPa s)
–	0W	–
22	5W	–
32	10W	–
46	15W	–
68	20W/20	2.6
100	30	2.9
150	40	2.9[a]; 3.7[b]
220	50	3.7
320	60	3.7

[a]0W40; 5W40; 10W40
[b]15W40; 20W40; 25W40

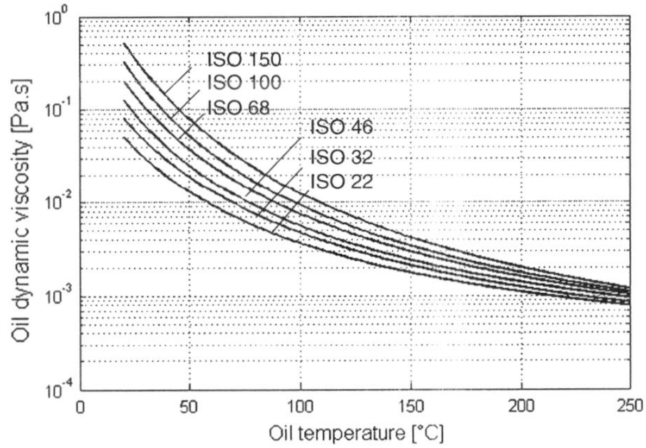

Fig. 9.1 Dynamic viscosities of lubricating oils versus temperatures

Table 9.2 Computed coefficients of oil dynamic viscosities

ISO VG grade	$\eta(T) = a\ \exp\left(\frac{b}{T(K)-c}\right)$		
	a (Pa s)	b (K)	c (K)
22	12.312e − 5	6.181e + 2	1.906e + 2
32	9.618e − 5	7.391e + 2	1.883e + 2
46	11.387e − 5	7.014e + 2	1.930e + 2
68	8.396e − 5	8.520e + 2	1.835e + 2
100	7.547e − 5	9.081e + 2	1.844e + 2
150	5.399e − 5	10.747e + 2	1.758e + 2

result from three points of the oil grade that are used to calculate of the dynamic viscosities η (Pa s) versus oil temperature T (K). The oil viscosities vary at each iteration step in the bearing dynamics and rotor dynamic computation.

9.3 HTHS Viscosity of Lubricating Oils

HTHS viscosity (high temperature high shear) is defined as the effective oil dynamic viscosity (mostly in mPa s) in the operating condition at the high oil temperature of 150 °C (302 °F) and large share rate of $10^6\ s^{-1}$.

The shear rate $\dot{\gamma}$ is the velocity gradient of the oil film that is defined as the change rate of the oil velocity to oil-film thickness.

$$\dot{\gamma} \equiv \left| \frac{\partial U}{\partial h} \right| \tag{9.2}$$

Fig. 9.2 Oil velocity profile and shear rate of the oil film

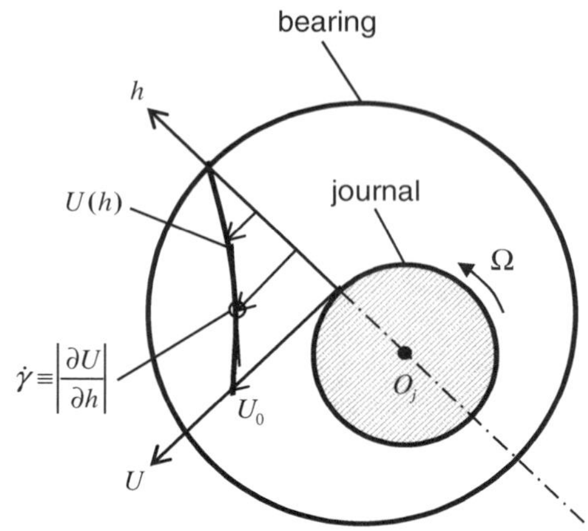

Figure 9.2 shows the velocity profile of the oil film in the bearing clearance of a fixed radial bearing with the journal circumferential velocity U_0. The shear stress τ of the oil film on the journal is proportional to the oil dynamic viscosity and shear rate.

$$\begin{aligned}\tau &= \frac{F_f}{A_S} = \eta \cdot \left|\frac{\partial U}{\partial h}\right| \\ &= \eta(T, \dot{\gamma}) \cdot \dot{\gamma} \equiv \tau(T, \dot{\gamma})\end{aligned} \tag{9.3}$$

where F_f is the friction force acting on the journal; A_S is the oil-lubricated surface of the journal; T is the oil temperature.

In case of a Newtonian fluid, such as single-grade oils (base oils) the dynamic viscosity depends only on the fluid temperature and not on the shear rate, as shown in Fig. 9.3a. Hence, the shear stress is linear to the oil shear rate (see Fig. 9.4).

$$\tau = \eta(T) \cdot \dot{\gamma} \tag{9.4}$$

In fact, multi-grade lubricating oils are generally non-Newtonian fluids due to long hydrocarbon chains and aromatic rings. As a result, the oil dynamic viscosity depends on not only the oil temperature but on oil shear rate, as shown in Fig. 9.3b. Hence, the shear stress is a nonlinear function of the oil temperature and shear rate (see Fig. 9.4).

$$\tau = \tau(T, \dot{\gamma}) \tag{9.5}$$

Figure 9.3 shows the dynamic viscosity of a Newtonian fluid is independent of the shear rate at a constant temperature; it reduces only with oil temperatures. On the contrary, the oil viscosity (mostly non-Newtonian oils) is nearly constant as long as the shear rate is below the critical shear rate of $10^4\ \mathrm{s}^{-1}$ and strongly decreases with

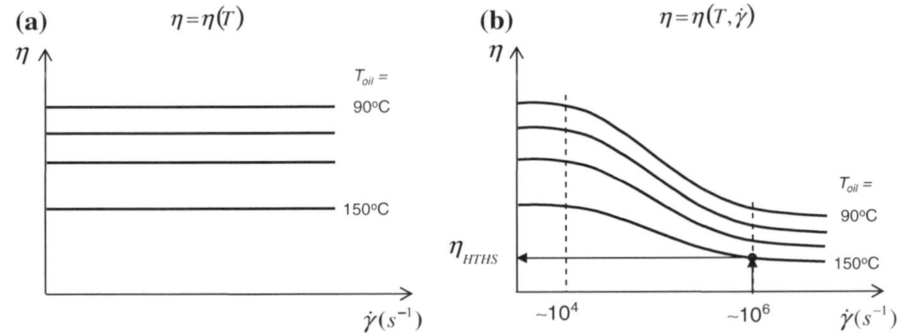

Fig. 9.3 Dynamic viscosity of lubricating oil versus shear rate and temperature shear rate. **a** Newtonian fluids. **b** Non-newtonian oils

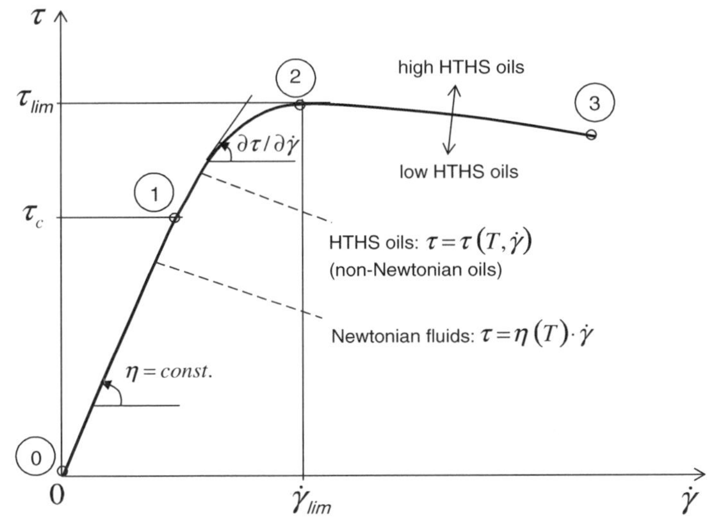

Fig. 9.4 Shear stress versus shear rate of lubricating oils

the oil shear rate up to $10^6\ s^{-1}$. From this shear rate, the oil viscosity remains unchanged at the lowest viscosity that is defined as the HTHS viscosity at the oil temperature of 150 °C. Generally, the oil shear rate in the bearing of the automotive turbochargers is very large due to their high rotor speeds in the small bearing clearances.

In the following section, the oil shear rate in the radial bearing of a typical turbocharger with a shaft diameter D = 7 mm supported on two rotating floating ring bearings (RFRBs) with a radial bearing clearance h = 10 μm is calculated.

Using Eq. (9.2), the shear rate of the oil film results in

$$\begin{aligned}\dot{\gamma} &\equiv \left|\frac{\partial U}{\partial h}\right| \approx \frac{U_{\text{eff}}}{h} \\ &= \frac{(1 - RSR)\,U_0}{h} = (1 - RSR)\frac{\pi ND}{60h}\end{aligned} \tag{9.6}$$

where

U_{eff} is the effective oil velocity in the RFRBs (m/s);
RSR is the ring speed ratio (−) in Eq. (6.92);
D is the shaft diameter (m);
N is the rotor speed (rpm);
h is the oil-film thickness (m).

According to Eq. (9.6), the oil shear rate is approximately $1.1 \times 10^6\ \text{s}^{-1}$ at the rotor speed from 40,000 rpm with a ring speed ratio of 25 %. Additionally, the effective oil temperature in the bearing clearance is mostly at 150 °C or higher at high rotor speeds. Therefore, the oil viscosity almost in the operating conditions of turbochargers is the HTHS viscosity (see Fig. 9.3b).

There are three operating areas of the shear stress versus shear rate of lubricating oils: (0)–(1), (1)–(2), and (2)–(3), cf. Fig. 9.4.

(0)–(1): *Newtonian Fluids* ($\tau \leq \tau_c$)

In this operating area, the lubricating oil is a Newtonian fluid (linear behavior) in which the oil viscosity only depends on the oil temperature and is independent of the shear rate. In case of constant temperature, the lubricant viscosity is unchanged until the shear stress reaches the critical stress τ_c. The shear stress is proportional to the shear rate with a constant viscosity η according to Eq. (9.4).

(1)–(2): *Non-newtonian Oils* ($\tau_c < \tau \leq \tau_{lim}$)

At increasing the shear rate, the lubricant behaves like a non-Newtonian fluid (nonlinear behavior). In this operating area, the lubricant viscosity depends on both oil temperature and shear rate. As a result, the shear stress is a nonlinear function of oil temperature and shear rate (see Fig. 9.4). The result shows that the shear stress gradient is smaller than in case of the Newtonian fluid; i.e., the shear stress slowly increases with the shear rate. On the one hand, the higher the shear rate is, the lower the oil viscosity becomes, as shown in Fig. 9.3b. On the other hand, the shear stress increases with the shear rate. Both opposite effects are responsible for the nonlinear behavior of the shear stress versus shear rate.

(2)–(3): *Non-newtonian Oils* ($\tau > \tau_{lim}$)

As the shear rate exceeds the limit shear rate, the shear stress begins slowly decreasing with the shear rate from (2) to (3). At high shear rates, the oil temperature rises due to friction in the bearing leading to a fall of the oil viscosity. It reduces the shear stress more than the increasing shear stress due to raising the shear rate. As a result, the shear stress slightly decreases with the shear rate in this operating area.

In Fig. 9.4, the curve (0)–(1)–(2)–(3) moves upwards for high HTHS oils and moves downwards for low HTHS oils. It is obvious that the shear stress of high HTHS is larger than the lower HTHS oils at the same shear rate. Hence, the lower HTHS the lubricating oil is, the less the friction power is generated in the engine and turbocharger as well leading to less fuel consumption, higher fuel mileage miles per gallon (MPG), and reduction of NO_x and CO_2.

Lubricating oils are generally divided into two HTHS levels: low HTHS viscosity with $\eta_{HTHS} < 3.5$ mPa s and high HTHS viscosity with $\eta_{HTHS} \geq 3.5$ mPa s, as given in Table 9.1. The low HTHS oils with the HTHS viscosity between 2.9 and 3.5 mPa s generates less friction in the engine (i.e., higher mileage MPG, lower NO_x, CO_2 emissions), but more wears could occur in the moving parts, such as cylinders and pistons of the engine, bearings in the turbochargers, etc. Thus, the lifetime of the engine becomes shorter due to wears. On the contrary, the high HTHS oils with the HTHS viscosity larger than 3.5 mPa s cause more friction power in the engine (i.e., lower mileage MPG, higher NO_x, CO_2 emissions), but the moving parts are protected from wears. As a result, the long lifetime is warranted for the engine even at high thermal and mechanical operating conditions.

Therefore, the automotive industry must take the compromise between the low HTHS for high mileage MPG and low NO_x, CO_2 emissions and the high HTHS for the wear protection and the long lifetime of the engines to find out which HTHS value is appropriate for the engines. Recently, in order to reduce NO_x and CO_2 emissions according to the new emission law, some car manufacturers have been thinking of using an ultra-low HTHS oil of 2.6 mPa s, such as lubricating oil SAE 0W20. To overcome the problem of wears, the materials of moving parts must be improved such as coatings or new endurable materials. As a result, the car costs a little more. Eventually, customers have to pay the price increase for the environment protection. In turn, they could save money at the fuel consumption.

At operating conditions with the high rotor speeds, the effective HTHS oil viscosity reduces; therefore, the bearing stiffness and damping coefficients decrease. To keep the rotor in balance with the external forces, the induced pressure in the bearing clearance has to increase itself at reducing the oil-film thickness. Hence, the journal eccentricity ε increases under the load F in order to raise the bearing stiffness and damping coefficients. If the oil-film thickness is below the limit, as given in Fig. 6.11, the mixed or boundary lubrication (BL) takes place in the bearing clearance. Note that the more the bearing friction is, the smaller the oil film thickness becomes due to the high oil temperature. Finally, the oil film breaks up in the radial direction because the boundary contact occurs between the journal and bearing leading to seizure of the journal in the bearing (see Fig. 9.5). As a result, the HTHS viscosity has somehow to do with the shear instability of the oil-film that is to blame for break up of the oil-film in the bearing. At high rotor speeds, the higher the shear rate is, the lower the oil viscosity becomes and the thinner the oil-film thickness will be (see Fig. 9.3).

Fig. 9.5 Oil-film break up in the bearing clearance

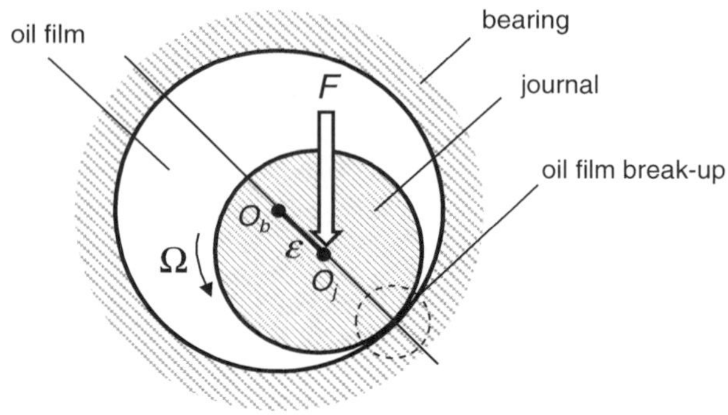

9.4 Viscosity Index of Lubricating Oils

To prevent wear, VI improvers (polymeric additives) are added in lubricating oils in order to modify the viscosity change with oil temperature. They are based on the VI index. The viscosity index (VI) relates to the change rate of the oil temperature to its dynamic viscosity.

$$VI \propto \left| \frac{\partial T}{\partial \eta} \right| \tag{9.7}$$

According to Eq. (9.7), the smaller the viscosity changes with oil temperature, the higher the viscosity index VI is leading to the better oil quality. The viscosities of lubricating oils with the high and low VI versus oil temperatures are shown in Fig. 9.6. The viscosity of the high VI oil decreases with temperature more slowly compared to the viscosity of the low VI oil. That means the absolute viscosity gradient of the high VI oil is larger than the low VI oil. The higher the viscosity index VI is, the less the viscosity changes with oil temperature and vice versa. Obviously, the high VI oils are better for wear protection than the low VI oils at a large operating range of oil temperatures.

Fig. 9.6 Dynamic viscosity versus temperature of low and high VI oils

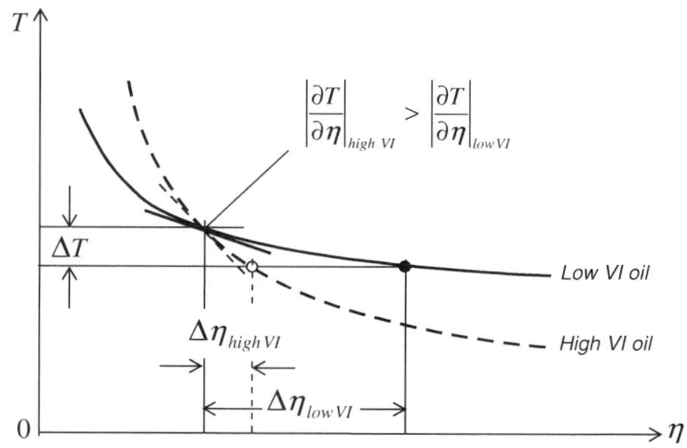

Synthetic oils (PAOs, esters) have mostly higher VI than the mineral oils because some chemical additives are added in the lubricants at the hydrocracking process. They have been produced and custom-tailored to optimize and simplify the molecular structures of the hydrocarbon chains and aromatic rings with predictably designed oil properties, instead of complex unpredictable molecular structures of the mineral oils with lower VI. As a result, synthetic lubricating oils are suitable for the extreme operating conditions with the high thermal and mechanical loads at high rotor speeds of automotive turbochargers.

Adding VI improvers (polymeric molecular additives) in the lubricating oils, the viscosity change with oil temperature reduces so that the HTHS effective viscosity is improved to prevent wears. The VI improvers regulate the oil viscosity with temperature in two ways. First, at low oil temperatures, the molecule chain contracts itself to generate more empty spaces for the oil molecules; therefore, the oil viscosity increases more slowly. Second, at high oil temperatures, the molecule chain expands itself to occupy more spaces in the oil molecules; hence, the oil viscosity reduces more slowly.

The viscosity index VI is a dimensionless number that is calculated by the ASTM D-2270 (American Society for Testing and Materials).

$$VI = 100 \times \left(\frac{L - U}{L - H} \right) \tag{9.8}$$

where

U is the viscosity of the test oil at 40 °C ($\sim$100 °F) in cSt;

L is the viscosity parameter at 40 °C of the first reference oil and is defined as

$$VI = 0 \; for \; L > H$$

H is the viscosity parameter at 40 °C of the second reference oil and is defined as

$$VI = 100$$

L and H are corresponding to the dynamic viscosity (cSt) of the test oil at 100 °C ($\sim$210 °F) are given in the ASTM D-2270;

To calculate the VI index, the oil viscosities at 40 and 100 °C are required. Their viscosity-related parameters L and H are given in the ASTM D-2270. The VI indexes of some commonly used lubricating oils are given in Table 9.3 [6].

Table 9.3 VI indexes of some automotive lubricating oils

Lubricating oils	Viscosity index (VI)
Mineral oils	80 … 120
Hydrocraking oils	125 … 150
Synthetic oils (PAOs)	140 … 160
Silicone oils (hydraulic fluids)	>200

9.5 Stribeck Curve

In the following section, the lubrication regions in the bearing clearance over the oil-film thickness are studied in the Stribeck curve. The oil-film thickness results from many rotordynamic and tribological influences, such as the acting force on the bearing, rotor speed, journal eccentricity, surface roughness, oil temperature, and as well as oil viscosity. If the current oil-film thickness is larger than the limit oil-film thickness, the friction in the bearing remains small; the bearing surface is protected from wear. In this case, the lubrication is fully hydrodynamic.

In order to look into the lubricating behaviors in the bearing, the dimensionless oil-film thickness λ is defined as the ratio of the minimum oil-film thickness to the root-mean-square (rms) combined surface roughness [1, 2, 4].

$$\lambda \equiv \frac{h_{\min}}{R_q} \tag{9.9}$$

where $h_{\min}$ is the minimum oil-film thickness, R_q is the rms combined surface roughness of the surfaces #1 and #2.

The rms-combined surface roughness is defined as

$$R_q \equiv \sqrt{R_{q1}^2 + R_{q2}^2} \tag{9.10}$$

within R_{q1} and R_{q2} are the rms roughness of the surfaces #1 and #2.

Using Gaussian distribution of the surface height, the rms roughness R_q is transformed into the arithmetic average roughness R_a by the factor of 1.25.

$$R_q = 1.25 R_a$$

Thus, the rms-combined surface roughness is written in the average roughness as

$$R_q = 1.25\sqrt{R_{a1}^2 + R_{a2}^2} \tag{9.11}$$

where R_{a1} and R_{a2} are the arithmetic average roughness of the surfaces #1 and #2.

The lubrication regions in the Stribeck curve are classified into four lubrication regions according to [4]:

- $\lambda \le 1$: boundary lubrication (BL)
- $1 < \lambda \le 5$: partial boundary lubrication (PBL)
- $3 < \lambda < 10$: mixed lubrication (ML) containing elasto-hydrodynamic lubrication (EHL: $5 \le \lambda \le 8$)
- $\lambda \ge 10$: fully hydrodynamic lubrication (HL: $\lambda \ge 8$–10)

In case of $\lambda \le 1$, the BL occurs between the surfaces in which the oil-film thickness is very small in the order of a few nanometers (10^{-9} m). In this oil film thickness, the molecules of hydrocarbon chains of the lubricating oil or polymer additives keep the moving surfaces separate from each other in the nanotribology consideration. Due to the very narrow oil film thickness, the friction coefficient strongly increases leading to seizure of materials between the moving surfaces under heavy loads.

As the dimensionless film thickness λ increases from 1 to 5, the partial BL occurs in which the surfaces partially contact each other between the BL (in the nanotribology $\sim 10^{-9}$ m) and ML (in the microtribology $\sim 10^{-6}$ m). The nanotribology is generally used to explain the tribological effects in the microtribology. At first, the friction coefficient remains nearly constant in the BL and then slightly decreases with the oil-film thickness in the ML. However, the friction coefficient is still high since the abrasive and adhesive frictions occur between the moving surfaces in the mixed BL.

The EHL occurs in the ML at $5 \leq \lambda \leq 8$, in which the asperities between the surfaces have been plastically deformed or removed due to abrasive and adhesive wears. As soon as the asperities of the surfaces disappear or do not touch with each other at increasing oil-film thickness, the friction coefficient drops significantly to the minimum where the fully HL begins (see Fig. 9.7).

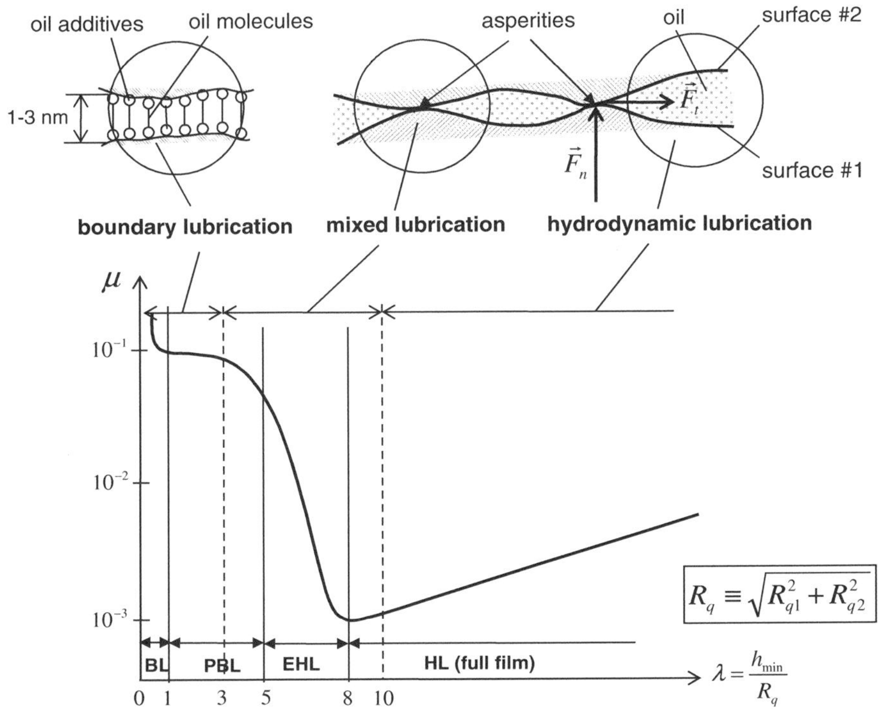

Fig. 9.7 Lubrication regions in the Stribeck curve μ-λ

In case of $\lambda \geq 10$, the moving surfaces are fully separated by the large oil-film thickness without contact of any asperity at less abrasive and adhesive frictions. This lubrication regime is called fully HL. The friction induced in this region is only the viscous friction of the oil film.

The friction coefficient in the fully HL is calculated as [9]

$$\mu_{\mathrm{HL}} = \frac{F_t}{F_n} \propto \frac{(1/h)}{1/h^2} = h \propto \lambda \tag{9.12}$$

where

h is the current minimum oil-film thickness;
λ is the dimensionless oil-film thickness;
F_t is the friction force proportional to inversely related to the film thickness ($1/h$);
F_n is the normal force proportional to $1/h^2$.

According to Eq. (9.12), the friction coefficient is proportional to the dimensionless oil-film thickness in the fully HL.

In the partial boundary or ML regions, the oil temperature in the bearing clearance is relatively high due to the large bearing friction. When the effective oil temperature in the bearing exceeds the flash points of the lubricating oil 210 °C (SAE 5W30) and 250 °C (SAE 20W30) [6], the oil film begins coking inside the bearing clearance and leaves the hard coked-oil layer on the rotor shaft.

The process of oil coking generates a hard black thin layer of the carbonaceous residue on the surface of the journal within the bearing clearance. The hard coked-oil layer continually increases after a long operating time; hence, the radial bearing clearance reduces. As a result, the effective oil temperature further increases as the bearing clearance reduces, and the coking process continuously takes place. Finally, it causes seizure of the shaft in the bearing and fatal damage of turbochargers.

9.6 Surface Texture Parameters

Surface tribological characteristics play a key role in rotor stability and for wear in the bearings as well. They strongly affect the oil-film thickness that depends on the lubrication regime in the bearings (see Fig. 9.7). The oil film thickness involves the rotordynamic stability, induced airborne noises, and prevention of wear in the bearings. The oil-film bearings work on the hydrodynamic principle in which the inner oil film supports the rotor against the unbalance excitation forces and keep it stable during the operation. Moreover, the outer oil film provides the rotor with a large damping to keep it in small amplitudes at resonances and prevent the rotor from the oil whirl instability. Therefore, the outer oil film reduces the induced airborne noises, such as the unbalance whistle and constant tone. Without or poorly lubricated oil supply, the bearings cannot fulfill their functions in the turbocharger.

That could cause damage in the bearings and furthermore in the compressor and turbine wheels due to seizure of the wheels in their housings. The surface texture parameters are discussed in the following section.

9.6.1 Surface Height Profile

The surface roughness characteristics of the bearings, journal in the radial bearing, and thrust rings in the thrust bearing are tribological analyzed. The roughness profile of the surface is defined as the trace profile about the waviness profile of the surface (see Fig. 9.8).

At sampling the surface, the surface profile consisting of the waviness and roughness height profile is measured by a stylus and digitalized. The peak-to-peak amplitude of the waviness profile is defined as the waviness height W_t. The measured signal of the surface profile is amplified and analyzed by band-filter technique, as displayed in Fig. 9.9a. Using the high-pass filter, one obtains the surface roughness profile; and the surface waviness profile results from the low-pass filter. The waviness profile shows the plateau shape of the surface; the roughness height profile describes the real roughness height of the surface that is measured from the reference line. The roughness height profile is the arithmetic average value of the surface height in an evaluation length. The roughness profile contains many peaks (asperities) and valleys of the measured surface [8].

The mean value of the surface height is defined in the evaluation length l_n as

$$\begin{aligned} \bar{z} &\equiv \frac{1}{l_n}\int_0^{l_n} z(x)dx \\ &\approx \frac{1}{N}\sum_{i=1}^{N} z_i(x_i) \end{aligned} \tag{9.13}$$

where N is the sampling number of the surface height z_i measured from the reference line z_{ref} within the evaluation length l_n.

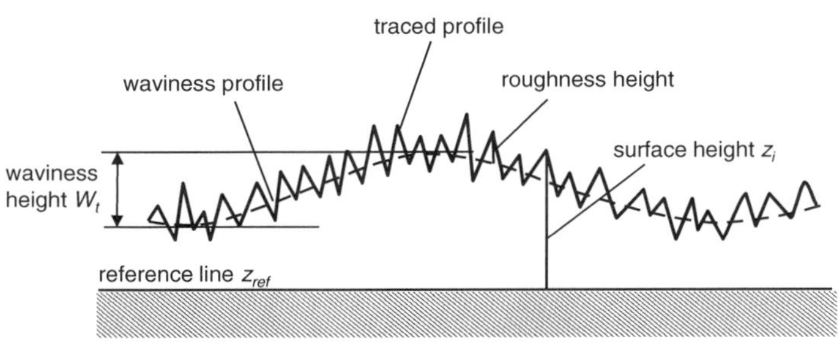

Fig. 9.8 Characteristics of the surface height profile

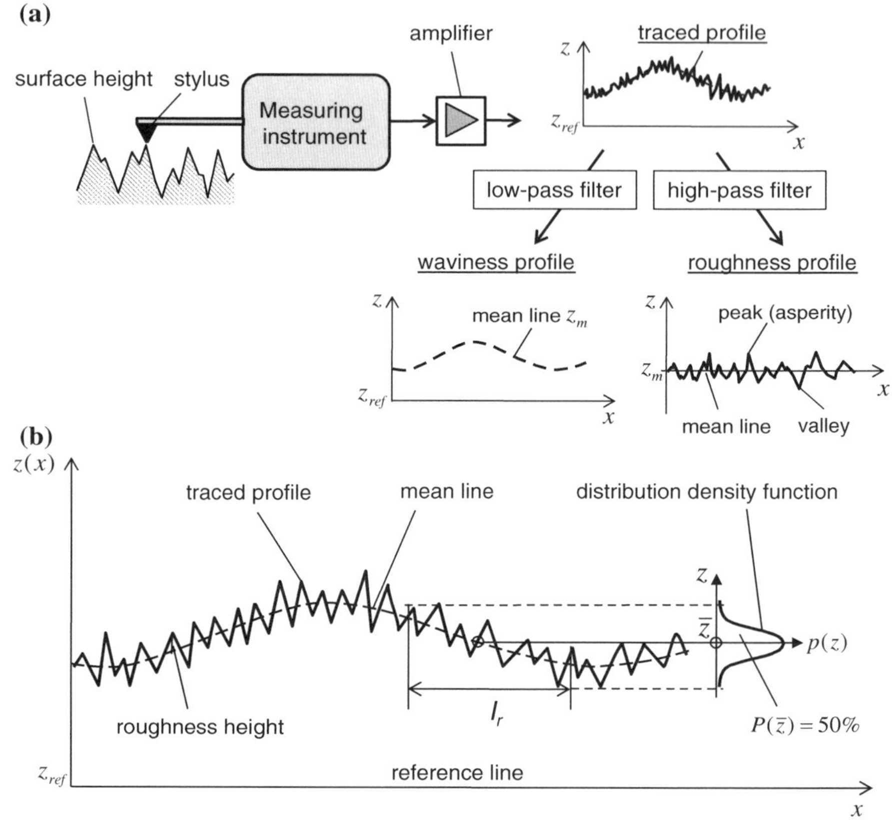

Fig. 9.9 **a** Analysis of the measured traced profile. **b** Distribution density function of the measured roughness height

In practice, the mean value of the surface height is determined by the Gaussian density function $p(z)$ within the sampling length l_r using the phase corrector filter. The mean value is the surface height value at the maximum density function whose probability function $P(z)$ equals 50 % (cf. Fig. 9.9b and Appendix D).

The variance s of the roughness height of N samplings in the sampling length l_r is calculated as

$$s \equiv \frac{1}{(N-1)} \sum_{i=1}^{N} (z_i - \bar{z})^2 \tag{9.14}$$

The standard deviation σ is defined as the square root of the variance s. It shows the variation of the measured values from the mean value of the surface roughness height in the normal distribution density function (see Appendix D). A small standard deviation indicates that the measured values tend to be close to the mean value; a large standard deviation shows that the measured values are spread out over a large range from the mean value.

Using Eq. (9.14), the standard deviation of N samplings in the sampling length l_r is written as

$$\sigma \equiv \sqrt{s} = \sqrt{\frac{1}{(N-1)}\sum_{i=1}^{N}(z_i - \bar{z})^2} \tag{9.15}$$

According to Appendix D, the smaller the standard deviation σ is, the better the production will be; e.g., a production with $\pm 3\sigma$ gives 99.7 % of its products that are within the given tolerances.

9.6.2 Surface Tribological Parameters

In the following section, we deal with some tribological parameters that are used to analyze the bearing surface roughness in the turbochargers [8].

- *Evaluation length* l_n is the length in which the values of the surface height are selected and evaluated. Its length is normally chosen by five times of the cutoff wavelength λ_c.

$$l_n \approx 5\lambda_c \tag{9.16}$$

- *Sampling length* l_r is the reference length for the roughness height evaluation and nearly equals the cutoff wavelength λ_c.

$$l_r \approx \lambda_c \tag{9.17}$$

- *Cutoff wavelength* λ_c is the profile filter that is used in analyzing the roughness height. It is generally chosen so that the amplitude of the roughness height with a sine function is reduced to about 50 % after filtering the measured signals. According to DIN EN ISO 4288:1998 and DIN EN ISO 3274:1998, some wavelengths are determined for profile filters as follows:

$$\lambda_c = 0.08;\ 0.25;\ 0.8;\ 2.5;\ 8.0\ \text{mm} \tag{9.18}$$

At decreasing the cutoff wavelength, the amplitude of the filtered surface roughness profile is also reduced; the amplitude of the filtered waviness profile of the surface roughness increases. Therefore, the short cutoff wavelength is rather preferred at small surface roughness of R_a and R_z.

- *Roughness* R_{sm} (DIN EN ISO 4287, ASME B46.1), see Fig. 9.10

It is called the *mean peak spacing* and defined by the arithmetic average value of five mean peak spacings of the roughness profile within the evaluation length l_n.

$$R_{sm} = \frac{1}{5}\sum_{i=1}^{5} s_{mi} \tag{9.19}$$

The mean peak spacing s_{mi} is the wavelength of the roughness profile that must contain at least one peak and one valley of the surface roughness.

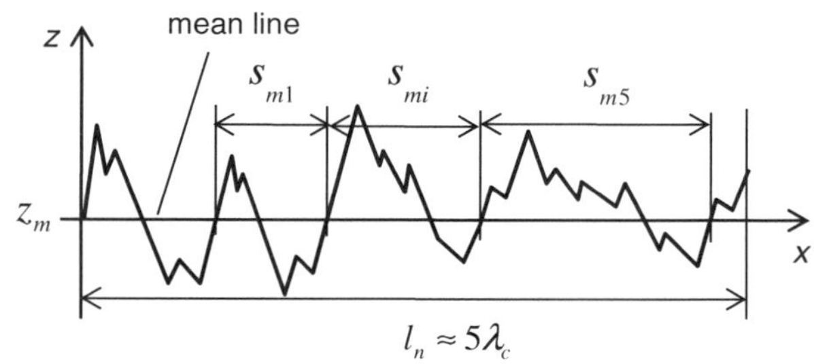

Fig. 9.10 Mean peak spacing R_{sm}

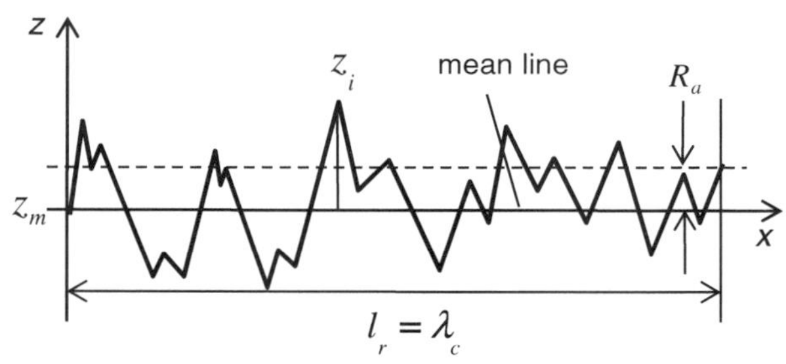

Fig. 9.11 Arithmetic mean roughness R_a

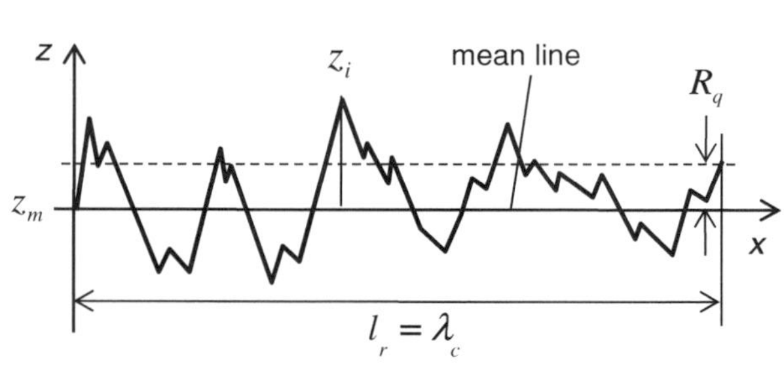

Fig. 9.12 Root mean square roughness R_q

- *Mean roughness R_a* (DIN EN ISO 4287, ASME B46.1), see Fig. 9.11

It is the arithmetic average roughness of N roughness heights z_i. For the simplicity, the roughness height z_i is measured from the mean line within the sampling length l_r.

$$R_a \equiv \frac{1}{l_r} \int_0^{l_r} |z(x)| dx \approx \frac{1}{N} \sum_{i=1}^{N} |z_i| \tag{9.20}$$

- *Root mean square (rms) roughness R_q* (DIN EN ISO 4287, ASME B46.1), see Fig. 9.12

The rms roughness height of N points within the sampling length l_r is defined as where the peak and valley heights are considered.

$$R_q \equiv \sqrt{\frac{1}{l_r} \int_0^{l_r} z^2(x) dx} \approx \sqrt{\frac{1}{N} \sum_{i=1}^{N} z_i^2} \tag{9.21}$$

- *Mean roughness depth R_z and maximum roughness depth $R_{\max}$* (DIN EN ISO 4287, ASME B46.1).

The mean roughness depth R_z is the arithmetic average value of the five single roughness depths $R_{z,i}$ of five consecutive sampling lengths l_r within the evaluation length l_n.

$$R_z \equiv \frac{1}{5} \sum_{i=1}^{5} R_{z,i} \tag{9.22}$$

The maximum roughness depth $R_{\max}$ is the largest single roughness depth of the five roughness depths within the evaluation length l_n.

$$R_{\max} \equiv \max_{i=1,..,5} (R_{z,i}) \tag{9.23}$$

where $R_{z,i}$ and $R_{\max}$ are shown in Fig. 9.13.

- *Material ratio R_{mr}* (DIN EN ISO 4287, ASME B46.1)

The material ratio is the ratio of the sum $l(c)$ of all material lengths l_i at the surface height z to the total material length l_n. It is also called the *bearing length ratio* according to the ASME B46.1.

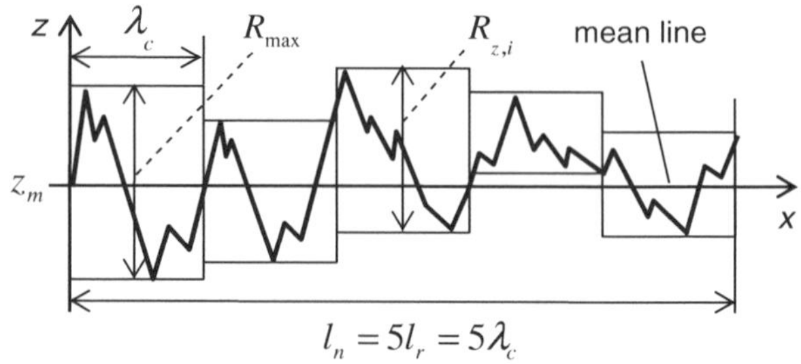

Fig. 9.13 Mean roughness depth R_z and maximum roughness depth R_{max}

$$R_{mr}(z) \equiv \frac{\sum l_i(z)}{l_n} 100\,[\%] \tag{9.24}$$

- *Abbott-Firestone curve (Abbott curve)*

The Abbott-Firestone curve is the most important surface roughness parameter that is additionally used to evaluate the surface roughness quality. The Abbott curve evaluates the tribological surface roughness quality of the samples that have the same mean roughness R_a and R_z. In this case, the shape of the Abbott curve decides which surface roughness is better.

According to the ASME B46.1, the Abbott curve is also called the *bearing area curve* (BAC). In the statistical analysis, the Abbott curve is in fact the probability distribution function of the surface roughness height resulting from its distribution density function (see Appendix D). At first, the construction of the Abbott curve is derived from the profile of the roughness height $z(x)$. At the cutting line with the surface height z, the material ratio R_{mr} is computed from the sum of all cutting lengths l_i to the total length l_n, as shown in Fig. 9.14.

The new dimensionless surface height c is defined as

$$c \equiv \frac{z - \bar{z}}{\sigma} \tag{9.25}$$

where σ is the standard deviation of the roughness height in Eq. (9.15).

The dimensionless surface height c versus the material ratio R_{mr} is plotted on the RHS of Fig. 9.14. The bell-shaped curve is called the *Abbott-Firestone curve*. Obviously, the material ratio equals 0 % at the maximum peak because no peak is cut; the material ratio arrives at 100 % at the minimum valley since all cutting lengths equal the total length. However, in practice, one moves the ordinate surface height c from the initial position of R_{mr} = 0–5 % (experience value) in the direction of the abscissa material ratio R_{mr}. Therefore, it makes sure that the reference line c_{ref} lies at the highest peak of the surface height because of initial wears after a short operating period.

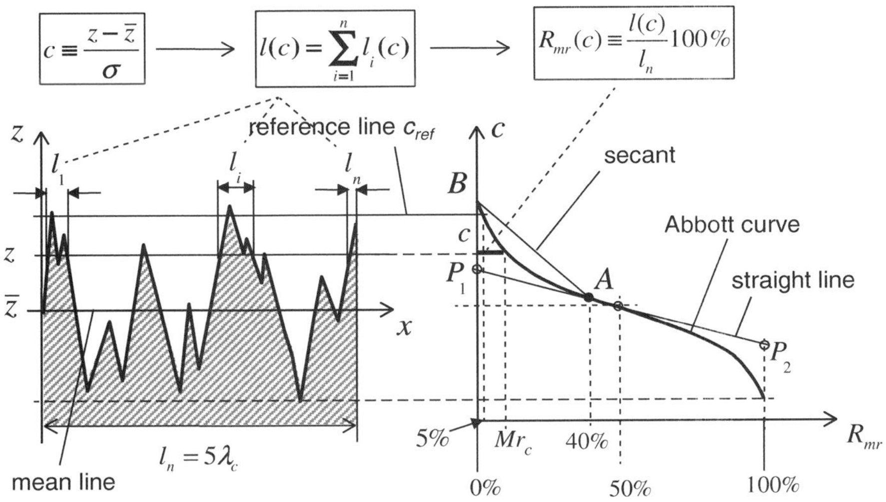

Fig. 9.14 Abbott curve and the material ratio R_{mr}

In another way, the Abbott curve can be derived using the amplitude density function (ADF) $p(c)$ of the surface roughness profile, as shown in Fig. 9.15 [2].

The ADF $p(c)$ is the number of surface heights between the two cutting heights z and $z + dz$. The cumulative distribution function $P(c)$ of the ADF $p(c)$ of the surface roughness height is in fact the material ratio at the dimensionless surface height c (see Appendix D).

$$P(c) = \int_{c}^{\infty} p(c)dc \equiv R_{mr}(c) \tag{9.26}$$

Thus,

R_{mr} $(c_{\max}) = P(c_{\max}) \approx 0$ % at the highest peak of the surface height;
R_{mr} $(0) = P(0) \approx 50$ % at the mean line ($c = 0$);
$\mathrm{R_{mr}}$ $(c_{\min}) = P(c_{\min}) \approx 100$ % at the lowest valley of the surface height.

- *Roughness R_{pk}, R_k, R_{vk}* (DIN EN ISO 13565-1 and -2)

R_{pk} is the reduced peak height in the Abbott curve in Fig. 9.16 that indicates the peak roughness of the surface; R_k is the core roughness depth indicating the plateau shape of the roughness surface; R_{vk} is the reduced valley height of the surface indicating the oil reservoir in the roughness surface.

The material ratios Mr_1 and Mr_2 are the smallest and largest material ratios at R_{pk} and R_{vk}, respectively. Both material ratios determine the shape of the Abbott curve that indicates the important tribological parameter besides the mean roughness R_a and mean roughness depth R_z.

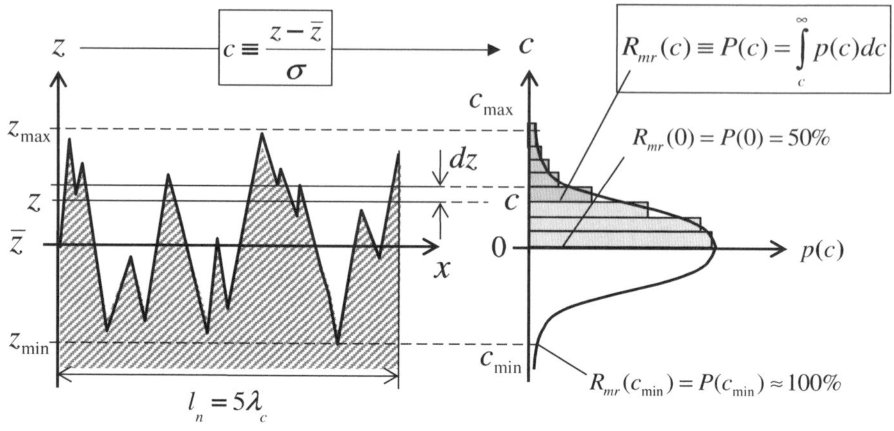

Fig. 9.15 Amplitude density function (ADF) of the surface roughness profile

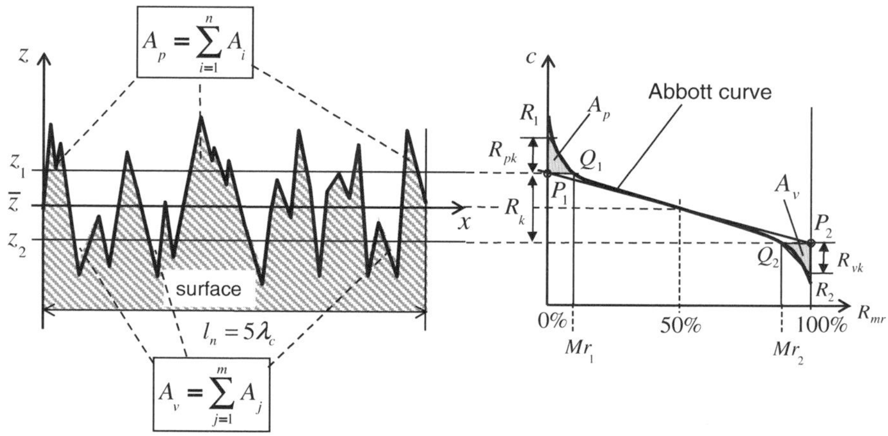

Fig. 9.16 Roughness R_{pk}, R_k, R_{vk} and material ratios Mr_1 and Mr_2

In the following section, the roughness R_{pk}, R_k, and R_{vk} and the material ratios Mr_1 and Mr_2 are determined in the Abbott curve in Figs. 9.14 and 9.16. At first, we create the secant *AB* with R_{mr} (*A*) of 40 % in Fig. 9.14. Then, we rotate the secant *AB* about point *A* until it is tangential to the left half-branch of the Abbott curve at the point *A*. The prolonged secant, called the straight line P_1AP_2, cuts the ordinates with R_{mr} = 0 and 100 % at P_1 and P_2, respectively, as plotted in Fig. 9.16. The corresponding surface heights z_1 and z_2 at P_1 and P_2 intersect the Abbott curve at Q_1 and Q_2, respectively. Therefore, the smallest and largest material ratios Mr_1 and Mr_2 are found at Q_1 and Q_2 (see Fig. 9.16).

The area A_p is the sum of all peak surfaces A_i above the cutting surface height z_1. In the Abbott curve, the triangle $P_1Q_1R_1$ is constructed so that its area equals A_p. The altitude P_1R_1 is defined by the reduced peak height R_{pk}. Similarly, the reduced valley height R_{vk} is equal to P_2R_2 that is derived from the triangle $P_2Q_2R_2$ whose area equals A_v of the total groove area below the cutting surface height z_2. Finally, the core roughness depth R_k is the surface roughness height of P_1P_2, as shown in Fig. 9.16.

The reduced peak and valley heights are calculated as

$$R_{pk} = \frac{2A_p}{l_n Mr_1} \times 100\,\% \tag{9.27}$$

$$R_{vk} = \frac{2A_v}{l_n(100 - Mr_2)} \times 100\,\% \tag{9.28}$$

Generally, the core roughness depth R_k ($< \sim 1$–2 μm) should be small in the bearings to increase the bearing load capability because the bearing surface has less plateau shape. Note that the smaller the reduced peak height R_{pk} ($< \sim 0.5$–1 μm) is, the better the surface quality becomes. On the contrary, the reduced valley height R_{vk} ($< \sim 1$–2 μm) should be larger than R_{pk} to maintain the reserved lubricating oil in the valley grooves. Hence, the bearing is well lubricated at the start-stop driving cycle.

Figure 9.17 shows an example of two surface roughness heights that have the same R_a but different roughness values of R_{pk}, R_k, and R_{vk}. Obviously, the first one has only a moderate tribological surface characteristic due to the higher value of $R_{pk,1}$ compared to $R_{pk,2}$ that could cause wear in the surface. On the contrary, the latter with smaller values of $R_{k,2}$ and $R_{pk,2}$ is tribological better because the surface has less plateau and the peaks of the surface roughness are reduced; its $R_{vk,2}$ is higher than the first one. Hence, the surface has more oil reservoir to lubricate the bearing surface.

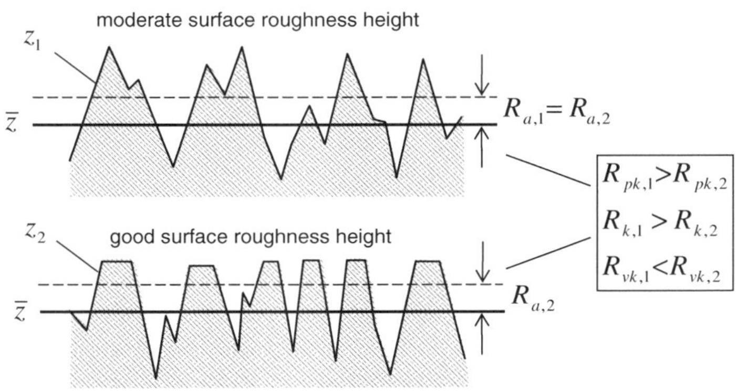

Fig. 9.17 Comparison of two surface roughness heights

Figure 9.18 shows three surface roughness heights that are finished by different methods of honing, eroding, and turning. Despite the nearly same roughness R_a and $R_{\max}$, they have different qualities of tribological surface characteristics. The honed surface is the best because all peak heights were removed and the groove depths are large enough to maintain sufficiently reserved oil for lubrication. The eroded surface has a moderate quality since many peaks remain and thus could cause wears at the asperities and reduces the bearing load capability. The turning surface with the largest R_z is the worst since it has many peaks with the largest reduced peak height R_{pk} due to the turning profile. It causes reduction of the bearing load capability and therefore induces the bearing friction due to abrasive wear at the peaks (asperities). In this case, we have to consider the other roughness parameters of R_{pk}, R_k, R_{vk}, and R_{mr}.

Besides the surface roughness depths R_z and R_a, the Abbott curve is additionally used to analyze the surface roughness quality. Generally, good tribological surface roughness characteristics are possibly smallest reduced peak heights R_{pk} and possibly largest reduced valley heights R_{vk}.

The Abbott curve is derived from the measured values of the surface roughness height with the lowest and largest material ratios of Mr_1 and Mr_2 according to Eqs. (9.27 and 9.28), as shown in Fig. 9.19.

They determine the shape of the Abbott curve of the roughness surface height. By experience, at the same roughness R_a or R_z, the Abbott curve with a *convex shape* indicates the best tribological quality of the surface roughness height (case *c*). Therefore, the best surface roughness characteristic is the third case with the small reduced peak heights R_{pk} (low peak surface roughness), the high reduced valley heights R_{vk} (more oil reservoir), and the small core roughness depth R_k with less plateau of the surface roughness. On the contrary, the surface with a *concave shape*

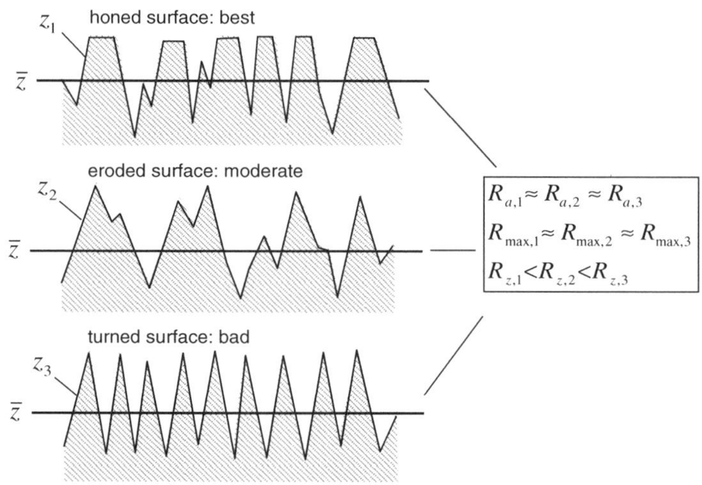

Fig. 9.18 Roughness heights at different finishing methods

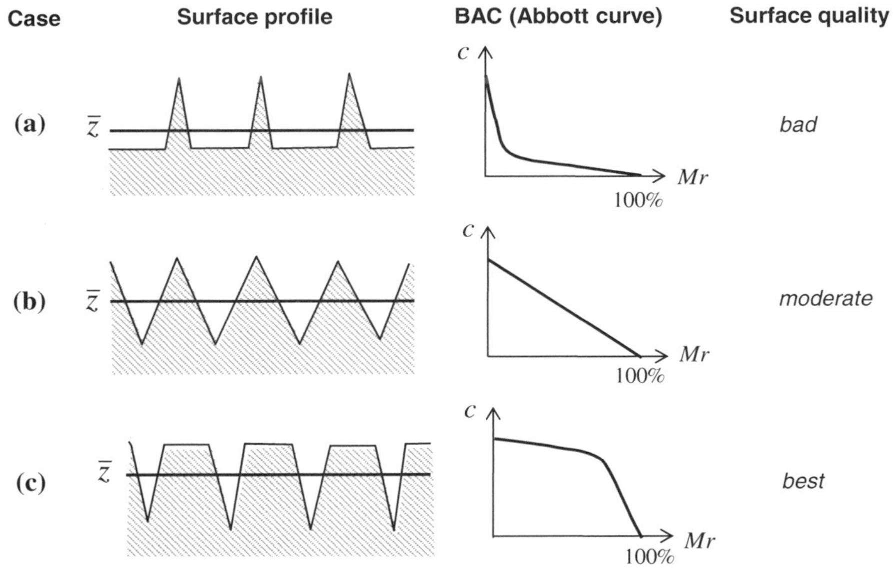

Fig. 9.19 Analysis of surface quality using Abbott curves

shows the bad tribological surface roughness quality (case *a*). They have many sharp asperities with the large reduced peak heights R_{pk} (high peak surface roughness) and the low reduced valley heights R_{vk} (less oil reservoir). The Abbott curve of the case *b* has a *linear shape* that is considered as moderate.

9.7 Elastic and Plastic Deformations in the Bearings

To study the wear mechanism in the bearings, some backgrounds of the elastic and plastic deformations of material are required [3, 5].

First, a small load acts upon the material that begins deforming; as soon as the load is removed, its form returns to the initial condition. This deformation is defined as *elastic deformation*. On the contrary, in case of *plastic deformation*, a heavy load acts upon the material. As a result, the resulting stress in the material exceeds the yield stress σ_o, the deformed material remains and does not return to the initial condition although the acting load is removed from it (see Fig. 9.20).

At further increasing the acting load at the plastic deformation, the material suddenly breaks up (material fracture) shortly after the ultimate tensile stress σ_u. Therefore, the wear process begins with the loss of the surface asperities (roughness peaks) that abrade the moving surfaces and cause wear paths on the surfaces.

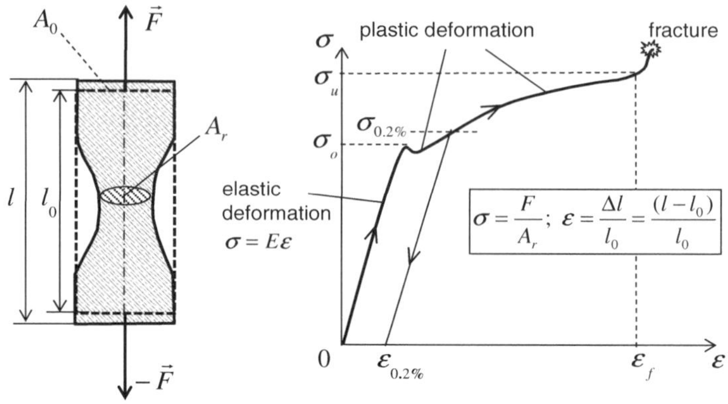

Fig. 9.20 Stress-strain diagram of a ductile material

9.7.1 Normal Stress

Applying the tensile force F on a cylindrical specimen with the initial cross-sectional area A_0 and length l_0, the body begins deforming in the axial direction.

The normal stress σ is defined as the ratio of the acting force to the real cross-sectional area.

$$\sigma = \frac{F}{A_r} \tag{9.29}$$

In the elastic deformation, the stress is proportional to the strain according to the Hooke's law.

$$\sigma = E\left(\frac{l - l_0}{l_0}\right) = E\frac{\Delta l}{l_0} \equiv E\varepsilon \tag{9.30}$$

where

E is the elasticity modulus (Young's modulus);
ε is the normal strain defined as the relative change of the length from l_0 to $l > l_0$.

The Hooke's law is valid as long as the stress is less than the *yield stress* σ_o; i.e., the deformation is elastic and returns to the initial position when the acting load is removed. At further increasing the acting force, the normal stress increases larger than the yield stress σ_o; hence, the deformation of the body becomes plastic where the Hooke's law has been no longer valid. With a plastic stress of 0.2 %, the strain remains at $\varepsilon = 0.2$ % after removing the acting force. The deformation is plastic up to the *ultimate tensile stress* σ_u. After exceeding the ultimate tensile stress, the normal stress sharply increases in a very short time, and the material breaks up. Shortly before the fracture occurs, the real cross-sectional area A_r significantly reduces due to material contraction at a constant load F, as shown in Fig. 9.20. Therefore, the normal stress sharply increases before the material fracture happens according to Eq. (9.29).

9.7.2 Shear Stress

Applying the force F acting on a specimen in the tangential direction, the body form deforms at a shear strain t. The *shear rate* $\dot{\gamma}$ is defined as the ratio of the shear strain t to the specimen height h (see Fig. 9.21).

The *shear stress* is written at the contact surface A as

$$\tau = \frac{F}{A} = G\left(\frac{t}{h}\right) \approx G\dot{\gamma} \tag{9.31}$$

where G is the *shear modulus* of the material.

The shear modulus is calculated from the *elasticity modulus* E and the *Poisson ratio* v as

$$E = 2G(1 + v) \tag{9.32}$$

where v equals 0.25–0.30 for most metals. Thus,

$$G = \frac{E}{2(1 + v)} \approx (0.385\ldots0.400)E \tag{9.33}$$

The elasticity modulus E can be used for

- low-alloy steels: $E \approx 212$ GPa;
- tool steels (highly alloyed steels): $E \approx 230$ GPa;
- brass, copper alloys, and bronze: $E \approx 96$ to 114 GPa.

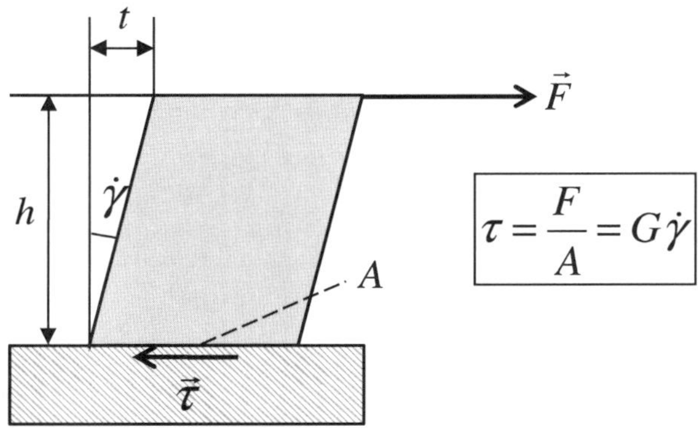

Fig. 9.21 Shear stress and shear rate in a testing specimen

9.7.3 Friction Force in the Bearings

The rotor unbalance induces the normal load F_n acting on the bearing surface. At the contact zone between the journal and bearing, the friction force F_t occurs on the journal surface opposite to the rotational direction of the rotor. The friction force is proportional to the normal load with a friction coefficient. Figure 9.22 shows the normal load and the friction force in the bearing. In fact, the adhesion and abrasion frictions between the bearing and journal induce the total friction force.

$$\begin{aligned} F_t &= \mu_f F_n \\ &= F_{\mathrm{adh}} + F_{\mathrm{abr}} = (\mu_{\mathrm{adh}} + \mu_{\mathrm{abr}}) F_n \end{aligned} \tag{9.34}$$

The total friction coefficient results from the adhesive and abrasive friction coefficients as

$$\mu_f = \mu_{\mathrm{adh}} + \mu_{\mathrm{abr}} \tag{9.35}$$

The adhesive friction force results from the shear stress and real contact area as

$$F_{\mathrm{adh}} = \tau A_r \tag{9.36}$$

within the real contact area A_r is determined as the ratio of the normal load to the hardness H of the softer material at the plastic deformation.

$$A_r = \frac{F_n}{H} \tag{9.37}$$

The hardness of material H is the mean contact pressure that results from the normal load and permanent indentation at the plastic deformation as

$$H \equiv p_{\mathrm{mean}} = \frac{F_n}{A_r} \tag{9.38}$$

The plastic deformation begins with $p_{\mathrm{mean}} \approx 1.07\text{–}1.1\sigma_o$ (σ_o = yield stress). At $p_{\mathrm{mean}} \approx 3\sigma_o$, it fully develops in the permanent indentation. According to [5], the hardness H is nearly 2.8–3 times of the yield stress for intenders with spherical, conical, pyramid, and flat-end shaped geometries.

$$H \approx (2.8\ldots3)\sigma_o \tag{9.39}$$

Substituting Eqs. (9.36 and 9.37), one obtains the adhesive friction coefficient.

$$\mu_{\mathrm{adh}} \equiv \frac{F_{\mathrm{adh}}}{F_n} = \frac{\tau}{H} \tag{9.40}$$

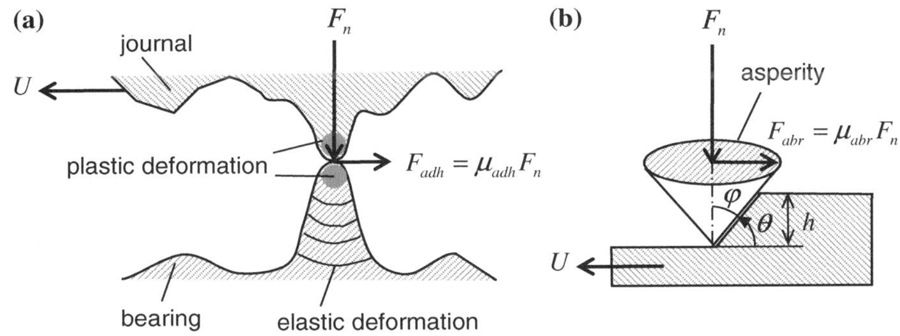

Fig. 9.22 Adhesive and abrasive friction forces in the bearing. **a** Adhesion friction. **b** Abrasion friction

According to [9], the abrasive friction coefficient is calculated as

$$\mu_{\text{abr}} = \frac{2}{\pi} \cot \varphi = \frac{2}{\pi} \tan \theta \tag{9.41}$$

where φ is the cone semi-angle of the asperity; θ is the abrasive angle of material (see Fig. 9.22).

Thus, the total friction coefficient results from Eqs. (9.35, 9.40, and 9.41).

$$\begin{aligned} \mu_f &= \mu_{\text{adh}} + \mu_{\text{abr}} \\ &= \frac{\tau}{H} + \frac{2}{\pi} \tan \theta \end{aligned} \tag{9.42}$$

Normally, the abrasive angle of most materials is less than 10°; hence, the abrasive friction coefficient is about 0.1. The adhesive friction coefficient is 0.17 to 0.2 for similar hardness of materials where the abrasive friction coefficient is negligible, and <0.3 for hard to softer materials. In fact, the total friction coefficient is quite larger than these theoretical values between 0.3 and 0.4.

The work hardening and junction growth in the contact zone is to blame for this discrepancy [9]. At the plastic deformation, the atomic dislocations are removed from the material grid structure. Therefore, the material becomes harder; the shear and yield stress, and the material hardness are strengthened. This process is called the *work hardening* during plastic deformation. Additionally, the real contact area increases at the plastic deformation, cf. Eq. (9.37). It is called the *junction growth* at the contact zone.

Straightforwardly, the plastic deformation occurs in the contact zone because the normal load and friction force concentrate on a very small contact area of the surface roughness asperities (high R_{pk} and R_z). The normal stress that is far away from the asperity is smaller than in the contact zone. Therefore, the elastic deformation occurs at far away from the contact zone (see Fig. 9.22). During plastic deformation at the contact zone, the shear stress increases faster than the hardness. As a result, the friction coefficient increases in the contact zone between the journal and the bearing. Therefore, the friction force increases in the plastic deformation due to the work hardening and junction growth, cf. Eqs. (9.34, 9.38, 9.40, and 9.42).

9.7.4 Friction Power in the Bearings

The following section deals with the friction power in the bearings that occurs in the boundary, mixed, and HLs in the Stribeck curve, as displayed in Fig. 9.7.

The friction power P_f in the bearings results from the friction force F_f and relative velocity U. It consists of the friction powers in the ML P_m and the HL P_h. At the large oil-film thickness $h \geq h_0$, the friction power in the HL P_h at $\varepsilon = 1$ dominates the bearing friction power. In other case $h \ll h_0$, the friction power in the boundary/ML P_m at $\varepsilon = 0$ is dominant in the bearing friction power at the small oil-film thickness.

Therefore, the bearing friction power can be expressed in terms of the hydrodynamic and mixed friction powers in the bearing.

$$\begin{aligned} P_f &= F_f U = (\tau \cdot A)U \\ &= (1-\varepsilon)P_m + \varepsilon P_h \\ &= [(1-\varepsilon)\mu_m + \varepsilon\mu_h]\sigma_N UA \end{aligned} \tag{9.43a}$$

where

ε is the weighting factor for the fully hydrodynamic ($\varepsilon = 1$) and ML ($\varepsilon = 0$);
μ_m is the friction coefficient in the ML;
μ_h is the friction coefficient in the HL;
σ_N is the normal stress on the bearing;
A is the bearing surface.

In case of a small bearing surface, the oil-film thickness reduces itself to increase the oil pressure in the bearing clearance against the bearing load. At further increasing the bearing load, the oil-film thickness becomes smaller than the oil-film thickness h_0 (see Fig. 9.23). Thus, the ML regime occurs in the bearing in which the friction coefficient μ_m significantly increases due to the adhesive and abrasive friction on the bearing surface. According to Eq. (9.43a), the mixed friction power P_m extremely increases much higher than the friction power P_h in the HL.

As a result, the effective friction power in the bearing increases drastically although the bearing surface is reduced. However, if we enlarge the bearing surface excessively, the oil-film thickness will increase in the fully HL regime. In this case, the hydrodynamic friction power P_h could also increase due to the large bearing

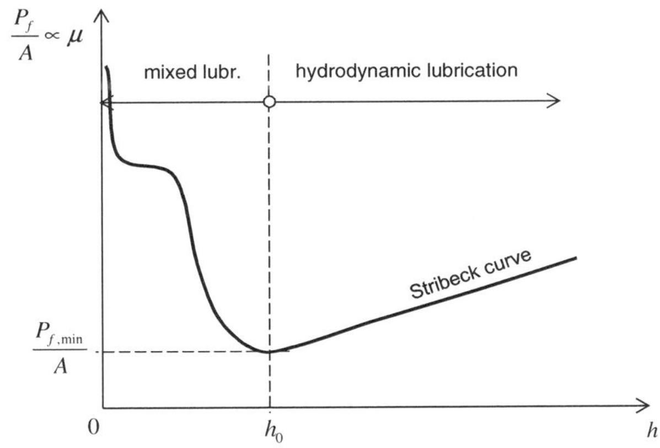

Fig. 9.23 Specific friction power in the bearing versus oil-film thickness

surface and high oil viscosity. Therefore, the effective bearing friction power also increases at enlarging the bearing surface in the fully HL. The friction power reaches a minimum value at the oil-film thickness h_0 from which the fully HL begins. The best way is to operate the bearing near the oil-film thickness h_0.

The specific friction power in the bearing is defined as

$$\begin{aligned}\frac{P_f}{A} &= [(1-\varepsilon)\mu_m + \varepsilon\mu_h]\sigma_N U \\ &\equiv \mu\sigma_N U \propto \mu\end{aligned} \tag{9.43b}$$

Figure 9.23 shows the behavior of the specific friction power of the bearing in the mixed and fully HL regimes in the Stribeck curve. The result indicates that the specific friction power strongly rises at $h \ll h_0$. Therefore, decreasing the bearing surface to reduce the bearing friction power is not always correct. In some cases, we obtain an increased friction power in the bearing due to the very large mixed friction coefficient μ_m.

The reduction of the bearing friction by moderately decreasing the bearing surface is correct when the operating condition is in the fully HL regime ($h > h_0$) with a smaller friction coefficient μ_h.

$$P_{f,h} = \mu_h \sigma_N U A$$

In the ML regime ($h \ll h_0$), the friction coefficient μ_m strongly increases leading to a high friction power in the bearing.

$$P_{f,m} = \mu_m \sigma_N U A \gg P_{f,h}$$

In this case, we must even enlarge the bearing surface or reduce the bearing load so that the oil-film thickness in the bearing increases to h_0.

9.7.5 Mohr's Circle Diagram

The normal force and bending moment caused by the friction force act upon the asperities of the bearing (see Fig. 9.24). The asperities begin deforming under the acting load and moment. At increasing external forces and moments, the plastic deformation occurs at the asperities. When the normal and shear stresses exceed their ultimate stresses, the asperities break up in the bearing clearance leading to abrasive wears.

In the following section, we compute the principle stresses using Mohr's circle method. The principle normal stresses result from solving the cubic stress equation of σ [3].

$$\begin{vmatrix} (\sigma_x - \sigma) & \tau_{xy} & \tau_{zx} \\ \tau_{xy} & (\sigma_y - \sigma) & \tau_{yz} \\ \tau_{zx} & \tau_{yz} & (\sigma_z - \sigma) \end{vmatrix} = 0 \tag{9.44}$$

In case of $\tau_{zx} = \tau_{yz} = 0$, and $\tau_{xy} \neq 0$ for simplicity, we have three solutions:

$$\sigma_1, \sigma_2 = \left(\frac{\sigma_x + \sigma_y}{2}\right) \pm \sqrt{\left(\frac{\sigma_x - \sigma_y}{2}\right)^2 + \tau_{xy}^2} \tag{9.45}$$

and

$$\sigma_3 = \sigma_z \tag{9.46}$$

The material failure takes place if the maximum tensile stress σ_1 in Eq. (9.45) due to bending moment exceeds the ultimate tensile stress σ_u, or the maximum shear stress $\tau_{\max}$ in Eq. (9.49) is larger than the critical shear stress τ_c.

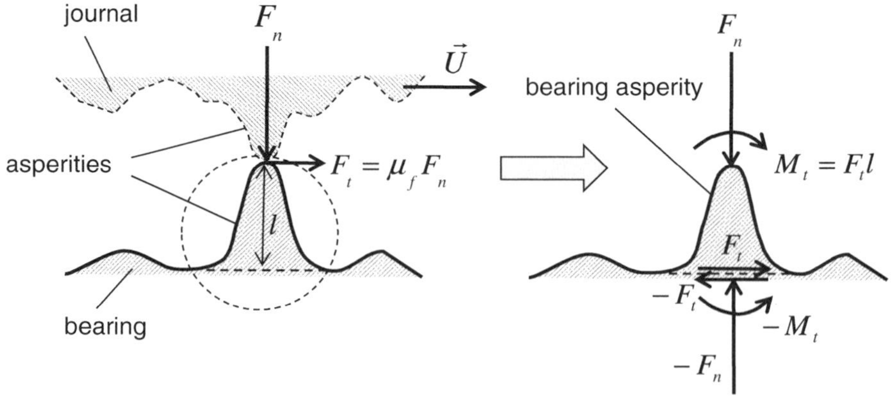

Fig. 9.24 Loads acting on the asperity of the bearing surfaces

In a three-dimensional case, all six components of normal and shear stresses in the directions of x, y, and z exist in the asperities. Using the coordinate transformation between the inertial coordinate system (x, y, z) and principle coordinate system (1, 2, 3), we obtain only the normal stresses σ_1, σ_2, and σ_3 in the directions 1, 2, and 3 in which the shear stresses τ_1, τ_2, and τ_3 equal zero.

Applying the Mohr's circle diagram, the related stresses in the principle coordinate system (1, 2, 3) are calculated graphically from the principle normal stresses σ_1, σ_2, and σ_3, as shown in Fig. 9.25.

The maximum principle shear stresses in the planes inclined 45° to the principle normal stress axes in the directions 1, 2, and 3 are calculated as

$$\tau_1 = \frac{|\sigma_2 - \sigma_3|}{2}; \quad \tau_2 = \frac{|\sigma_1 - \sigma_3|}{2}; \quad \tau_3 = \frac{|\sigma_1 - \sigma_2|}{2} \tag{9.47}$$

The normal stresses perpendicular to the principle shear stresses result as

$$\sigma_{\tau 1} = \frac{\sigma_2 + \sigma_3}{2}; \quad \sigma_{\tau 2} = \frac{\sigma_1 + \sigma_3}{2}; \quad \sigma_{\tau 3} = \frac{\sigma_1 + \sigma_2}{2} \tag{9.48}$$

The maximal shear stress for any plane in the asperity is the largest shear stress of the principle shear stresses from Eq. (9.47).

$$\tau_{max} = \max(\tau_1, \tau_2, \tau_3) \tag{9.49}$$

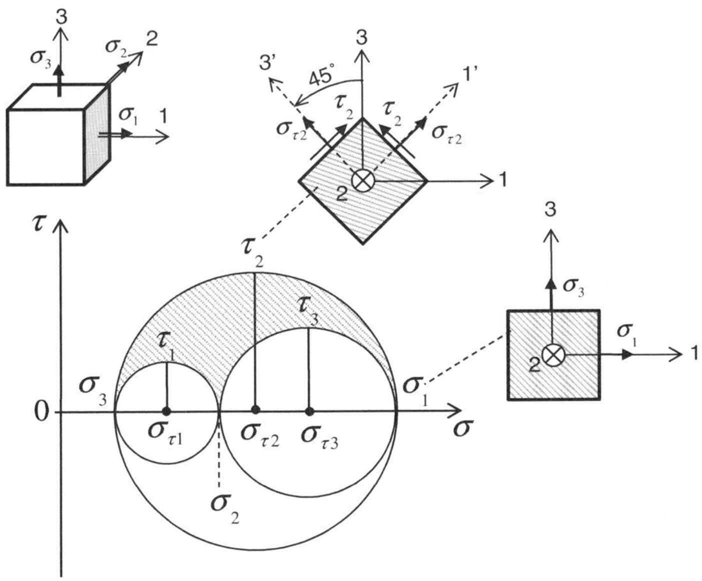

Fig. 9.25 Calculation of principle stresses using Mohr's circle diagram

All states of stress (σ, τ) in the asperities occur in the shaded area, which lies in the upper half plane outside the two small circles and inside the large one displayed in the Mohr's circle diagram in Fig. 9.25.

9.8 Wear Mechanisms in the Oil-Film Bearings

Oil-film bearings work on the hydrodynamic principle in the various lubrication regimes, such as fully hydrodynamic, mixed, and BLs in the Stribeck curve (see Fig. 9.7).

Theoretically, wear in the bearing does not occur or is negligibly small in the fully HL because the oil film is thick enough to prevent wear in the bearing. In case of clean oils, only the hydrodynamic friction occurs in the oil film, as discussed in Sect. 9.3. However, contaminated oil with hard particles causes wear in the bearing due to the adhesion and abrasion forces between the bearing and journal.

If the oil-film thickness is smaller than the limit oil-film thickness, the mixed and BLs occur in the bearing clearance. The asperities of the bearing and journal surfaces slide to each other leading to the plastic deformation due to thermomechanical loads acting on the asperities. As a result, it causes at first the *adhesive wear* at the asperities due to plastic shearing, then losses the material because their asperities break up. In the abrasive wear, the hard particles in the contaminated oil, broken bits of the asperities, and still remaining asperities of the surfaces abrade the surface of the softer material. Therefore, the material is removed from the bearing surface with time leading to wear. In the mixed and partial BLs, the wear process in the bearing begins with the adhesive friction at the asperities of the roughness surfaces. Wears continue further by loss of the asperities, and it is eventually intensified by the abrasive wear of the hard particles in the BL leading to seizure and damage of the bearing.

Figure 9.26 shows the adhesive wear mechanism where the asperities in the softer material bearing (brass) contact the other asperities of the moving journal of

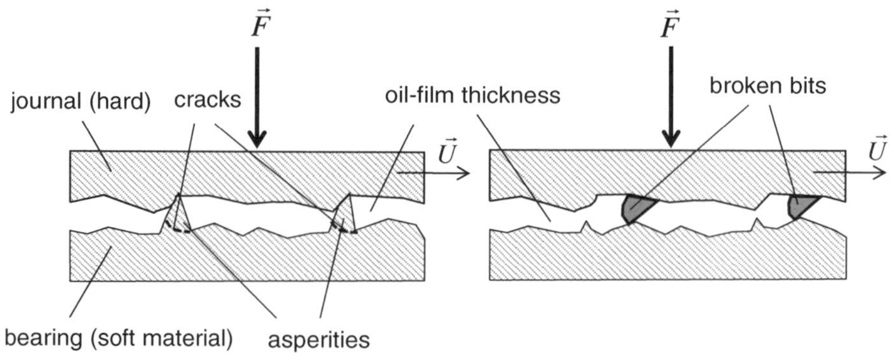

Fig. 9.26 Adhesive and abrasive wear in the bearing clearance

hard materials (highly alloyed steels) under the unbalance force and bending moment that act upon the bearing asperities with the relative circumferential velocity U of the journal to the bearing.

According to the deformation mechanism, as shown in Fig. 9.24, the asperities of the bearing surface begin deforming plastically under the acting loads. When the tensile and shear stresses in the asperities exceed the ultimate tensile and critical shear stresses, some asperities of the bearing and journal break off. Then, the broken bits together with the hard particles in lubricating oil cause the abrasive wear in the bearing.

The theoretical critical shear stress of a perfect material without any atomic dislocations in the material grid structure is given in [3] and [9].

$$\begin{aligned}\tau_{c,th} &= \frac{Gb}{2\pi h} \\ &\approx \frac{G}{10} \approx 10^2 \sigma_o\end{aligned} \tag{9.50}$$

where

G is the shear modulus;
b, h are the distance between the atoms in horizontal and vertical direction with ($b/h \sim 0.5$–1);
σ_o is the yield stress of material.

However, the real critical shear stress in an imperfect pure metal can be reduced by a factor of 300–10,000 of its theoretical value.

The theoretical tensile stress to break chemical bonds between the atoms in a perfect metal grid is estimated at

$$\sigma_{u,th} \approx \frac{E}{10} \tag{9.51}$$

where E is the elasticity modulus of material, see Eq. (9.33).

In fact, the real ultimate tensile stress of the common metals could be lower than the theoretical value by a factor of 10 to 100.

The contact types of wears are normally classified into *sliding, rolling, erosive, fretting,* and *slurry wear* [1].

Lubricating contaminated oil with hard particles supplies the two-oil-film bearing. The hard particles enter the bearing convergent wedge due to the wedge-velocity effect of the rotating journal and cause abrasive wear in the bearing that is classified into three different types of wears A, B, and C (see Figs. 9.27, 9.28 and 9.31).

- *Sliding wear* occurs when the hard and soft surfaces move relative to each other; the hard asperities slide over the soft surface. Hence, the soft material is removed due to plastic deformation and fracture of asperities.

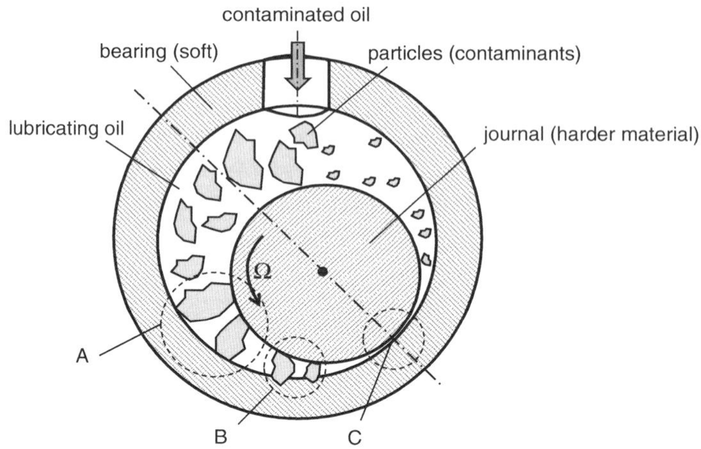

Fig. 9.27 Abrasive wear mechanism in a radial bearing

Fig. 9.28 Wear paths in a rotating floating ring bearing (RFRB)

- R*olling wear* happens when the hard particles floating in lubricating oil roll over two moving surfaces in the small gap, such as the convergent clearance of the radial bearing and axial clearance of the thrust bearing (see Figs. 9.27, 9.28 and 9.29, 9.21). The asperities of the surfaces break up and cause abrasive wear on their surfaces in the rotating direction.
- *Erosive wear* is induced by the impact of the particles in lubricating oil or hard asperities against the surface. The impact kinetic energy of the hard particles and moving broken asperities deforms the asperities at the contact zones and causes material fracture when the tensile and shear stresses exceed the ultimate and critical values.
- *Fretting wear* is caused by the repeated cyclical microvibration of two contact surfaces under load, especially at a still stand. Due to the periodic rubbing, the bonding force between the atoms is weakened after repeating many vibration

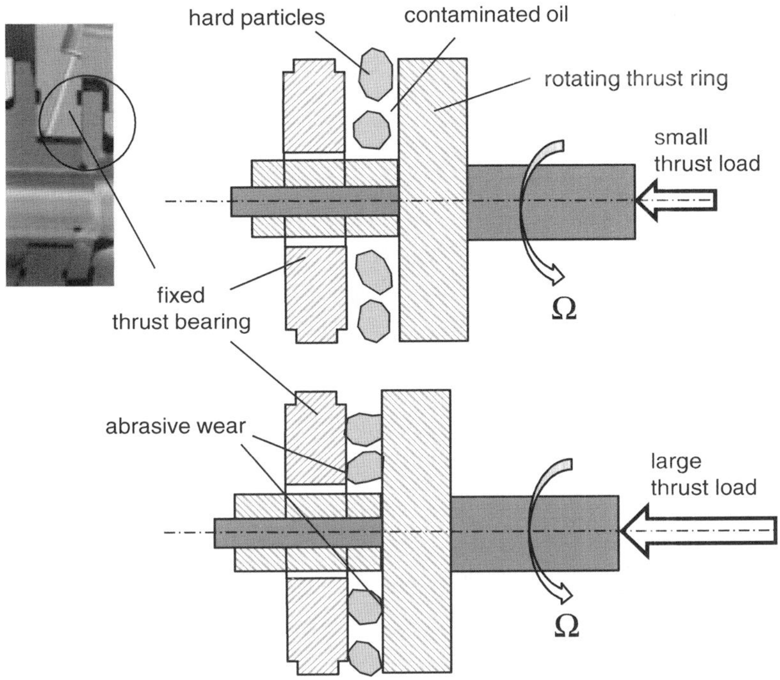

Fig. 9.29 Abrasive wear mechanism in a thrust bearing

cycles over a long operating time. As a result, adhesive wear occurs on the surfaces and is called the fretting wear (*false Brinelling*).

- *Slurry wear* occurs when the abrasive particles in lubricating oil float in the bearing clearance and abrade the asperities and surface of the bearing.

Figures 9.27, 9.28, 9.29 and 9.30 show the most typical types of the abrasive wear occurring in the radial and thrust oil bearings of the automotive turbochargers.

Figures 9.29 and 9.30 describe the wear mechanism in the thrust bearing. The oil-film thickness between the bearing and thrust ring becomes large at a small thrust load; hence, the hard particles in lubricating oil enter the active bearing clearance that is between the bearing and thrust ring. At increasing thrust load, the oil-film thickness decreases to the minimum of the axial clearance. The large thrust load induces the friction force between the particles and surfaces; the friction force causes abrasive wear on the surfaces between the bearing and thrust ring, especially in the soft bearing surface. Finally, abrasive wear leaves deep wear grooves on the bearing and disk surfaces. The wear grooves do not cause the failure of the bearing at once. However, the bearing damage could happen if the abrasive wear is continuously fortified leading to the mixed and BLs in the bearing.

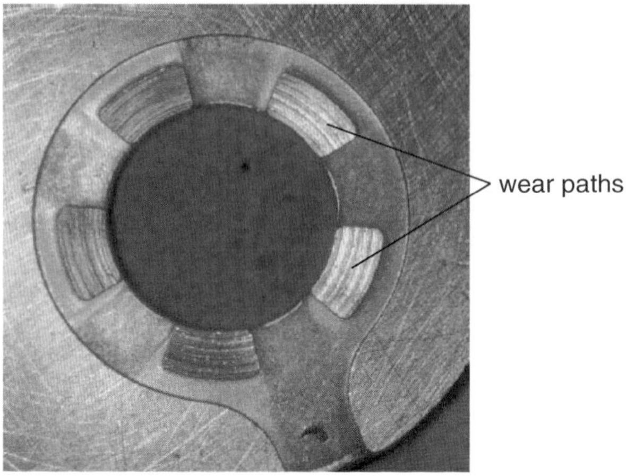

Fig. 9.30 Wear paths on the surface of a thrust bearing

The abrasive wear is classified into three different types A, B, and C according to [9], as shown in Fig. 9.31 at which the journal material of highly alloyed steels is much harder than the bearing material of brass.

- Type A called the *three-body abrasive wear* shows the hard particles slide and roll on the bearing and journal surfaces, touch the asperities, deform them plastically, and finally remove them from the bearing surface.
- Type B called the *two-body abrasive wear* shows the hard particles and broken bits of the asperities are embedded in the softer bearing surface. Due to rotation, they abrade the journal surface as if a sand paper slides on it at a high speed. To prevent the journal from the abrasive wear, its surface is treated by nitriding. Nitriding is a heat-treating process of diffusing nitrogen (N_2) into the surface to create a hard coating layer on the surface, such as boron nitride (BN), titanium nitride (TiN), and silicon nitride (Si_3N_4).
- Type C called the *surface abrasive wear* shows the hard asperities of the journal surface abrade the softer bearing surface in case of poorly lubricated oil film in the partly and BLs. It mostly occurs at the minimum oil-film thickness in the bearing clearance, as displayed in Figs. 9.27, 9.28, 9.29, 9.30, and 9.31. The abrasive wear leaves wear paths on the bearing surface.

In the point of view of the tribological mechanism of wears, *fatigue and corrosive wears* should be taken into account besides the adhesive and abrasive wears (see Fig. 9.32).

- *Fatigue wear* is generated by the repeated cyclical friction between two moving surfaces after a certain number of rubbing cycles. It is caused by the fatigue fracture where the yield stress of material strongly reduces at increasing the number of rubbing cycles according to the Wöhler curve. There are two kinds of fatigue wear, the high-cycle fatigue wear (HCFW) occurs after a high number of rubbing cycles; the low-cycle fatigue wear (LCFW), after a low number of rubbing cycles.

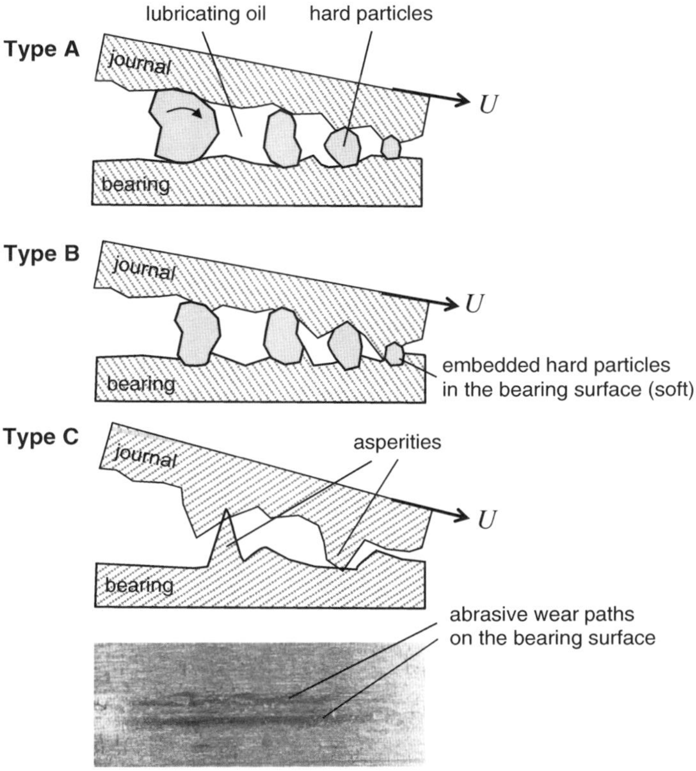

Fig. 9.31 Types of abrasive wear in the convergent wedge of a radial bearing

- *Corrosive wear* happens when the materials of the surfaces contact a corrosive substance (liquid or gas), such as dissolved water, fuel, and diffused air in lubricating oil that induces some tribochemical reactions (chemical and electrochemical) on the surfaces of the journal and bearing. Therefore, the materials of the surfaces are removed by the tribochemical corrosion.

In case of the ball bearings, two most failures of flaking (spalling) and false Brinelling due to contact fatigue often occurs after a long operating time. Both mechanisms reduce the lifetime of the bearing and cause the total failure of the machine.

The flaking-related failure occurs on the ball/races and generates the formation of large and deep cavities by a heavy load at the contact surface between the ball and races of a ball bearing. It induces intensive noises (NVH) and drastically reduces the lifetime of the bearing leading to a total failure in a few hundred km after noises occur.

Two root causes for the flaking failure are the *subsurface initial* and *surface initial microcracks* on the ball and race surfaces.

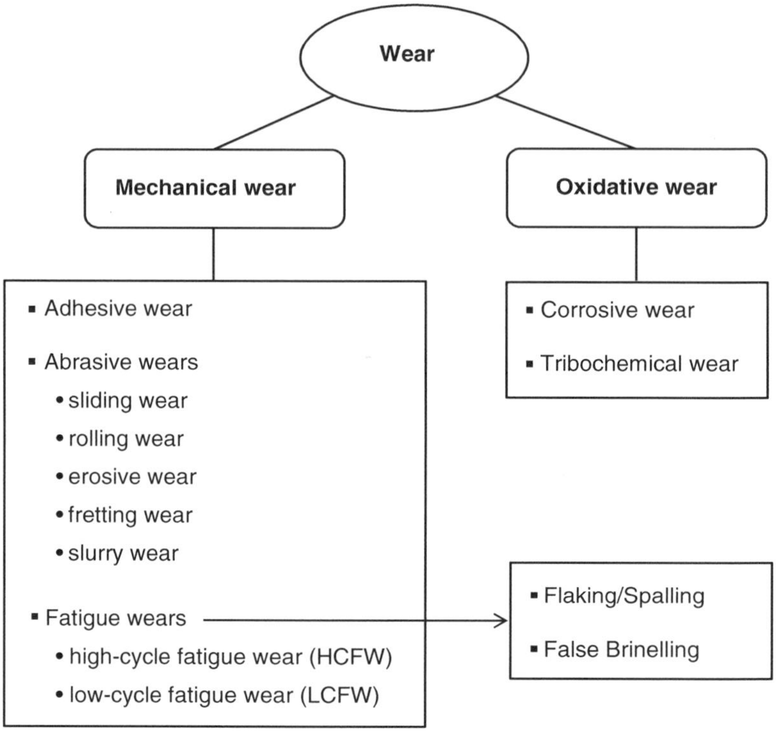

Fig. 9.32 Classification of wear mechanisms

First, the initial subsurface microcracks occurring about 100–150 μm under the surface are developed into the macrocracks that are caused by the high Hertzian stress at the contact surface in many repeated cyclical driving cycles. The developed subsurface macrocracks prolong in the upward direction to the surface. Due to contact fatigue, they lead to the material break up at the contact surface between the ball and races.

Second, the initial surface microcracks are induced on the surface due to the break up of the asperities of the surface at the small oil film thickness in the mixed/BL (see Fig. 9.7). Then, due to the elasto-hydrodynamic (EHD) effect, the microcracks turn into the macrocracks in the rotating direction of the ball. Further development of the macrocracks continues by the high shear stress of the oil film acting upon the surface in the repeated cyclical driving cycles. As a result, the developed macrocracks cause the material break up of the surface that is called flaking or spalling, as shown in Fig. 9.33. Note that the smaller the oil-film thickness is, the higher the shear stress acts upon the surface leading to the surface distress at the dimensionless oil-film thickness λ less than 3.

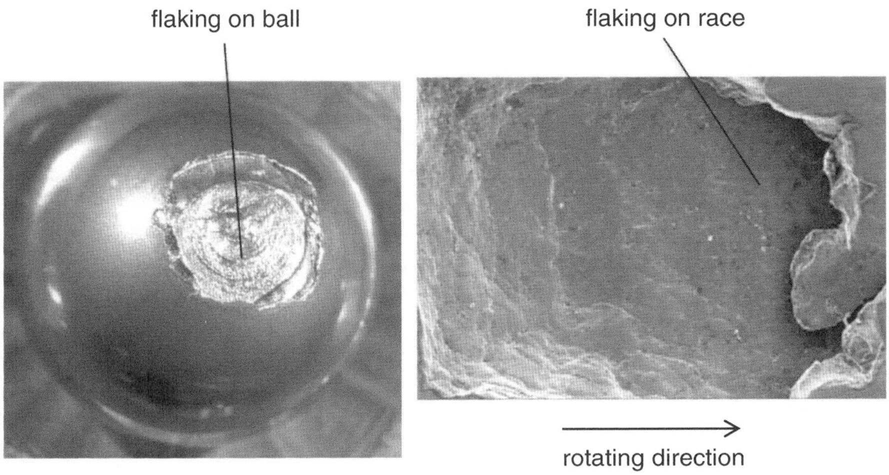

Fig. 9.33 Flaking in a ball bearing

Figure 9.34 shows the *false-Brinelling* (fretting mark) failure occurring in a ball bearing. False Brinelling is caused by high-frequency microvibrations in the axial direction of the bearing at the contact surfaces between the balls and races. It occurs at the transport or idle operating conditions under the bearing load, such as the rotor weight or radial preload of the bearing. In practice, it is known that false Brinelling has two negative effects of inducing bearing noises (NVH) and reducing the lifetime of the bearing. In order to prevent the false-Brinelling-related failures, the bearing should be preloaded in the axial direction with an available thrust load using a flat spring (wave washer).

Fig. 9.34 False Brinelling in a ball bearing

Adhesive, abrasive, and fatigue wears are followed by the plastic deformation and fracture of materials; hence, they are called “mechanical wear”. On the contrary, the corrosive wear is caused by tribochemical reactions of corrosive mediums; therefore, it is called “oxidative wear”.

In practice, used and old lubricating oils after a long operating time in the automotive turbochargers contain about 5 % volumetric dissolved water and approximately 10 % volumetric dissolved fuel besides contaminated hard particles. In this case, the contaminated used oil is very thin and looses the necessary viscosity for the bearing lubrication. Under heavy loads, the oil-film thickness in the bearings becomes small and breaks up leading to the bearing wear and failure. Therefore, to prevent wears in the bearings, both lubricating oil and oil filter must be regularly changed in every period of 12–18 months at an average driving rate of 10,000 km/year according to the norm ISO 4406-1999: 24/23/18.

References

1. Bhushan, B.: Modern Tribology Handbook—Two-Volume Set. CRC Press Inc, Boca Raton (2000)
2. Bhushan, B.: Introduction to Tribology. Wiley, New York (2002)
3. Dowling, N.E.: Mechanical Behavior of Materials, 3rd edn. Pearson-Prentice Hall (2007)
4. Hamrock, B., Schmid, S.R., Jacobson, B.O.: Fundamentals of Fluid Film Lubrication, 2nd edn. Marcel Dekker Inc, New York (2004)
5. Johnson, K.L.: Contact Mechanics. Cambridge University Press, Cambridge (1985)
6. Kennedy F.E., et al.: Tribology, Lubrication, and Bearing Design—The CRC Handbook of Mechanical Engineers. CRC Press, Boca Raton (1988)
7. Khonsari, M., Booser, E.: Applied Tribology and Bearing Design and Lubrication, 2nd edn. Wiley, New York (2008)
8. Mahr: Surface Texture Parameters. Mahr GmbH, Germany (1999)
9. Mate, C.M.: Tribology on the Small Scale. Oxford University Press, Oxford (2008)

Chapter 10
Design of Turbocharger Platforms

10.1 Introduction

After learning how to deal with rotordynamics, bearing systems, balancing of turbochargers, and tribology in the bearings, we will learn how to design a platform of automotive turbochargers so that most current and future customer requirements should be covered in the platform. Therefore, we do not need to design a specific turbocharger for each customer requirement. Instead, we just take a suitable CHRA (Center Housing and Rotating Assembly) and insert it into the required housings of compressor and turbine for the customer applications. Additionally, we cut short the development time, time to market, and save money as well.

Furthermore, the common parts of turbochargers, such as compressor and turbine wheels, seal rings, thrust bearings, etc., could be used in different type series in the platform. As a result, much money can be saved if we order a huge number of the same part instead of different parts with a small number at the suppliers.

All the above considered are the reasons why the platform is necessary and important. Note that it is very difficult, sometimes impossible and costs a lot of time and money to correct the already unsuitable platform of turbochargers.

10.2 Market Analyses of Combustion Engines

Before we design the platforms of turbochargers for passenger (PV) and commercial vehicles (CV), two questions need to be answered. First, which combustion engines are required for the worldwide market in the future? Second, how many type series of turbochargers in the platform could cover most applications of customers currently and in the future?

In the following section, a market analysis of diesel engines for passenger vehicles (PV) is chosen as an example of designing turbocharger platforms for

H. Nguyen-Schäfer, *Rotordynamics of Automotive Turbochargers*,
Springer Tracts in Mechanical Engineering, DOI 10.1007/978-3-319-17644-4_10

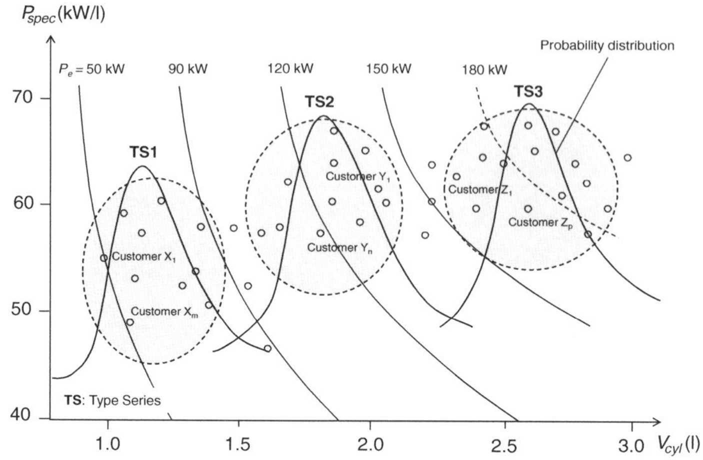

Fig. 10.1 Schematic market analysis for diesel passenger vehicles

passenger vehicles (PV). Using the same method, the turbocharger platform for commercial vehicles (CV) is designed.

The specific and total engine powers versus engine cylinder volume are displayed in Fig. 10.1. The density probability distribution of the engines in the market concentrates on three main areas of V_{cyl} from 0.9 to 1.5 l, 1.6 to 2.3 l and 2.5 to 3.0 l. The type series of the turbocharger platform should cover the required engines.

As a result, three *type series* (TS) of turbochargers TS1, TS2, and TS3 are designed for the platform of passenger vehicles respectively. Note that some engines in the requirement are intentionally not considered in the platform of turbochargers because they are only used in a small number of very special applications. Therefore, they are not commonly used in the platform. If they had been taken into account in the platform, it would lead to high cost for the platform development. Therefore, we concentrate on the essential thing and exclude such applications in order to save development time and cost.

The first type series TS1 for small engines with maximum engine powers from 45 to 90 kW, the second TS2 for middle engines (95 to 145 kW), and the third TS3 for large engines (155 to 200 kW) are summarized with maximum specific powers (Table 10.1). The engine specific power is defined as an engine power per unit of cylinder volume (kW/l).

Table 10.1 Type series for a diesel turbocharger platform of PV

Type series	Cylinder Volume *V*cyl (l)	Maximum engine power P_e (kW)	Maximum specific power *P*spec (kW/l)
TS1	0.9–1.5	45–90	50–60
TS2	1.6–2.3	95–145	60–63
TS3	2.5–3.0	155–200	62–67

Generally, the platform of turbochargers must be defined at the beginning of the development. Unfortunately, neither the geometrical sizes of the compressor and turbine wheels nor the shaft diameters of rotor of each type series are known in the platform.

In the following sections, we apply the physical similarity laws to designing the new platform. The physical similarity laws are based on many theories of combustion engines, turbomachinery, technical thermodynamics, rotordynamics, bearing dynamics, and classical mechanics.

10.3 Calculating Sizes of Compressor and Turbine Wheels

The geometrical sizes of the compressor and turbine wheels are calculated for each type series in the platform using the physical similarity law. It is obvious that the engine power is proportional to the mass flow rate of the charge air, as discussed in Chap. 1. In principle, the larger the mass flow rate of air, the higher the engine power will be.

The engine power is calculated according to Eq. (1.6a) as [1–3]

$$P_e = \eta_f Q_f \cdot \left(\eta_{vol}\rho_a V_{cyl}\frac{N}{n_R}\right) \cdot \frac{1}{AFR} \propto \dot{m}_C \tag{10.1}$$

where $\dot{m}_C$, the mass flow rate of the charge air for the engine, is defined as

$$\dot{m}_C = \eta_{vol}\rho_a V_{cyl}\frac{N}{n_R}.$$

Using thermodynamics of turbomachinery according to Eq. (2.14), the real compressor power is calculated [1, 3] as

$$P_C = \frac{\dot{m}_C c_{p,a} T_1}{\eta_C}\left[\left(\frac{p_2}{p_1}\right)^{\left(\frac{\kappa-1}{\kappa}\right)_a} - 1\right] \propto \dot{m}_C \tag{10.2}$$

The mass flow rate of the charge air from the compressor is proportional to squared inflow diameter D_1 of the compressor wheel (cf. Fig. 10.2).

$$\dot{m}_C \propto D_1^2 \tag{10.3a}$$

Using the definition of the trim ratio *TR* of the compressor wheel ($TR = D_1/D_2$), the inflow diameter D_1 (*inducer diameter*) of the compressor wheel is written in its outflow diameter D_2 (*exducer diameter*).

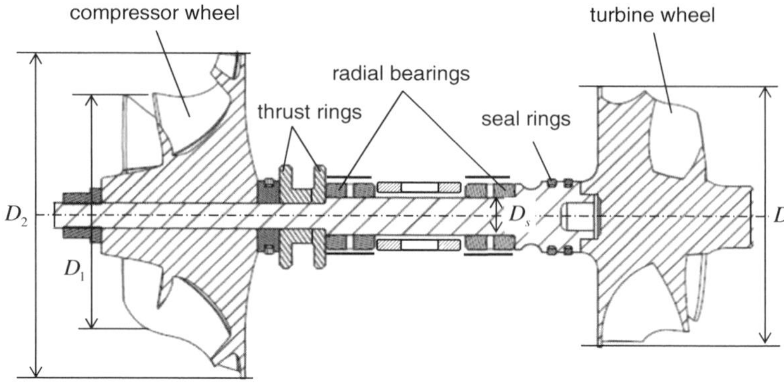

Fig. 10.2 Setup of the CHRA of a turbocharger

$$D_1 = D_2 \cdot TR \tag{10.3b}$$

Substituting Eq. (10.3b) into Eq. (10.3a), one obtains the mass flow rate of the charge air from the compressor.

$$\dot{m}_C \propto D_2^2 \cdot TR^2 \tag{10.4}$$

Using Eq. (10.1), the physical similarity law between the engine and compressor powers gives the relation of the engine power and outflow diameter of the compressor wheel [4].

$$\begin{aligned} P_e &\propto \dot{m}_C \propto D_2^2 \cdot TR^2 \\ \Rightarrow D_2 &\propto \frac{\sqrt{P_e}}{TR} = f \cdot \frac{\sqrt{P_e}}{TR} \end{aligned} \tag{10.5}$$

The result of Eq. (10.5) indicates that the exducer diameter D_2 of the compressor wheel is proportional to the square root of the engine power and inversion of the trim ratio of the compressor wheel. However, we do not know the proportional factor f for the turbocharger type series. Now, the question is how to determine D_2 corresponding to the engine power for the type series in Table 10.1.

Usually, a few designed turbochargers must be built to measure the proportional factor. This work would take a long time and cost a lot of money. Fortunately, the competitors had already done this task. However, we cannot use it one-to-one in the designed platform because the type series are different from the competitor type series. Notwithstanding, we can use their results with subtle skill to achieve the target in a very short time and at minimum cost.

In order to save the development time, time to market, and cost, we use the similarity between the designed turbocharger and the commercial turbochargers on the market (marked *) to find out the proportional factor f for calculating the exducer diameter D_2 of the designed turbocharger type series in the platform.

The similarity of the engine powers gives the relation of related parameters.

$$P_e \propto D_2^2 \cdot TR^2 \Leftrightarrow P_e^* \propto (D_2^*)^2 \cdot TR^{*2}$$

Thus, the exducer diameter D_2 of the designed compressor wheel results in [4]

$$\frac{P_e}{P_e^*} = \frac{D_2^2 \cdot TR^2}{\left(D_2^*\right)^2 \cdot TR^{*2}}$$
$$\Rightarrow D_2 = D_2^* \cdot \left(\frac{TR^*}{TR}\right)\sqrt{\frac{P_e}{P_e^*}} \tag{10.6}$$

In this case, all starred parameters of the commercial turbochargers on the market have been known; they are used to calibrate the proportional factor for the platform. At the given engine power P_e and the trim ratio TR (normally, $TR \approx 0.76$ for compressor wheels), the exducer diameters D_2 of the compressor wheel are calculated for each type series of the platform using Eq. (10.6).

The wheel ratio of D_2/D_3 between the exducer diameter of compressor wheel and the inducer diameter of turbine wheel is chosen between 1.05 and 1.25 [4].

$$\frac{D_2}{D_3} = 1.05\text{–}1.25 \tag{10.7}$$

In Table 10.2, the wheel ratios of D_2/D_3 are given in the diagonal band of the platform matrix. Outside this range, the turbochargers are not suitable for automotive applications. If the wheel ratio is too small, the turbine wheel is larger than the compressor wheel. Thus, the mass center of the CHRA lies nearer to the turbine radial bearing on

Table 10.2 Ratios of compressor to turbine wheels in the platform

Turbine wheel diameter D_3 (mm)	Compressor wheel diameter D_2 (mm)								
	D_{21}	D_{22}	D_{23}	D_{24}	D_{25}	D_{26}	D_{27}	D_{28}	D_{29}
D_{31}	1.10	1.15							
D_{32}	1.05	1.10	1.25						
D_{33}		1.05	1.15						
D_{34}			1.10	1.20	1.25				
D_{35}			1.06	1.15	1.20				
D_{36}					1.10	1.20			
D_{37}						1.12	1.20	1.25	
D_{38}						1.06	1.15	1.20	
D_{39}								1.08	1.20
CHRA Platform:	Type series 1 (45–90 kW)			Type series 2 (95–145 kW)			Type series 3 (155–200 kW)		

the RHS; it leads to a large load acting upon the turbine bearing. This could cause rotordynamic instability and bearing failure at the turbine side. Otherwise, the wheel ratio is too large; i.e., the turbine wheel is much smaller than the compressor wheel. It leads to large axial load on the thrust bearing that is between the thrust rings. This could cause a bad transient behavior of the rotor (large turbolag) and possibly failure of the thrust bearing due to overload in the thrust bearing.

By experience, the ratio should be chosen between 1.05 and 1.25 in order to compromise the acting loads on the turbine bearing and the thrust bearing. As a result, the inducer diameters D_3 of turbine wheels are calculated using Eq. (10.7) for each type series in the platform.

Until now, the rotor shaft diameter D_s has not been dealt with. In the following section, we will learn how to compute it for the designed platform.

10.4 Calculating Diameters of the Rotor Shaft

Besides the key function of connecting the compressor and turbine wheels, the rotor shaft has two additional functions: supporting the radial bearings and keeping the compressor wheel fixed on the rotor. This is against the aerodynamic torque M_{CW} acting on the compressor wheel and keeps the compressor wheel rotating.

The friction torque of the compressor wheel on the rotor shaft is used to determine the rotor shaft diameter. The rotor shaft diameter D_s is defined as the journal diameter inside the radial bearing, as shown in Figs. 10.2 and 10.3.

The aerodynamic torque acting on the compressor wheel (compressor wheel torque) is calculated using the theory of turbomachinery at a rotating speed ω [1, 2].

$$\begin{aligned} P_C &= M_{CW}\omega \propto (D_2)^2 \cdot TR^2 \\ \Rightarrow M_{CW} &= \frac{P_C}{\omega} \propto \frac{(D_2)^2 \cdot TR^2}{\omega} \end{aligned} \tag{10.8}$$

The maximum circumferential velocity of the compressor wheel is assumed as constant due to stability of the compressor wheel.

$$U_2 = \frac{\omega D_2}{2} = const. \tag{10.9a}$$

Using Eq. (10.9a), the rotor speed is proportional to the inversion of the compressor wheel diameter.

$$\omega \propto \frac{1}{D_2} \tag{10.9b}$$

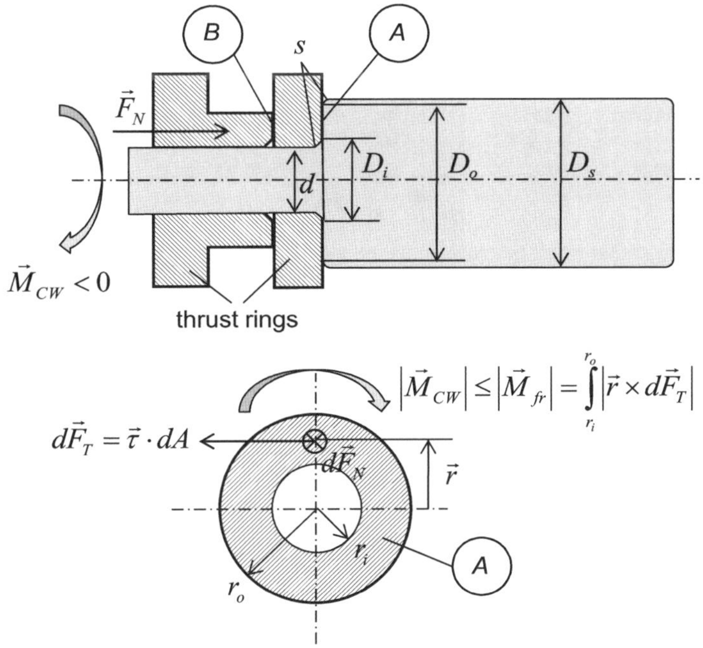

Fig. 10.3 Friction torque versus compressor wheel torque

Substituting Eq. (10.9b) into Eq. (10.9a), one obtains the compressor wheel torque

$$M_{CW} \propto \frac{(D_2)^2 \cdot TR^2}{\omega} = (D_2)^3 \cdot TR^2 \tag{10.10}$$

The friction torque acting on the contact areas A and B between the rotor shaft and thrust rings must be at least larger than the compressor wheel torque in order to keep the compressor wheel fixed on the rotor.

At first, integrating the tangential force F_T over the rotor radius r, the friction torque is calculated at the contact area A.

$$\begin{aligned} M_{fr} &= \int r dF_T = \int_A r\tau dA \\ &= \int_{r_i}^{r_o} r(2\pi r\tau)dr = \frac{\pi}{12}\tau\left(D_o^3 - D_i^3\right) \\ &\propto \left(D_o^3 - D_i^3\right) \end{aligned} \tag{10.11}$$

where τ is the shear stress in the contact area A.

The aerodynamic torque on the compressor wheel must be smaller than the friction torque in order to keep the compressor wheel from not slipping over the rotor shaft. Hence, the shaft diameter D_s of the rotor results from the necessary condition [4].

$$\begin{aligned} M_{CW} \propto (D_2)^3 \cdot TR^2 \le M_{fr} \propto \left(D_o^3 - D_i^3\right) \\ \Rightarrow (D_2)^3 \cdot TR^2 \le \left(D_o^3 - D_i^3\right) \end{aligned} \tag{10.12}$$

The inner and outer diameters in Eq. (10.12) are calculated as

$$\begin{aligned} D_i &= d + 2s; \\ D_o &= D_s - 2s \end{aligned} \tag{10.13}$$

where

d is the bore diameter of the compressor wheel;
D_s is the rotor shaft diameter;
s is the fillet radius ($s \approx 0.300$–0.350 mm).

Substituting Eqs. (10.12 and 10.13), one obtains the function of the rotor shaft diameter in other parameters [4].

$$M_{CW} \propto (D_2)^3 \cdot TR^2 \le M_{fr} \propto (D_s - 2s)^3 - (d + 2s)^3 \tag{10.14}$$

The similarity law between the designed and commercial turbochargers on the market (marked *) gives the required rotor shaft diameter D_s [4].

$$\begin{aligned} M_{CW} &: (D_2)^3 \cdot TR^2 \le (D_s - 2s)^3 - (d + 2s)^3 \\ \Leftrightarrow M_{CW}^* &: (D_2^*)^3 \cdot TR^{*2} \le (D_s^* - 2s^*)^3 - (d^* + 2s^*)^3 \\ \Rightarrow & \frac{(D_s - 2s)^3 - (d + 2s)^3}{(D_s^* - 2s^*)^3 - (d^* + 2s^*)^3} \ge \frac{(D_2)^3 \cdot TR^2}{(D_2^*)^3 \cdot TR^{*2}} \end{aligned}$$

Therefore, the rotor shaft diameter results as

$$D_s \ge 2s + \left[(d + 2s)^3 + \left(\frac{D_2}{D_2^*}\right)^3 \left(\frac{TR}{TR^*}\right)^2 \left[(D_s^* - 2s^*)^3 - (d^* + 2s^*)^3\right]\right]^{\frac{1}{3}} \tag{10.15}$$

Likewise, the required starred parameters in Eq. (10.15) of the commercial turbochargers have been already known. Therefore, it is easy to calculate the rotor shaft diameter D_s for the platform. To determine the rotor shaft diameter of the type

Table 10.3 An examplary platform of automotive turbochargers (Diesel PV)

	Compressor wheel diameter D_2 (mm) →								
Turbine wheel diameter D_3 (mm)	D_{21}	D_{22}	D_{23}	D_{24}	D_{25}	D_{26}	D_{27}	D_{28}	D_{29}
D_{31}	1.10	1.15		Shaft diameter $D_{s2} > D_{s1}$			Shaft diameter $D_{s3} > D_{s2}$		
D_{32}	1.05	1.10	1.25						
D_{33}		1.05	1.15						
D_{34}			1.10	1.20	1.25				
D_{35}			1.06	1.15	1.20				
D_{36}					1.10	1.20			
D_{37}						1.12	1.20	1.25	
D_{38}	Shaft diameter D_{s1}					1.06	1.15	1.20	
D_{39}								1.08	1.20
	Type series 1 (45–90 kW)			Type series 2 (95–145 kW)			Type series 3 (155–200 kW)		
CHRA Platform:									

series, its high-end diameter D_2 (i.e., the maximum diameter) of the compressor wheel should be employed in Eq. (10.15) because the largest compressor wheel induces the highest aerodynamic torque acting on the compressor wheel.

The rotor shaft diameter D_{si} in the platform should be at least equal to the RHS of Eq. (10.15). Note that if the fillet radius s is larger due to some production processes, the RHS also increases with the fillet radius s. The rotor shaft diameter D_{si} should be checked for whether the condition for D_s in Eq. (10.15) is satisfied for the application. If it is not the case, some measures of surface coating between the thrust rings and rotor shaft should be additionally used to increase the surface friction coefficient. As a result, the condition in Eq. (10.12) of $M_{fr} \geq M_{CW}$ is always satisfied.

In general, the higher the engine power, the larger the rotor shaft diameter will be, as shown in Table 10.3. It is obvious that the higher engine power requires more mass flow rate of the charge air. Therefore, the compressor wheel must be larger (increased D_2) to provide the engine with more mass flow rate of the charge air. This requires a larger shaft diameter according to Eq. (10.15).

Analogously, the same computation at the contact surface B is carried out to design the geometry of the thrust rings.

Some applications overlap at the boundary between the type series (See Table 10.3). For this case, the condition in Eq. (10.12) between the compressor wheel and friction torques must be checked by FEM computations.

10.5 Design of CHRA Geometry for the Platform

In this section, the overall geometry of CHRA is calculated for each type series in the platform. Normally, the geometrical computation begins from inside to outside of the rotor center in both axial and radial directions (cf. Fig. 10.4).

At first, the bearing distance l_0 between two radial bearing centers is chosen by experience so that the ratio of l_0 to the shaft diameter D_s lies between about 3.2 and 3.5 for turbochargers of passenger vehicles (PV) and between approximately 3.7 and 4.0 for turbochargers of commercial vehicles (CV) [4].

$$\frac{l_0}{D_s} = \begin{cases} 3.2\text{–}3.5 \text{ for } PV \\ 3.7\text{–}4.0 \text{ for } CV \end{cases} \tag{10.16}$$

If the ratio is too small, the rotor mass center is nearer to the turbine bearing leading to a large radial load acting on the turbine bearing and possibly to the rotor instability. Otherwise, the ratio is too large, the total length of the CHRA increases so that the turbocharger total length does not fit in the given space under the car hood. Note that the longer and slender the rotor is, the more easily unstable the rotor is with conical vibration mode.

The next step is to design the radial bearings discussed in the earlier chapters. Both bearings at the compressor and turbine sides have the same geometry that avoids confusing different radial bearings at the assembly line.

The length between the bearing ends l_b results from the bearing width W and bearing distance l_0. It is written as a function of D_s and W using Eq. (10.16).

$$l_b = l_0 + W = f(D_s, W) \tag{10.17}$$

Furthermore, the thrust bearing and seal rings are designed, as discussed in Chap. 6. Using aerodynamics of the turbomachinery, both compressor and turbine

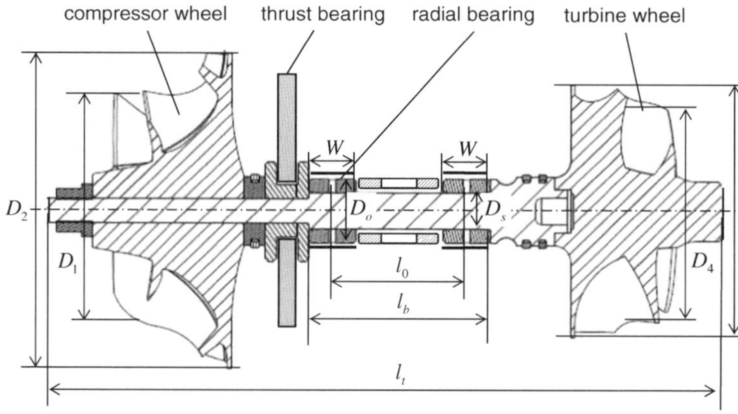

Fig. 10.4 Geometry of the CHRA of a type series

geometries are also computed. Finally, the total length l_t of the CHRA results from l_b and all related lengths l_i.

$$l_t = l_b + \sum_i l_i \tag{10.18}$$

Likewise, the CHRA geometry in radial direction begins from the shaft diameter, the outer diameters of the radial bearing, thrust rings, thrust bearing, and the outflow diameters of the compressor and turbine wheels.

Having known the CHRA geometry in both axial and radial directions, the bearing housing and the housings of the compressor and turbine wheels with all connections of oil, cooling water, exhaust gas, and charge air to the engine could be designed in the given space under the car hood.

References

1. Baines, N.C.: Fundamentals of turbocharging. Concepts ETI, Inc, Boston (2005)
2. Heywood, J.B.: Internal combustion engine fundamentals. McGraw-Hill, New York (1988)
3. Golloch, R.: Downsizing bei verbrennungsmotoren. Springer, Heidelberg (2005)
4. Nguyen-Schäfer, H.: Design of turbocharger platforms, Personal documents (2012)

Appendix A
Transformation Between the Precessing and Inertial Coordinates

To calculate the gyroscopic and unbalance moments acting upon the rotor, the chain rule of differentiation of the angular momentum theorem is used. The resulting moments in the precessing coordinate system (x', y', z') are transformed into the inertial coordinate system (x, y, z) for the vibration equations. Therefore, the transformation of the basis vectors between the coordinates is required (see Fig. A.1).

The two-dimensional angular position vector θ is written in the precessing coordinate system (x', y', z') with $\theta_z = 0$.

$$\begin{aligned}\boldsymbol{\theta} &= \theta_x \mathbf{i}' + \theta_y \mathbf{j}' + \theta_z \mathbf{k}' \\ &= \theta_x \mathbf{i}' + \theta_y \mathbf{j}'\end{aligned} \tag{A.1}$$

where θ_x, θ_y, and θ_z are the proper Euler angles in the directions x', y', and z'.

The transformation between the basis vectors (**i'**, **j'**, **k'**) and (**i**, **j**, **k**) of the processing and inertial coordinate systems is written as [1]

$$\begin{pmatrix} \mathbf{i}' \\ \mathbf{j}' \\ \mathbf{k}' \end{pmatrix} = \mathbf{I}\mathbf{R}_{\boldsymbol{\theta}x}\mathbf{R}_{\boldsymbol{\theta}y} \begin{pmatrix} \mathbf{i} \\ \mathbf{j} \\ \mathbf{k} \end{pmatrix} = \mathbf{R} \begin{pmatrix} \mathbf{i} \\ \mathbf{j} \\ \mathbf{k} \end{pmatrix} \tag{A.2}$$

where **I** is the unit matrix; $\mathbf{R}_{\theta x}$ and $\mathbf{R}_{\theta y}$ are the transformation matrices in functions of θ_x and θ_y that are written as

$$\mathbf{R}_{\boldsymbol{\theta}x} = \begin{pmatrix} 1 & 0 & 0 \\ 0 & \cos\theta_x & \sin\theta_x \\ 0 & -\sin\theta_x & \cos\theta_x \end{pmatrix}; \ \mathbf{R}_{\boldsymbol{\theta}y} = \begin{pmatrix} \cos\theta_y & 0 & -\sin\theta_y \\ 0 & 1 & 0 \\ \sin\theta_y & 0 & \cos\theta_y \end{pmatrix} \tag{A.3}$$

H. Nguyen-Schäfer, *Rotordynamics of Automotive Turbochargers*,
Springer Tracts in Mechanical Engineering, DOI 10.1007/978-3-319-17644-4

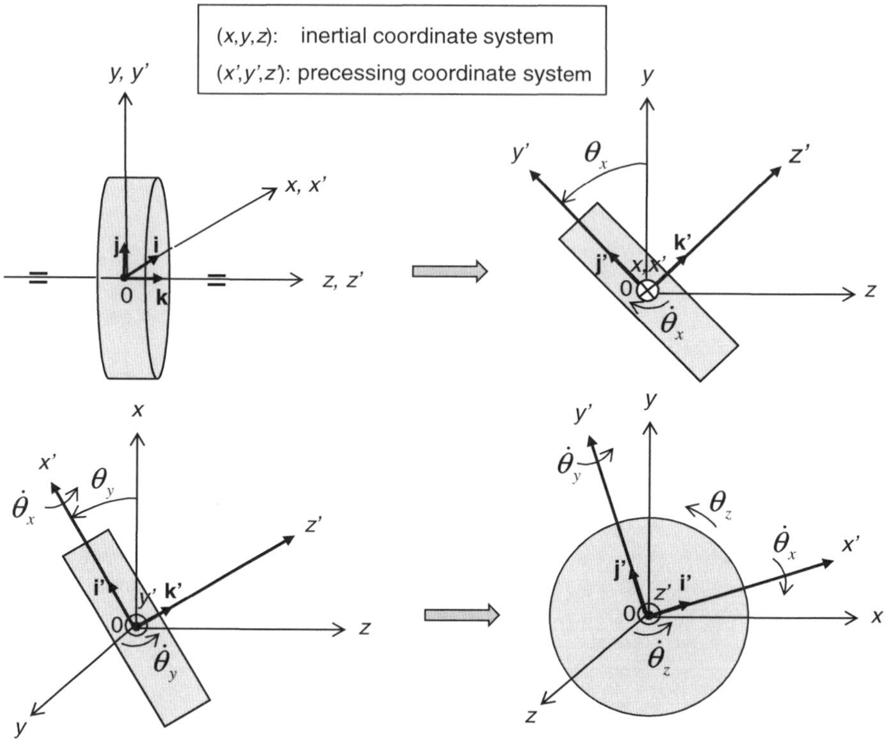

Fig. A.1 Transformation between the precessing and inertial coordinates

The transformation matrix **R** is calculated from Eqs (A.2 and A.3) as

$$\mathbf{R} = \mathbf{I}\mathbf{R}_{\theta x}\mathbf{R}_{\theta y} = \begin{pmatrix} \cos\theta_y & 0 & -\sin\theta_y \\ \sin\theta_x \sin\theta_y & \cos\theta_x & \sin\theta_x \cos\theta_y \\ \cos\theta_x \sin\theta_y & -\sin\theta_x & \cos\theta_x \cos\theta_y \end{pmatrix} \tag{A.4}$$

Therefore,

$$\begin{aligned} \mathbf{i}' &= \cos\theta_y\,\mathbf{i} - \sin\theta_y\,\mathbf{k} \\ \mathbf{j}' &= (\sin\theta_x \sin\theta_y)\mathbf{i} + \cos\theta_x\,\mathbf{j} + (\sin\theta_x \cos\theta_y)\mathbf{k} \\ \mathbf{k}' &= (\cos\theta_x \sin\theta_y)\mathbf{i} - \sin\theta_x\,\mathbf{j} + (\cos\theta_x \cos\theta_y)\mathbf{k} \end{aligned} \tag{A.5}$$

For small angles θ_x, $\theta_y \ll 1$, thus $\cos\theta \approx 1$ and $\sin\theta \approx \theta$, one obtains from Eq. (A.5)

$$\begin{aligned} \mathbf{i}' &\approx \mathbf{i} - \theta_y \mathbf{k} \approx \mathbf{i} \\ \mathbf{j}' &\approx \mathbf{j} + \theta_x \mathbf{k} \approx \mathbf{j} \\ \mathbf{k}' &\approx \theta_y \mathbf{i} - \theta_x \mathbf{j} + \mathbf{k} \approx \mathbf{k} \end{aligned} \tag{A.6}$$

In general, the three-dimensional angular position vector $\boldsymbol{\theta}$ is written in the precessing coordinate system (x', y', z') as

$$\boldsymbol{\theta} = \theta_x \mathbf{i}' + \theta_y \mathbf{j}' + \theta_z \mathbf{k}' \tag{A.7}$$

The transformation between the basis vectors of two coordinate systems can be expressed as

$$\begin{pmatrix} \mathbf{i}' \\ \mathbf{j}' \\ \mathbf{k}' \end{pmatrix} = \mathbf{R}_{\boldsymbol{\theta}z} \mathbf{R}_{\boldsymbol{\theta}x} \mathbf{R}_{\boldsymbol{\theta}y} \begin{pmatrix} \mathbf{i} \\ \mathbf{j} \\ \mathbf{k} \end{pmatrix} \equiv \mathbf{R}^* \begin{pmatrix} \mathbf{i} \\ \mathbf{j} \\ \mathbf{k} \end{pmatrix} \tag{A.8}$$

The transform matrices $\mathbf{R}_{\theta x}$ and $\mathbf{R}_{\theta y}$ are given in Eq. (A.3) and $\mathbf{R}_{\theta z}$ is written as

$$\mathbf{R}_{\boldsymbol{\theta}z} = \begin{pmatrix} \cos\theta_z & \sin\theta_z & 0 \\ -\sin\theta_z & \cos\theta_z & 0 \\ 0 & 0 & 1 \end{pmatrix} \tag{A.9}$$

Using the matrix product of $\mathbf{R}_{\theta z}$, $\mathbf{R}_{\theta x}$, and $\mathbf{R}_{\theta y}$, the transformation matrix $\mathbf{R}^*$ results as

$$\mathbf{R}^* = \begin{pmatrix} \cos\theta_z\cos\theta_y + \sin\theta_z\sin\theta_x\sin\theta_y & \sin\theta_z\cos\theta_x & -\cos\theta_z\sin\theta_y + \sin\theta_z\sin\theta_x\cos\theta_y \\ -\sin\theta_z\cos\theta_y + \cos\theta_z\sin\theta_x\sin\theta_y & \cos\theta_z\cos\theta_x & \sin\theta_z\sin\theta_y + \cos\theta_z\sin\theta_x\cos\theta_y \\ \cos\theta_x\sin\theta_y & -\sin\theta_x & \cos\theta_x\cos\theta_y \end{pmatrix} \tag{A.10}$$

For small angles θ_x, θ_y, $\theta_z << 1$, thus $\cos\theta \approx 1$ and $\sin\theta \approx \theta$, one obtains the transformation between the basis vectors of two coordinate systems.

$$\begin{aligned} \mathbf{i}' &\approx \mathbf{i} + \theta_z \mathbf{j} - \theta_y \mathbf{k} \approx \mathbf{i} \\ \mathbf{j}' &\approx -\theta_z \mathbf{i} + \mathbf{j} + \theta_x \mathbf{k} \approx \mathbf{j} \\ \mathbf{k}' &\approx \theta_y \mathbf{i} - \theta_x \mathbf{j} + \mathbf{k} \approx \mathbf{k} \end{aligned} \tag{A.11}$$

Appendix B
Calculation of the Value x from the Value X in the $\log_{10}$ Scale

To read a value X in the $\log_{10}$ scale, it is needed the transformation from the value X into the value x. The corresponding value x is interpolated from the measured values X in the $\log_{10}$ scale at an arbitrary ratio ξ (see Fig. B.1).

The intervals ΔX and ΔX_{i+1} in the $\log_{10}$ scale are calculated as

$$\Delta X = X - X_i = \log_{10}\left(\frac{x}{x_i}\right); \tag{B.1}$$

$$\Delta X_{i+1} = X_{i+1} - X_i = \log_{10}\left(\frac{x_{i+1}}{x_i}\right) \tag{B.2}$$

Dividing Eq. (B.1) by Eq. (B.2), one obtains

$$\begin{aligned} \log_{10}\left(\frac{x}{x_i}\right) &= \left(\frac{\Delta X}{\Delta X_{i+1}}\right) \log_{10}\left(\frac{x_{i+1}}{x_i}\right) \\ &= \log_{10}\left(\frac{x_{i+1}}{x_i}\right)^{\left(\frac{\Delta X}{\Delta X_{i+1}}\right)} \end{aligned} \tag{B.3}$$

Thus,

$$\left(\frac{x}{x_i}\right) = \left(\frac{x_{i+1}}{x_i}\right)^{\left(\frac{\Delta X}{\Delta X_{i+1}}\right)} = \left(\frac{x_{i+1}}{x_i}\right)^{\xi} = 10^{\xi} \tag{B.4}$$

where

$$\xi \equiv \frac{\Delta X}{\Delta X_{i+1}} = \frac{X - X_i}{X_{i+1} - X_i};\ 0 < \xi < 1 \tag{B.5}$$

H. Nguyen-Schäfer, *Rotordynamics of Automotive Turbochargers*,
Springer Tracts in Mechanical Engineering, DOI 10.1007/978-3-319-17644-4

Fig. B.1 Calculated value x from X of the $\log_{10}$ scale

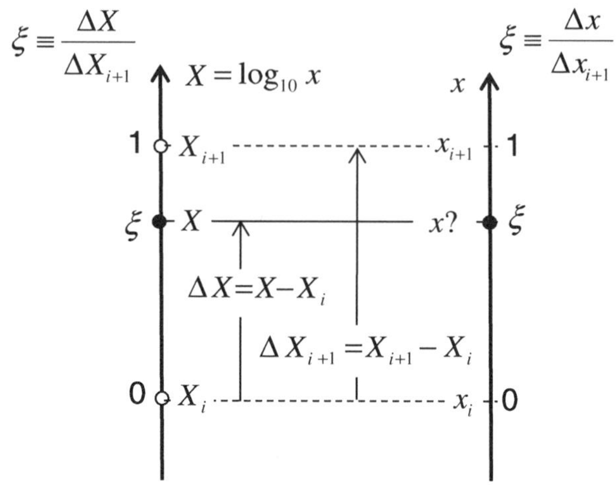

The corresponding value x of X in the $\log_{10}$ scale is computed from the ratio ξ in Eq. (B.4) where ξ is the relative distance of X between X_{i+1} and X_i (see Table B.1). The values x are displayed for various values ξ in the case $x_i = 10^1$ (see Fig. B.2). The results help us to calculate the corresponding value x from the measured value X in the $\log_{10}$ scale at any ratio ξ.

Table B.1 Computed values x from ξ of the $\log_{10}$ scale

ξ	x/x_i	$x = x_i \cdot 10^{\xi}$		
		$x_i = 10^0$	$x_i = 10^1$	$x_i = 10^2$
0.00	1	1.00	10.00	100.00
0.05	1.12	1.12	11.22	112.20
0.10	1.26	1.26	12.59	125.89
0.15	1.41	1.41	14.13	141.25
0.20	1.58	1.58	15.85	158.49
0.25	1.78	1.78	17.78	177.83
0.30	2.00	2.00	19.95	199.53
0.35	2.24	2.24	22.39	223.87
0.40	2.51	2.51	25.12	251.19
0.45	2.82	2.82	28.18	281.84
0.50	3.16	3.16	31.62	316.23
0.55	3.55	3.55	35.48	354.81
0.60	3.98	3.98	39.81	398.11
0.65	4.47	4.47	44.67	446.68
0.70	5.01	5.01	50.12	501.19
0.75	5.62	5.62	56.23	562.34
0.80	6.31	6.31	63.10	630.96
0.85	7.08	7.08	70.79	707.95
0.90	7.94	7.94	79.43	794.33
0.95	8.91	8.91	89.13	891.25
1.00	10	10.00	100.00	1000.00

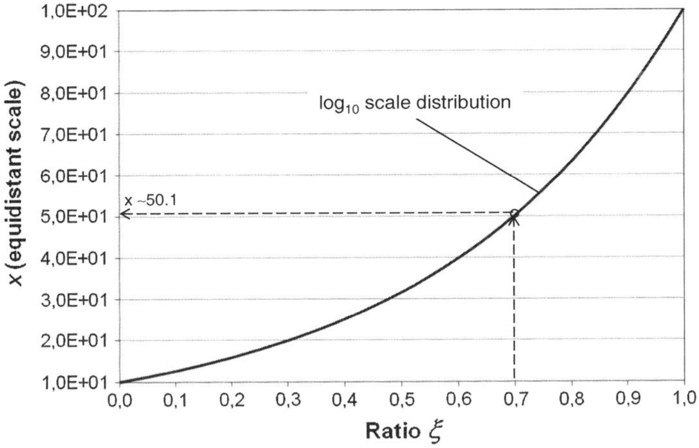

Fig. B.2 Computed values x from X of the $\log_{10}$ scale at $\xi = 0.7$

Appendix C
Solutions of the Characteristic Equation with Complex Coefficients

The characteristic equation with complex coefficients is derived from the vibration equation [C.2].

$$D(s) \equiv s^2 + (a + jb)s + (c + jd) = 0 \tag{C.1}$$

where;

$s = (\alpha \pm j\omega_n) \in \mathbf{C}$	is the complex eigenvalue;
$a, b, c, d \in \mathbf{R}$	are the real numbers.

Solving the characteristic equation $D(s) = 0$, the eigenvalue s results in

$$\begin{aligned} s_{1,2} &= -\left(\frac{a + jb}{2}\right) \pm \sqrt{\left(\frac{a + jb}{2}\right)^2 - (c + jd)} \\ &= \alpha \pm j\omega_n \end{aligned} \tag{C.2}$$

Calculating and rearranging the real and imaginary terms of the eigenvalue, one obtains

$$\alpha = -\frac{a}{2} \pm \frac{1}{\sqrt{2}}\left[\left(\frac{a^2 - b^2}{4} - c\right) + \sqrt{\left(\frac{a^2 - b^2}{4} - c\right)^2 + \left(\frac{ab}{2} - d\right)^2}\right]^{\frac{1}{2}} \tag{C.3}$$

and

$$\omega_n = \frac{b}{2} \pm \frac{1}{\sqrt{2}}\left[-\left(\frac{a^2 - b^2}{4} - c\right) + \sqrt{\left(\frac{a^2 - b^2}{4} - c\right)^2 + \left(\frac{ab}{2} - d\right)^2}\right]^{\frac{1}{2}} \tag{C.4}$$

H. Nguyen-Schäfer, *Rotordynamics of Automotive Turbochargers*,
Springer Tracts in Mechanical Engineering, DOI 10.1007/978-3-319-17644-4

The necessary condition for the rotordynamic stability is that the real term α of the eigenvalue must be negative (see Sect. 4.2).

$$\alpha = -\frac{a}{2} \pm \frac{1}{\sqrt{2}} \left[-E + \sqrt{E^2 + F^2} \right]^{\frac{1}{2}} < 0 \tag{C.5}$$

within

$$E \equiv -\left(\frac{a^2 - b^2}{4} - c \right) = -\left(\frac{a^2 - b^2}{4} \right) + c; \tag{C.6a}$$

$$F \equiv \left(\frac{ab}{2} - d \right) \tag{C.6b}$$

Thus, the stability condition ($\alpha < 0$) for the rotor given in Eq. (C.5) results as

$$a(ac + bd) - d^2 \geq 0. \tag{C.7}$$

Appendix D
Normal Distribution Density Function and Probability Function

The normal distribution density function is defined using the parameters $\bar{z}$ and σ as

$$p(z) = \frac{1}{\sigma\sqrt{2\pi}} \exp\left[-\frac{1}{2}\left(\frac{z-\bar{z}}{\sigma}\right)^2\right] \tag{D.1}$$

where

$\bar{z}$ is the mean value of the sampling values;
σ is the standard deviation resulted from the sampling values.

$$\sigma = \sqrt{\frac{1}{(N-1)}\sum_{i=1}^{N}(z_i - \bar{z})^2} \tag{D.2}$$

within the mean value is

$$\bar{z} = \frac{1}{N}\sum_{i=1}^{N} z_i \tag{D.3}$$

The normal distribution density function is also called the *Gaussian density function* with a bell shape. Its probability distribution function is calculated integrating $p(z)$ from $-\infty$ to ζ. The probability function $P(\zeta)$ is the area under the bell curve (see Fig. D.1).

$$P(\zeta) = \frac{1}{\sigma\sqrt{2\pi}} \int_{-\infty}^{\zeta} \exp\left[-\frac{1}{2}\left(\frac{z-\bar{z}}{\sigma}\right)^2\right] dz \tag{D.4}$$

H. Nguyen-Schäfer, *Rotordynamics of Automotive Turbochargers*,
Springer Tracts in Mechanical Engineering, DOI 10.1007/978-3-319-17644-4

Changing z with the new dimensionless variable

$$c \equiv \frac{z - \bar{z}}{\sigma} \tag{D.5}$$

the normal distribution function $p(z)$ is written as

$$p(c) = \frac{1}{\sqrt{2\pi}} \exp\left[-\frac{c^2}{2}\right] \tag{D.6}$$

The probability function at the new variable c is written as

$$P(c) = \frac{1}{\sqrt{2\pi}} \int_{-\infty}^{c} \exp\left[-\frac{1}{2}c^2\right] dc \tag{D.7}$$

Therefore,

$$P(-1 \le c \le +1) = \frac{1}{\sqrt{2\pi}} \int_{-1}^{1} \exp\left[-\frac{1}{2}c^2\right] dc \tag{D.8}$$

The probability distribution values for various parameters are calculated from Eq. (D.8), see Fig. D.2.

$$\begin{aligned} P(-1 \le c \le +1) &= 68.3\,\% \\ P(-2 \le c \le +2) &= 95.4\,\% \\ P(-3 \le c \le +3) &= 99.7\,\% \\ P(-4 \le c \le +4) &= 99.99\,\% \end{aligned} \tag{D.9}$$

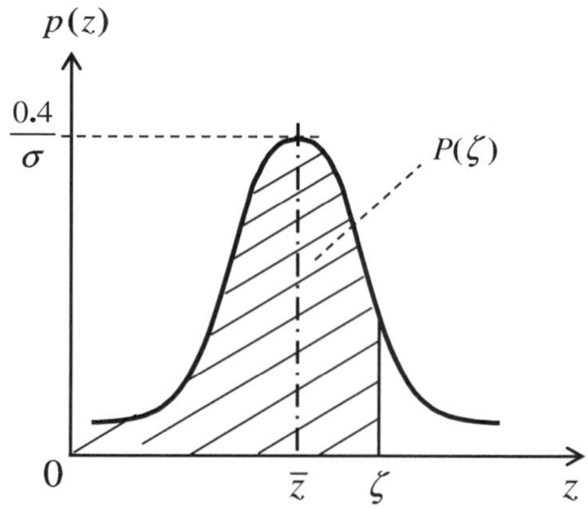

Fig. D.1 Distribution density function $p(z)$ and probability function $P(\zeta)$

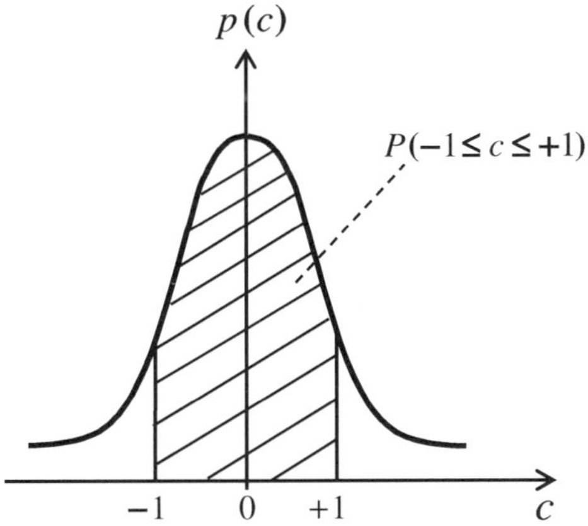

Fig. D.2 Distribution density function $p(c)$ and probability function $P(c)$

This result indicates that a production with $\pm 3\sigma$ delivers 99.7% of the products that fulfill the lowest and highest tolerances.

References

1. Adams Jr, M.L.: Rotating Machinery Vibration. Taylor and Francis, CRC (2001)
2. Muszýnska, A.: Rotordynamics. Taylor and Francis, CRC (2005)
3. Taylor, J.R.: An Introduction to Error Analysis. University Science Books (2005)

Further Readings

1. Adams Jr, M.L.: Rotating Machinery Vibration. Taylor and Francis, CRC (2001)
2. Bently, D.E. and Hatch, C.: Fundamentals of Rotating Machinery Diagnostics. Bently Pressurized Bearing Press (2002)
3. Childs, D.: Turbomachinery Rotordynamics. Wiley (1993)
4. Den Hartog, J.P.: Mechanical Vibrations. McGraw-Hill (1956)
5. Ehrich, F.: Handbook of Rotordynamics. Krieger Publishing Company (2004)
6. Gasch, R., Nordmann, R., Pfuetzner, H.: Rotordynamik, 2 Auflage. Springer (2006)
7. Genta, G.: Dynamics of Rotating Systems. Springer (2005)
8. Goodwin, M. J.: Dynamics of Rotor Bearing Systems. Thomson Learning (1989)
9. Gunter, E.J.: Dynamic Stability of Rotor-Bearing Systems. NASA SP-113 (1966)
10. Hori, Y.: Hydrodynamic Lubrication. Springer (2006)
11. Kraemer, E.: Rotordynamics of Rotors and Foundations. Springer (1993)
12. Lee, C.W.: Vibration Analysis of Rotors. Kluwer Academic (1993)
13. Muszýnska, A.: Rotordynamics. Taylor and Francis, CRC (2005)
14. Neville, F., Rieger, N.F.: Balancing of Rigid and Flexible Rotors. Shock and Vibration Information Center, N.F. (1986)
15. Rao, J.S.: Rotor Dynamics, 3rd edn. New Age Intl Publishersi (2007)
16. Rieger, N.F.: Rotordynamics 2—Problems in Turbomachinery. CISM Courses and Lectures No. 297 (1988)
17. Rieger, N.F.: Balancing of Rigid and Flexible Rotors. DoD, U.S. (1986)
18. Rieger, N.F. and Crofoot, J.F.: Vibrations of Rotating Machinery. Vibration Institute (1977)
19. Schweitzer, G. et al.: Magnetic Bearings. Springer (2009)
20. Vance, J.: Rotordynamics of Turbomachinery. Wiley (1988)
21. Vance, J., Zeidan, and Murphy, B.: Machinery Vibration and Rotordynamics. Wiley (2010)
22. Tondl, A.: Some problems of Rotor Dynamics. Chapman & Hall (1965)
23. Yamamoto, T. and Ishida, Y.: Linear and Nonlinear Rotordynamics. Wiley (2001)
24. Wen, J.C. and Gunter, J.: Introduction to Dynamics of Rotor-Bearing Systems. Eigen Technologies ETI (2001)
25. Wowk, V.: Machinery Vibration and Balancing. McGraw-Hill (1995)

H. Nguyen-Schäfer, *Rotordynamics of Automotive Turbochargers*,
Springer Tracts in Mechanical Engineering, DOI 10.1007/978-3-319-17644-4

Index

H. Nguyen-Schäfer, *Rotordynamics of Automotive Turbochargers*,
Springer Tracts in Mechanical Engineering, DOI 10.1007/978-3-319-17644-4